C0-AOB-779

HANDBOOK OF

SATELLITE TELECOMMUNICATION AND BROADCASTING

The Artech House Telecommunication Library

Telecommunications: An Interdisciplinary Text, Leonard Lewin, ed.
Telecommunications in the U.S.: Trends and Policies, Leonard Lewin, ed.
The ITU in a Changing World by George A. Codding, Jr. and Anthony M. Rutkowski
Integrated Services Digital Networks by Anthony M. Rutkowski
The ISDN Workshop: INTUG Proceedings, G. Russell Pipe, ed.
Introduction to Satellite Communication by Bruce R. Elbert
Handbook of Satellite Telecommunications and Broadcasting, L. Ya. Kantor, ed.
Telecommunication Systems by Pierre-Gerard Fontolliet
The Law and Regulation of International Space Communications by Harold M. White, Jr. and Rita Lauria
Introduction to Telephones and Telephone Systems by A. Michael Noll
The Deregulation of International Telecommunications by Ronald S. Eward
The Competition for Markets in International Telecommunications by Ronald S. Eward
Machine Cryptography and Modern Cryptanalysis by Cipher A. Deavours and Louis Kruh
Cryptology Yesterday, Today, and Tomorrow, Cipher A. Deavours et al., eds.
Teleconferencing Technology and Applications by Christine H. Olgren and Lorne A. Parker
The Executive Guide to Video Teleconferencing by Ronald J. Bohm and Lee B. Templeton
World-Traded Services: The Challenge for the Eighties by Raymond J. Krommenacker
The Telecommunications Deregulation Sourcebook, Stuart N. Brotman, ed.
Communications by Manus Egan
Preparing and Delivering Effective Technical Briefings by David L. Adamy
Writing and Managing Winning Technical Proposals by Timothy Whalen
World Atlas of Satellites, Donald M. Jansky, ed.
Communication Satellites in the Geostationary Orbit by Donald M. Jansky and Michel C. Jeruchim
New Directions in Satellite Communications: Challenges for North and South, Heather E. Hudson, ed.
Television Programming across National Boundaries: The EBU and OIRT Experience by Ernest Eugster
Evaluating Telecommunications Technology in Medicine by David Conrath, Earl Dunn, and Christopher Higgins
Optimization of Digital Transmission Systems by K. Tröndle and G. Söder
Optical Fiber Transmission Systems by Siegfried Geckeler
Traffic Flow in Switching Systems by Gerard Hebuterne
Measurements in Optical Fibers and Devices: Theory and Experiments by Giovanni Cancellieri and Umberto Ravaioli
Fiber Optics Communications, Henry F. Taylor, ed.
The Cable Television Technical Handbook by Bobby Harrell

HANDBOOK OF

SATELLITE TELECOMMUNICATION AND BROADCASTING

L. Ya. Kantor, ed.

English translation edited
by Donald M. Jansky

Artech House
BOSTON • LONDON

Library of Congress Cataloging-in-Publication Data

Spravochnik po sputnikovoĭ svi͡azi i veshchanii͡u. English.
Handbook of satellite telecommunication and broadcasting.

Translation of: Spravochnik po sputnikovoĭ svi͡azi i veshchanii͡u.
Bibliography: p.
Includes index.
1. Artificial satellites in telecommunication.
I. Kantor, Lev I͡Akovlevich. II. Title
TK5104.S6513 1987 621.38′0422 87-19578
ISBN 0-89006-220-X

ARTECH HOUSE, INC.
685 Canton Street
Norwood, MA 02062

International Standard Book Number: 0-89006-220-X
Library of Congress Catalog Card Number: 87-19578

Translation from the Russian of *Spravochnik po Sputnikovoi Svyazi i Veshchaniyu,* copyright 1983 by Radio i Svyaz, Moscow.

10 9 8 7 6 5 4 3 2 1

Contents

The Authors

G.B. Askinazi
V.L. Bykov
G.V. Vodop'yanov
M.N. D'yachkova
L.Ya. Kantor, Editor
A.M. Model'
A.M. Pokras
V.V. Timofeev
V.M. Tsirlin
I.S. Tsirlin

Preface

The rapid expansion in the USSR of the network of earth satellite communications stations is a most characteristic feature of recent years. New geostationary satellites of the Statsionar and Statsionar-T (Ehkran) series have appeared; the capacity of telephone communications systems has been increased; systems for transmitting new kinds of information (wire photos and audio broadcast programs) have been developed; and the Moskva and Ehkran mass distribution systems, containing hundreds of thousands of earth stations, have been built. The number of experts engaged in or preparing for the development and operation of satellite communications and broadcasting systems has increased sharply in this connection. This handbook is addressed basically to these people.

The authors' hope that the book will also be beneficial to experts who are not engaged directly in satellite communications, but who are interested in this new and rapidly developing industry and are educated in the field of radio engineering. Engineering solutions, used for developing non-Soviet satellite communications systems, international and national, are taken into consideration in the manual for the purpose of giving the reader broader familiarity with the state of the art, but the authors based their work mainly on experience in the development of Soviet systems.

The proposed handbook, in keeping with the long tradition of similar publications at the Radio i Svyaz' publishing house, is a systematized presentation of satellite communications broadcasting principles and technology. This book is written in handbook form and the material is presented in the most systemized form possible. The derivations of formulas and detailed substantiation of various solutions are not presented; however, the following are included:

- examination of all typical elements and methods of the design and calculation of satellite communications systems, as well as examples of the calculations;
- explanations of the values used in the formulas, and the coefficients necessary for the calculations;
- tables of values necessary for calculations;
- numerous illustrations to enhance the text;
- the most concise definitions of the concepts that are utilized.

Basic concepts and terms are defined, and a detailed analysis of and a procedure for calculating the energy parameters of satellite radio links and crosstalk interference between satellite systems are presented. Much attention is devoted to multiple access techniques. Soviet and some non-Soviet communications and broadcasting satellite systems, the main types of ground stations, receiving-transmitting equipment, multiple access equipment, antenna feeds, antenna systems and airborne repeaters are described. Aspects of reliability, geostationary orbit utilization, and the principles by which the performance characteristics of satellite links are normalized are examined.

The book is intended for engineering technicians, engaged in the operation, design and development of satellite communications systems, and also for students at colleges and universities specializing in radio communications.

Chapter 1

Design Principles and Functions of Satellite Communications Systems

1.1 SATELLITE COMMUNICATIONS SYSTEMS: BASIC DEFINITIONS, STRUCTURE, AND PURPOSE

The main idea of building satellite communications systems is simple—to put an intermediate repeater of a communications system on an artificial earth satellite. The satellite moves in a rather high orbit for a long time without using energy for its motion. Power is supplied to the satellite repeater and other systems of the artificial earth satellite from solar batteries, almost always operating on rays from the sun, which nothing shades.

In a high enough orbit the artificial earth satellite "sees" a very large territory (about one-third of the earth's surface), and for this reason any station located in that territory can be connected directly through its satellite repeater. Three artificial earth satellites may be enough to build a nearly global communications system. At the same time, modern technology is capable of producing a satellite signal narrrow enough that the energy of the transmitter of a satellite can be focused, if necessary, on a limited area, for example, on a small country. This provides an opportunity to utilize an artificial earth satellite effectively for servicing small areas.

It should also be mentioned that the path of the satellite signal between the artificial earth satellite and an earth station usually has high elevation angle to the earth's surface. This shortens its time in the earth's atmosphere, thereby reducing signal loss due to atmospheric moisture and noise.

Satellite communications began to be developed in the middle 1960s after the appearance of the Soviet artificial earth satellite Molniya and the American Telstar. Since then they have been undergoing rapid development all over the world. Many satellite communications and broadcasting systems have been built for different functions.

We will give definitions of the basic concepts, discussed in the handbook, using the traditional practice of the use of terms. *Space radio communications* are radio communications between artificial earth satellites and earth stations.

An *earth station* is a radio communications station, located on the earth's surface (or in the main part of the earth's atmosphere) and intended for communications with space or with other earth stations through space stations or other objects in space, for example, passive (reflector) artificial earth satellites. In contrast to this, stations of ground radio communications systems, which have nothing to do with space communications systems or radio astronomy, are called *ground stations.*

Satellite communications are communications between earth stations through space stations or through passive artificial earth satellites. Thus, satellite communications are a particular case of space communications.

A *satellite link* is a communications line between earth stations through one satellite and consists of an earth-satellite link (Figure 1.1, the *up link*) and satellite-earth link (the *down link*). Earth stations are connected to switching centers of a communications network (for instance, with a long-distance telephone exchange) through connecting terrestrial links (see Figure 1.1).

Key: **(1) Space station**
(2) Long-distance telephone exchange
(3) Earth station

FIGURE 1.1. Satellite link (hypothetical reference circuit).

Satellite broadcasting is the transmission of radio broadcast programs (television and audio) from earth transmitting stations to receiving stations through a space station—an active repeater. Thus, satellite broadcasting is a particular case of satellite communications, in which a particular class of one-way (simplex) messages is transmitted, and these messages are received all at once by several earth stations or by many receiving stations (circular transmission).

The following are kinds of radio communications services, depending on the type of earth stations and the purpose of a system [1]:

- fixed satellite service—between earth stations, located at certain fixed points;
- mobile satellite service—between mobile earth stations using one or several space stations (there are land, sea, and air mobile satellite services, depending on the location of a mobile earth station);
- satellite radio broadcasting service—a radio communications service in which the signals of space stations are intended for direct reception by the population. Here both individual and collective reception is considered direct reception; in the latter case a broadcast program is delivered to individual subscribers through a terrestrial distribution system (cable or air waves) by a small transmitter. It should be pointed out that the term "radio broadcasting" includes both television and audio broadcasting. A satellite radio broadcasting service, thus defined, does not use all kinds of satellite broadcasting systems, but only those that are intended for reception with comparatively simple receiver installations.

The Soviet satellite broadcasting system, Orbita, during the first years of its existence, was intended only for the distribution of audio broadcasting programs and TV programs with audio accompaniment in the USSR. According to the CCIR Radio Regulations [1] it is not a radio broadcasting system, but rather a fixed satellite service, because programs are not delivered to telecenters. This is not a strict limitation, of course, but it should be kept in mind, because different frequency bands are allocated to services (see Chapter 3).

Satellite communications systems are used for transmitting the following different kinds of information:

- TV programs: in this case there are systems for exchanging TV programs between equal earth stations, and systems for one-way distribution or programs from a transmitting station to many receiving earth stations;
- different kinds of one-way messages, most often of a one-way nature: wire photos, audio broadcasting programs;
- telephone messages, two-way in nature: audio frequency channels or groups of them can be used for exchanging different kinds of information such as telegraph, discrete information from electronic computers and other sources.

Depending on the kinds of information being transmitted, there are general-purpose multifunctional systems with earth stations that exchange different kinds of information (examples are the Canadian Telesat space communications system and the Soviet Orbita system), and special systems for transmitting one kind or several uniform kinds of information (the Ehkran satellite broadcasting system for the one-way distribution of TV and audio broadcasting is an example).

Depending on the territory that is served, the location and affiliation of earth stations, and the structure of control, satellite communications systems may be classified as:

- international, in which there are stations of different countries; systems of this kind can be global (world coverage), like Intelsat, Intersputnik, or regional, like Eutelsat;
- national, with earth stations located within the boundaries of one country including zonal, with earth stations located within one of the areas (regions) of a country and departmental (commercial and business) systems, with earth stations belonging to a particular department (organization, company), and transmitting only business information and data in the interests of the department.

In all satellite communications systems, whatever their differences, there are several elements, used for the same purpose: earth stations of different types; and space stations, used for retransmission (receiving-transmitting) facilities are located on artificial earth satellites, with antennas for receiving and transmitting radio signals and with support systems, such as power sources, systems for aiming antennas (at the earth) and solar batteries (at the sun), satellite orbital correction systems, *et cetera.* Types of earth stations are examined in greater detail.

Receiving earth stations of distribution systems (satellite broadcasting systems) are the simplest type of station. They are used simply for receiving TV programs or other broadcast information (see Figure 1.2), for example audio broadcasting or wire photos; receiving earth stations usually are equipped with a small antenna to reduce costs, and there are usually many of these earth stations in a system.

Transmitting earth stations of satellite broadcasting systems are stations that perform transmission in an earth-satellite link of broadcast information, intended for distribution through a network of receiving stations; if the transmitting earth station is located in the service area and signals radiated by the artificial earth satellite of this system can be received by it, then this kind of reception usually is accomplished for listening purposes, and it is still called a transmitting station. There can be several transmitting stations in a system.

Receiving-transmitting earth stations, operating in a two-way telephone communications network (including a network in which different kinds of messages (telegraph, data, audio broadcasting programs, *et cetera*)

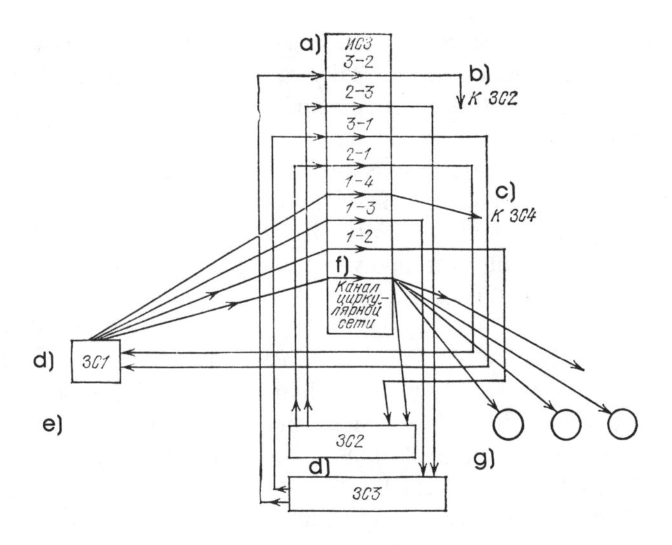

Key: (*a*) **Artificial earth satellite**
(*b*) **To earth station 2**
(*c*) **To earth station 4**
(*d*) **Earth station**
(*e*) **Transmission of broadcast information and telephone traffic**
(*f*) **One-way network channel**
(*g*) **Receiving earth stations of one-way network**

FIGURE 1.2. Diagram of one- and two-way channels through an artificial earth satellite: space stations 1–3—earth stations for receiving broadcast information and telephone traffic.

can be transmitted on telephone channels or groups of channels), and also in a TV program network. Stations of this kind often are outfitted with equipment for operating through several satellite transponders at the same time. Sometimes receiving-transmitting stations of a telephone system are also receiving stations of a broadcasting system; the Orbita earth stations are such (see Figure 1.2).

Monitor earth stations are stations that keep track of the operating mode of a communications satellite, of the observance by earth stations of a network of parameters that are important for the operation of an entire network—radiated power, transmission frequencies, emission with a certain polarization (if polarization division is used in a system), the quality of the modulating signal, *et cetera.* The role of monitor earth stations in keeping a system functioning normally is exceedingly great. Often the functions of a monitor station are turned over to one of the receiving-transmitting stations of a network.

Monitor and central stations of a network usually can exchange information with the stations of a network through a special service communications subsystem. This kind of subsystem usually can be set up through the same satellite, through which the main network operates, but in some cases it is necessary to use service communications terrestrial links.

Earth stations of a satellite control and monitor system are stations that control the operation of a space station and all the other subsystems of the artificial earth satellite and monitor their condition. These stations play a very important role in the insertion of satellites into orbit, in initial tests and in the placement of space stations in operation. During the operation of modern long-life artificial earth satellites, especially those in a geostationary orbit, where a space station usually operates 24 hours a day, the earth stations of the control system have almost nothing to do with the operation of the system, and the functions of a satellite monitor and control station in this case can be turned over to one of the receiving-transmitting stations of a network.

Terrestrial links are used for connecting earth stations to the sources and users of transmitted information, because the earth stations usually are separated from them for reducing interference, antenna coverage angles, *et cetera.* Examples are connecting lines between a receiving-transmitting earth station and a long-distance telephone exchange or other switching center of a telephone network, between a receiving earth station and TV transmitter, typography facility or radio broadcasting station.

Outside equipment is the part of satellite communications equipment that is not installed at satellite communications stations, but at other facilities. For instance, echo suppressors, and sometimes multiplexing channeling and modulation equipment with output signal passing through a connecting terrestrial link (usually a radio relay line), goes straight into the circuit of a satellite communications link, necessary for the operation of satellite channels, can be installed at long-distance telephone exchanges.

The control center of a communications system is the organization that manages the operation of a system and its development, i.e., the delivery of new earth stations and satellites, the scheduling of their operation, delivery of transponders to consumers, maintenance and repair work, *et cetera.* The control center usually is connected to the stations of a network by service communications channels. The control center sometimes can be combined with the transmitting station of a satellite broadcasting system or with a monitor earth station.

1.2 THE BASIC CHARACTERISTICS OF SATELLITE COMMUNICATIONS SYSTEMS

1.2.1 The Most Important Characteristics of Earth Stations

The receiving and transmitting bands, in which the equipment of a station is intended for operation—the antenna, receiving and transmitting equipment (most earth stations operate in the 4/11 GHz band (down link) for receiving and 6/14 GHz (up link) band for transmissions (see Chapter 3)).

The Q-factor G/T of a receiving station is the ratio of the gain of the antenna (in dB on the receiving frequency) to the equivalent noise temperature of the station (in dB relative to 1 K); it ranges up to 42 dB for the largest antennas that are used in practice (with a diameter of 32 m) and is 20–31.7 dB for the earth stations of most national and regional systems, stations of the Intersputnik system, and class B stations of the Intelsat system.

The diameter of an antenna is important, even when the G/T is known, because it is specifically the diameter of an antenna that determines the size and cost of an earth station. If polarization is used in a system, it is necessary to know the cross-polarization characteristics of the antenna and the polarization with which a station is operating when transmitting and receiving.

An antenna is also distinguished by the characteristics of the mount used to aim the antenna at a satellite; it is important to distinguish between 360° antennas, capable of being aimed at any point in the heavens, and antennas with a limited range of working angles, within which an antenna can be aimed at a signal source. Antenna steering systems also are characterized by the rate of angular motion. In recent years slow-moving antennas, with more and more limited range, suitable for operation only with geostationary satellites, have begun to be used.

The effective isotropic radiated power (EIRP) is the product of the transmitter power, multiplied by the antenna gain (in the transmission band). It usually ranges from 50–95 dB. The maximum spectral density of the power current (in $W/m^2 \cdot Hz$), radiated by an earth station, often is specified for simplifying the calculation of crosstalk interference, although the exact calculation of crosstalk interference requires a knowledge of the structure of the signals that are used in a system (the kind and parameters of modulation, *et cetera*) [2]. It is also important to characterize a transponder of a space station in terms of its ability to be tuned in the working frequency band, of linear and nonlinear distortions of messages and of the bandwidth of each transponder.

1.2.2 The Basic Characteristics of a Space Station

A space station is distinguished basically by the same characteristics as an earth station: The G/T of the receiver, the EIRP of each transmitter, and the polarization of the radiated and received signals. However, some of the parameters are much different from those mentioned above for earth stations. For instance, the G/T of the receiver of a space station usually is -20 dB to $+3$ dB, not only because a smaller antenna is used, but also because of the use of a simpler (and with a higher noise temperature) input amplifier. The EIRP of a space station usually does not exceed 23–45 dB. The number of transponders is an important characteristic of the satellite communications space station.

A transponder of a space station is produced when radio signals pass through a common amplifier (a common transmitter output stage) in a certain common frequency band. Now the fact that this definition is to some extent arbitrary becomes obvious, in any case, for an earth station. For instance, several transponders can share common elements—the antenna, waveguide signal path, and input amplifier. On the other hand, the bandwidth of one transponder at an earth station can be separated by filters for the purpose of separating and detecting signals from different earth stations, passing through a common satellite transponder (in the case of frequency division multiple access (FDMA), see Chapter 7).

The definition of transponder is more concise for a communications satellite. The entire frequency band in which a communications system operates is divided into a number of portions (with a bandwidth of 35–40 MHz, 80–82 MHz, or 120 MHz), in which the signals are amplified by a transponder.

Usually six to twelve transponders operate simultaneously on a satellite, and up to 27 operate on the largest satellites (the Intelsat-V). The signals of these transponders are separated in terms of frequency, space, and polarization. The number of transponders, their bandwidth, and EIRP determine (basically!) the most important summary characteristic of a satellite—its carrying capacity, i.e., the number of telephone and TV channels that can be transmitted through a satellite, or in more general form the number of binary units per second that can be transmitted through a given satellite. The definition of the carrying capacity of a satellite is, of course, only conventional, because it depends on the G/T of the earth stations that are used in a system and on the kind of radio signals that are used. The carrying capacity is essentially a characteristic of a system, not of a satellite. Nevertheless the concept of the carrying capacity of a satellite is used often in the literature.

It should be pointed out that the carrying capacity of a satellite transponder depends not only on the basic characteristics (the bandwidth and EIRP), but also to some extent on other characteristics: the linearity of the amplitude response, the amount of AM-PM conversion, the nonuniformity of the group delay time in the IF band of the transponder, *et cetera.* These parameters influence the crosstalk interference between the signals of different earth stations, the reliability with which the signals are received and therefore the losses due to the satellite.

Depending on the character of the radiation pattern of the satellite antennas, a satellite (or its individual transponders, if there are several antennas on-board and they are different) is characterized by its coverage area. This is the part of the surface of the globe, within which the satellite signal level, necessary for reception with a given quality at an earth station with a certain G/T, is received. The capability of receiving signals from earth stations with a certain EIRP at the input of a satellite is also obtained in this coverage area. The coverage area of the satellite obviously is a characteristic of a satellite communications system, and not just of a satellite itself.

The coverage area is determined not only by the character of the radiation pattern of the antenna of a satellite, but also by features of geometric plots that appear when the earth's surface is sectioned by the cone of the antenna beam [3, 5]. The shape of this cross section depends on the point where a satellite is located. The "sighting point" is the point where the axis of the main lobe of the antenna of a satellite intersects the earth's surface, and also is the point of the positional instability of the satellite and the orientation of its antennas. The concept of guaranteed service area, in which the above-mentioned reception and transmission conditions are specified, regardless of the combinations of deviations of the satellite and the antenna of the satellite from the mean position, is introduced [1.2] in connection with the latter circumstance—instability (more information on this will be given in Section 1.4).

The position of a satellite in orbit, the point on the earth's surface where its antenna is aimed, and the instabilities of these parameters are not only important for calculating service areas, but also for calculating crosstalk interference between satellite communications systems (see Chapter 3). The maximum spectral density of the power radiated by a satellite (in $W/m^2 \cdot Hz$) also is often specified for the purpose of simplifying the calculation of crosstalk interference.

Finally, the most important characteristic of a satellite, which determines not only the reliability and continuity of communications, but most importantly determines economic characteristics of an entire communications system, is the service life of a satellite—the accrued operating time to the complete failure of even one (or of several, depending on technical requirements) of the transponders of a space station. This can be determined with a high probability, usually 0.9 and higher. The service life of modern satellites is three to seven years or longer, due to the high reliability of components, and flexible and extensive redundancy.

1.2.3 The Basic Characteristics of Satellite Communications Systems

The service area of a system is the combination (union) of the service areas of individual satellites in a system (Figure 1.3); the determination of the service area of a satellite is explained in Section 1.4, and it is a little different from the concept of coverage area, already discussed.

The term union (and not sum) is used because the areas of individual satellites usually overlap (this is unavoidable if continuous coverage is to be achieved, and it is useful for organizing communications between earth stations in different areas), and therefore the overall area of a system is smaller than the sum of the individual areas.

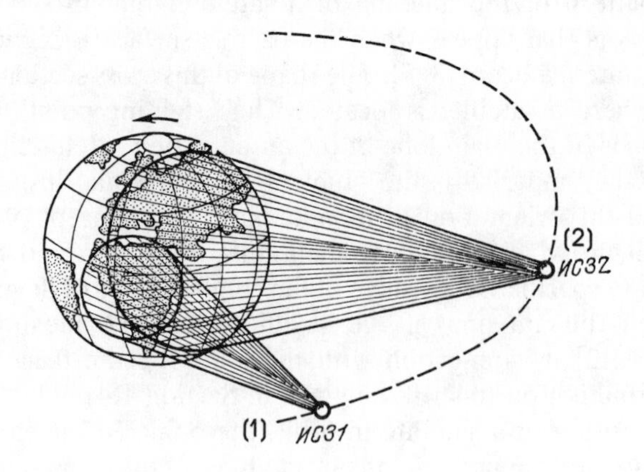

Key: **(1) Satellite 1**
(2) Satellite 2

FIGURE 1.3. Determination of the service area of a satellite communications system.

The carrying capacity of a system is the union of the carrying capacities of the satellites in a system. In this case we will define the term union as follows. The carrying capacity of a system is less than the sum of the carrying capacities of the individual satellites, because for communications between stations operating through different satellites, some of the channels are transferred successively by two space stations by means of double-hop links (earth-satellite-earth-satellite-earth) or of direct satellite to satellite links (earth-satellite-satellite-earth). When only one satellite is used in a satellite communications system the service area and carrying capacity of the system and of the satellites coincide.

Now a satellite communications system often is characterized by the number and location of the earth stations, the number of satellites, and the type of orbit and location in a geostationary orbit. Other characteristics of a system are the number of transponders in a satellite, their bandwidth and the central frequencies of the transponders in earth-satellite and satellite-earth links.

One of the most important characteristics of a system is the multiple access modulation method—the method by which signals radiated by different earth stations are combined, for their passage through a common transponder of the satellite repeater of a space station. Multiple access is used because it is usually impractical or impossible to create the same number of transponders on a satellite as at an earth station of a system. Multiple access with frequency, waveform, and time division is discussed in Chapter 7. Any multiple access method involves losses of up to 3–6 dB of the carrying capacity of a transponder (in comparison with the single-signal mode). However, losses in the most modern systems (using time division) may not exceed 0.5–2 dB.

The modulation method that is used has a significant effect on the energy characteristics of a communications system, the necessary frequency band, and its electromagnetic compatibility with other systems; frequency modulation (FM) for transmitting messages in analog form, and phase-shift keying (PSK) for transmitting messages in discrete form are used extensively. The most important of the modulation parameters for FM is the frequency deviation, and for PSK it is the number of carrier phases (the modulation factor). For the case of the transmission of TV programs and also the method used to transmit audio accompaniment (time or frequency, subcarrier frequency, *et cetera*) see Chapter 14.

Another most important characteristic of a system is the quality of its message transmission channels—TV, telephone, *et cetera.* We will not examine in detail here the entire list of the performance characteristics of channels. (More will be said about them in Chapter 2.) It is enough to simply mention that a satellite communications system usually is used for establishing international or long-distance communications channels of great length, and the quality of these channels meets the requirements, formulated for such channels in the recommendations of the International

Radio Consultative Committee (CCIR) or of national standards. However, in some satellite communications systems, depending on the specific purpose, or because of economic considerations, less rigid quality standards are acceptable.

Just as in TV broadcasting systems, in which signals are received by simple sets, a lower signal-to-noise ratio often is permitted. This is recommended, in particular, by the plan of satellite broadcasting systems, adopted by the World Administrative Conference on Radio in 1977 [4, 6]; a similar decision was made in the Soviet Ehkran system [4]. The reasons for the lower signal-to-noise ratio are both the desire to reduce the cost of a receiving station and the capability of still offering the subscriber adequate reception quality. Actually, a receiving station of such a system is close to the subscriber, and the satellite link replaces not only a long-distance terrestrial link, but also a part of the distribution network, and a TV transmitting center on the receiving end of the link is simplified or eliminated altogether.

A somewhat lower signal-to-noise ratio or a bandwidth that is narrower than the bandwidth recommended for long-distance channels also is established sometimes in telephone channels. This is usually permitted if a satellite communications system is intended for special purposes, because in specialized systems of this kind, simplified earth stations are close to the subscriber and the quality of a subscriber's channel remains acceptable.

In certain non-Soviet satellite communications systems, built on the basis of frequency division multiple access (FDMA) and the transmission of every channel on a separate carrier (see Chapters 7 and 18), companders are used; these operate on the basis of features of perception of noise in an acoustic signal. The use of companders makes it possible to lower the audibility of noises by 10–20 dB and accordingly to achieve an increase in the carrying capacity of a communications system. This strips the channels of their versatility, because the above-mentioned gain is not achieved when messages, data, *et cetera,* are transmitted on audio frequency channels.

1.3 THE STRUCTURE OF EARTH AND SPACE STATIONS

The equipment of stations is described in detail in several subsequent chapters. Only the definitions of some of the most important elements of stations are given here.

The simplest earth station is intended only for one-way reception (see Figure 1.2)—a receiving earth station for receiving signals from one transponder of the space repeater of a satellite.

The signals that are radiated by a satellite are received (Figure 1.4(a)) by antenna 1 of an earth station (most often a reflector). This signal is amplified by system 2, which contains an amplifier and a mixer for intermediate-frequency (IF) preamplification. The oscillations necessary for frequency conversion are generated by heterodyne 3. Most of the amplification of the signal takes place on an intermediate frequency (or on several intermediate frequencies in the case of multiple conversion) in IF amplifier 4, which contains a filter (or filters) that determine the bandwidth that is optimum for receiving a signal. The bandwidth would be optimum if it is close to the passband for reception of TV programs, multichannel telephone messages with time multiple access, *et cetera,* or if it is only a fraction of the transponder bandwidth, for example for receiving telephone signals with frequency multiple access. The amplifier is followed by demodulator 5, which extracts the transmitted message, and by terminal channeling equipment 6. For example, in the case of the reception of TV programs, the regeneration of the synchromixture, the selection of audio accompaniment channel, *et cetera,* can be accomplished in equipment 6. The information that is received passes through connecting terrestrial link 7 to the program consumer. Complex 8 is used for aiming the antenna at the satellite, and it includes a drive that moves the antenna, and steering equipment, which controls its motion.

More sophisticated earth stations, intended for two-way communications and operating on several satellite trunks, are constructed in accordance with the more general diagram in Figure 1.4(b), where 1 is an antenna with a steering complex, usually used for receiving and transmitting at the same time; 2 is a receiving and transmitting dividing network; 3 is a low-noise amplifier; 4 is a combiner (a combining network) for the transmitters of different transponders; 5 is a dividing network (a dividing filter), which separates the received signals of different trunks; 6 is a transponder transmitter; 7 is a transponder receiver; 8 is transponder channeling equipment; and 9 is connecting line equipment. Also possible, and often used, are earth stations, in which transmit-receive communications is combined with a receive-only capability. Many stations of the Orbita network are presently designed in this way (see Figure 1.2).

The following describes the basic elements of the radio engineering complex of a space station which is a part of a satellite communications system. This complex consists of two main components—an antenna and a transponder. In contrast to earth stations that have one antenna, several receiving and transmitting antennas usually are installed on modern communications satellites. This is explained by the need to provide for different service areas and to focus the radiation of the antennas with the sites of the earth stations on the earth's surface so that the energy will not

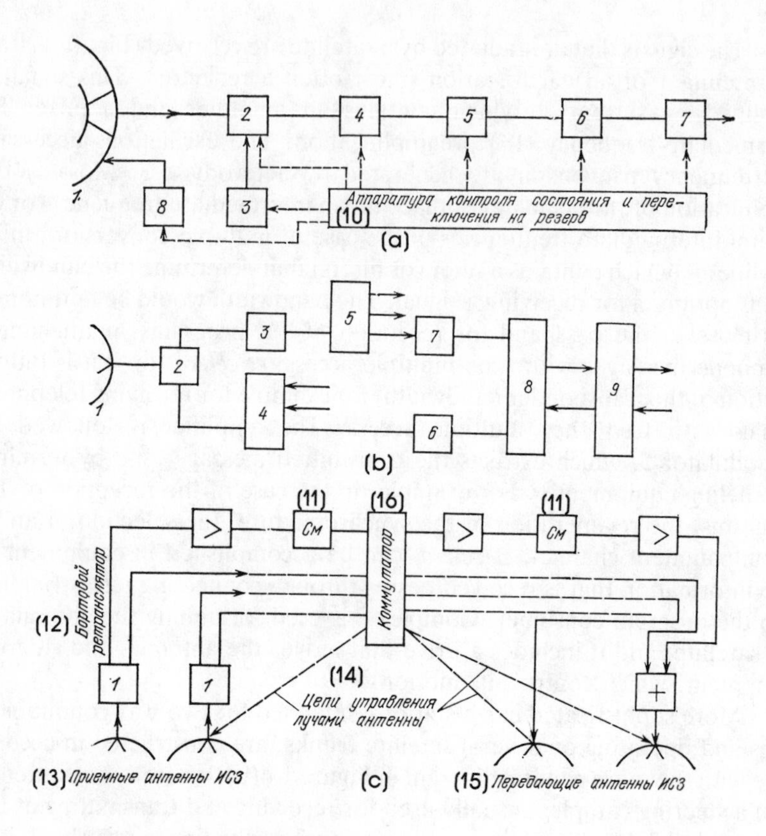

Key: **(10) Condition monitoring and standby switching equipment**
(11) Mixer
(12) Transponder
(13) Satellite receiving antennas
(14) Antenna steering circuits
(15) Satellite transmitting antennas
(16) Switchboard

FIGURE 1.4 Simplified schematic diagrams of single-transponder earth station, receiving broadcast information (a); a multitransponder earth station (b); and the satellite communications space station (c).

be scattered uselessly to areas where it is not used. The high directivity of the receiving and transmitting antennas of a satellite also reduces crosstalk interference with other communications systems (satellite and ground) and increases the utilization efficiency of the geostationary orbit (see Chapters 5 and 6).

The signal received by the antenna of a space station goes to input 1 (Figure 1.4(c)), which on a satellite consists of mixers and amplifiers, utilizing low-noise traveling wave tubes (TWT) or transistors, and only rarely uncooled parametric amplifiers. The received signal is amplified further on the receiving frequency, intermediate frequency, and transmission frequency. In some cases, single frequency conversion from input to output is used instead of double conversion, and in this case there is no IF amplifier.

Signal separation, switching and combining networks (the switchboard in Figure 1.4(c)) may be used for sending the signals, addressed to different earth stations, to the transmitting antennas with the appropriate service area. Signals can be switched both in one or several transponders. Systems with a high-speed steerable narrow-beam antenna (beam switching) are promising, because they make it possible to conduct communications with many earth stations through pencil-beam antennas without increasing the number of antennas on-board a satellite (see Chapter 7).

Standby elements and standby switching equipment are not shown in Figure 1.4(c); they are usually quite sophisticated systems, because the extent of redundancy varies from element to element of a circuit, depending on their reliability, importance for the viability of a satellite, and necessary service life, which is seven years and longer. In some cases signals may be processed in a more complicated manner in a space station, for example, conversion by modulation and the regeneration of signals, transmitted in discrete form.

1.4 THE ORBITS OF COMMUNICATIONS SATELLITES: SERVICE AREAS

1.4.1 The Orbits of Satellites

After being inserted in the necessary orbit, a satellite obeys the laws of celestial mechanics under the influence of inertia and the earth's gravity. Other heavenly bodies exert some action that distorts the orbit and changes its parameters, so that, in order to maintain the necessary orbital parameters, it is necessary from time to time to activate a small jet motor on a satellite. Therefore, a complicated and expensive measure, the insertion of a satellite in a high-altitude near-earth orbit by a multistage rocket, is justified by the low cost of keeping a satellite in the proper position in orbit for a long period of time, determined by the reliability of all the systems of the satellite.

According to Kepler's first law any orbit (motion trajectory) of an earth satellite lies in a fixed plane, passing through the center of the earth, and it is an ellipse, at one focus of which the earth is located. The orbit of a satellite usually is characterized by its inclination [angle i between the orbital plane and the equatorial plane* (Figure 1.5)], by altitude of its apogee (the farthest) and altitude of its perigee (the point of the orbit that is nearest the earth's surface), and by its rotation period (the time it takes the satellite to complete one revolution around the earth, returning to the starting point). Because a circle is a special case of an ellipse, circular orbits also are possible, in which the foci merge with the center, i.e., the earth is at the center of an orbit.

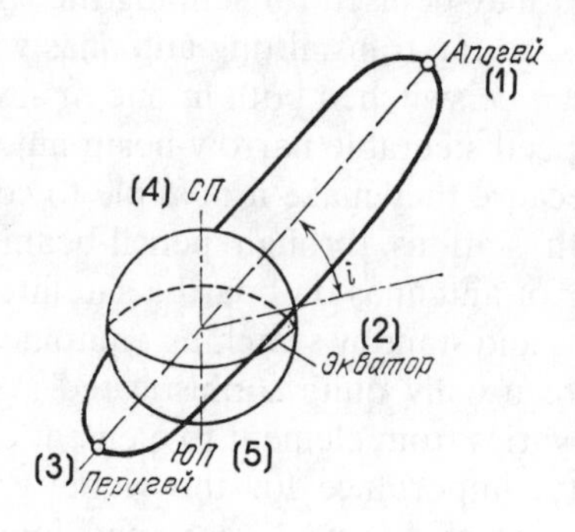

Key: **(1) Apogee**
(2) Equator
(3) Perigee
(4) North Pole
(5) South Pole

FIGURE 1.5. An elliptical satellite orbit.

When selecting the type of orbit for the satellites of communications and broadcasting systems it is necessary, first of all, to consider that the satellite has to have communications with an earth station, situated in a given service area. It must be able to see a satellite at least a few hours, and for TV broadcasting systems at least 12–16 hours; and a communications

---

*As is known, the angle between straight lines, lying in these planes and perpendicular to the line of intersection of the planes, is called angle α between the planes.

system, as a rule, must operate around the clock. If a communications window is not 24 hours, then it is desirable that it be repeated the same time of day. So-called synchronous or subsynchronous orbits are used for this purpose. The rotation period of this type of satellite is precisely equal to the length of the day, or to the length of a day divided by an integer. If one satellite does not provide the necessary 24-hour a day communications between earth stations in a service area, then several satellites are used in the very same orbit, and these satellites are separated so that around the clock (*albeit* with brief interruptions) operations of a communications system can be achieved as all the earth stations pass simultaneously and rapidly to a new satellite.

A geostationary orbit is most often used; it is a circular orbit, located in the plane of the equator at an altitude of 35,786 km. A satellite in such an orbit makes one revolution around the earth in exactly one day, i.e., in exactly the same time that it takes the earth to complete a revolution around its axis. Therefore, a satellite in a geostationary orbit (it is called a geostationary satellite) is as though parked over some equatorial point on the earth's surface. This offers numerous advantages: no interruptions of communications, a communications session becomes a 24-hour session, and it becomes easier to aim the antennas of earth stations at the satellite. In some systems the antennas of the earth stations, after initial adjustment, can be nearly or completely locked. By virtue of the high altitude of a geostationary orbit, its area of visibility on the earth's surface is large, about one-third of the earth's surface (Figure 1.6). Geostationary orbits are used very extensively because of these advantages. It is hard to insert new satellites in certain segments of such an orbit due to crosstalk interference with satellite communications systems that already exist, and the problem of the efficient utilization of geostationary orbits acquired independent significance (see Chapter 6).

As can be seen in Figure 1.6, the polar regions receive poor service from a geostationary orbit, because the satellite is seen at a small angle of elevation to the earth's surface, if at all. The small angle of elevation causes an increase in receiver noise (due to the influence of the earth's thermal emission), and it also opens the possibility that the received signal will be weakened due to the screening action of the local features, buildings, trees, *et cetera.* The angles of elevation decrease toward the poles, and also as an earth station gets farther in terms of longitude from the point where the satellite is parked (Figures 1.6 and 1.7).

It is also necessary to consider that due to the influence of various disturbances, including those of the earth and the gravitational forces of the moon and sun, the position of a satellite becomes unstationary, and its mean position shifts (drifts) slowly with a 24-hour cycle. These disturb-

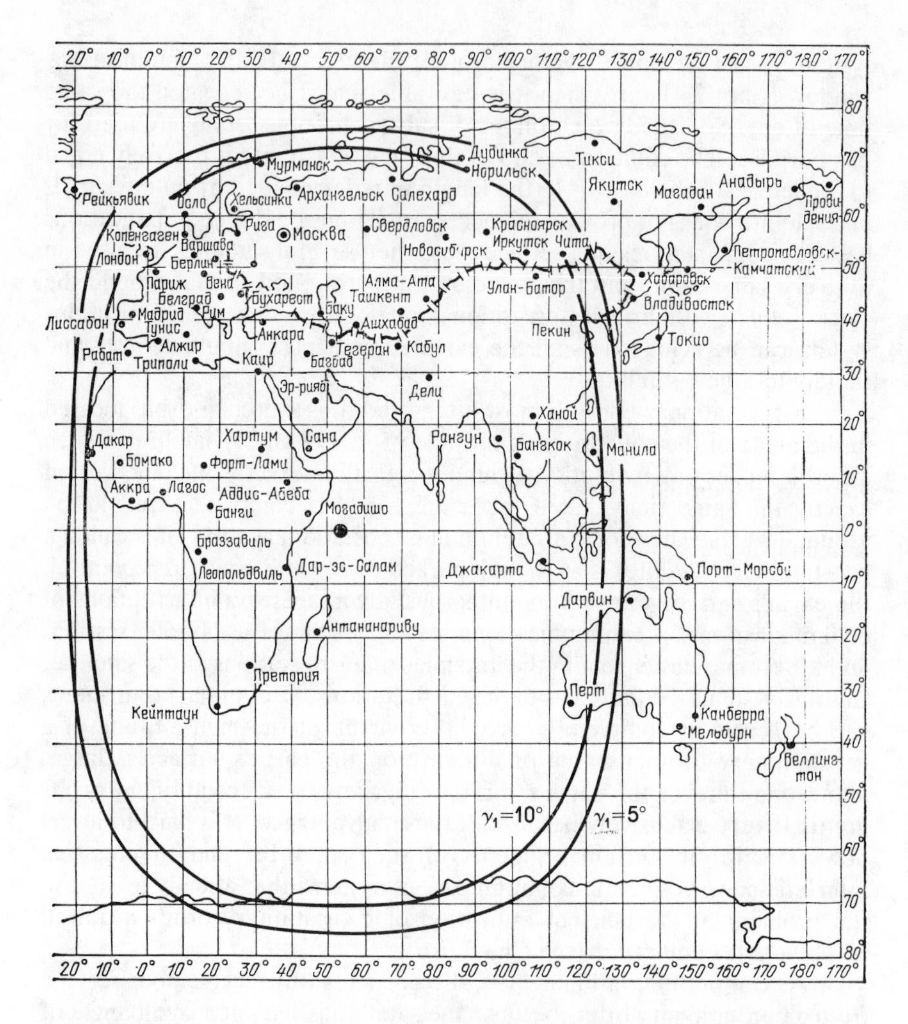

FIGURE 1.6. Area of visibility of a geostationary satellite (γ is the angle of elevation of the antenna of an earth station, aimed at the satellite).

ances also cause the plane of inclination of an orbit to change by approximately 0.92° a year. This means that a satellite, inserted in an orbit that is precisely in the equatorial plane, after just one year will describe every day in the heavens a shape (a figure eight or an ellipse projected upon the earth) with a north to south amplitude of 0.92°. For these reasons a more general definition of a geostationary orbit is used in official documents [1]: a geostationary orbit is a part of space, in which geostationary satellites are deployed.

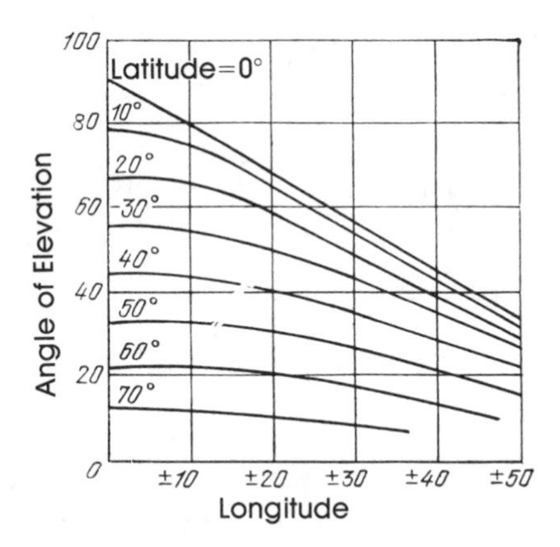

FIGURE 1.7. Angles of elevation γ for steering earth antennas at geostationary satellite as functions of the longitude and latitude of the earth antenna site.

Orbital disturbances are corrected by periodically activating corrective rockets on a satellite. It takes considerably more energy to correct the plane of inclination than to correct the longitude of the orbit position. But if these disturbances are not corrected, then it becomes necessary to track a satellite with the antennas of the earth stations or to use earth-station antennas with a wide radiation pattern (and consequently with low gain); failure to correct perturbations of the position of a satellite or to do it imprecisely obviously detracts from the use of the satellite (see Chapter 6).

Because of the heavy use of the satellite geostationary orbits, the use of subsynchronous circular and elliptical orbits with periods of twelve, eight, six, and three hours is being examined. In this case the satellite is inserted at a lower altitude and can service a smaller territory in periods that are repeated several times a day at the very same time (see Table 1.1) [5, 1.4].

A high-altitude elliptical orbit with a rotation period of 12 hours is used extensively today, in addition to a geostationary orbit. Molniya satellites, used by the USSR communications and broadcasting system, are being placed in this orbit. The inclination of the orbital plane of a Molniya satellite is 63.4°, the perigee is 500 km, and the apogee is 40,000 km. According to Kepler's second law, the motion of a satellite at high altitude (in the vicinity of the apogee) is slow, and a satellite passes very quickly through the vicinity of the perigee, located over the Southern Hemisphere of the earth.

TABLE 1.1

Rotation period, hours	*Number of turns per day*	*Altitude of circular orbit, km*	**Altitude of elliptical orbit, km**	
			perigee	*apogee*
4	6	6750	500	13000
6	4	10750	500	21000
8	3	14250	500	28000
12	2	20375	500	40250
24	1	35875	500	71250

The area of visibility of a satellite in a Molniya orbit, because of its considerable altitude for much of the time, is about the same as for a geostationary satellite. The area of visibility is located in the Northern Hemisphere and therefore is suitable for northern countries (Figure 1.8). The entire USSR can be serviced by one satellite for at least eight hours, and therefore three satellites are enough for around-the-clock operation (when the longitude of the apogee of one of the orbits is over the middle of the USSR), even if the second, the apogee of which is over the Western Hemisphere, is not used. Actually, these orbits can be used some of the time for servicing Soviet territory. As can be seen in Figure 1.8, a satellite is visible for six hours (from $t = 3$ hours after the perigee to $t = 9$ hours) to all points north of 53° north latitude in the worst regions, shifted 180° relative to the longitude of the apogee of the satellite. The orbits of a satellite whose apogee was in the Western Hemisphere was used for a long time by the Intersputnik system, because both Cuba and the socialist countries of Europe were in its area of visibility at the same time. Now, however, the Intersputnik system, to eliminate interruptions of communications and to simplify the user earth stations, and for other operational advantages, is using two geostationary satellites, the Statsionar-4 and the Statsionar-5, located at 14° west longitude and 53° east longitude.

1.4.2 Areas of Visibility, Coverage, and Service

The concept of the area of visibility of a satellite should be construed as the part of the earth's surface from which a satellite is visible at an angle of elevation greater than some minimum tolerable angle (for example 5°) for a given communications session; instantaneous area of visibility (see, for example, Figure 1.8) is defined as the area of visibility at a certain time,

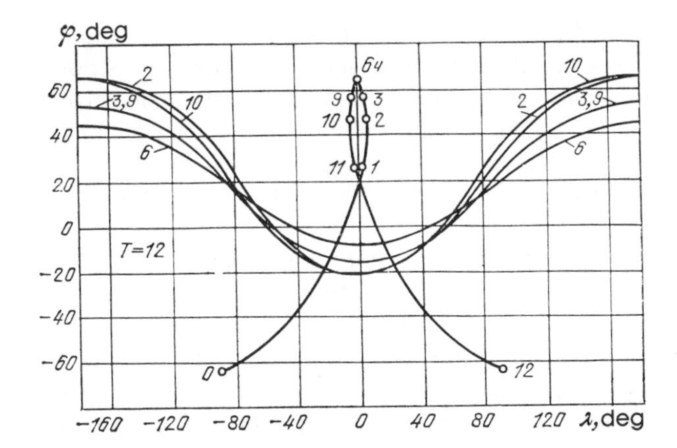

FIGURE 1.8. The trajectory of the subsatellite point and the southern boundary of instantaneous areas of radio visibility for satellites of the Molniya type.

i.e., for a communications session of zero duration. As a satellite travels, the instantaneous area shifts, and therefore the area of visibility for a certain period of time is always smaller than the instantaneous area, because it represents the inside envelope of the instantaneous areas.

The most important characteristic of a satellite communications system is the coverage area, the part of the area of visibility in which necessary energy relations are guaranteed in a communications link at an earth station with specific energy parameters (also see Section 1.2). This characteristic is important in the design of satellite communications systems and in the analysis of the interaction between them, and therefore is examined in greater detail.

If the radiation patterns of the on-board receiving and transmitting antennas of a satellite are wide enough to cover all of the earth that is visible from a satellite with uniform gain, then the coverage area coincides with the area of visibility and is called the global coverge area. All of the first communications satellites had these areas, and now many satellites, such as Intelsat, Gorizont, and Molniya, are providing these areas to maximize the coverage of the earth's surface. However, smaller coverage areas, as close as possible to the boundaries of the territory being serviced, be it a region, a country or even a part of it, are being created more and more often for the purpose of improving the energy characteristics of communications links.

Often (for instance, [5]) a coverage area is defined as the territory, where at every point the angle of elevation with the antenna of an earth station is aimed at the satellite and is not less than the minimum tolerable, and the power flux density of the satellite transmitter is not less than the specification. This definition is no different from the one given above if the energy relations in the earth-satellite communications link are ignored, i.e., if the possible influence of the radiation pattern of the receiving antenna of the satellite is ignored. In practice, the earth-satellite link often is not crucial, and in any case the necessary input signal level of a receiving satellite can be achieved by appropriately increasing the transmitter power of an earth station. The latter definition of service area is perfectly valid in this situation. However, in a satellite communications system with an extensive network of earth stations, the transmitters of the earth stations are one of the most expensive parts of a system, and the characteristics of the up link must be taken into consideration. Therefore the definition above is more general. Thus, the coverage area is drawn on a map in four steps.

1. *Determination of the visibility area.* For this we solve the purely geometric problem of determining the angle of elevation for an earth station at some point on the earth's surface—the angle between a line from this point to the satellite and the earth's surface.

It is helpful to draw on a map lines of equal angles of elevation, corresponding to small angles of elevation, for example 10°, 5°, *et cetera.* Lines of equal elevation angles for a satellite in a geostationary orbit are shown in Figure 1.9 [5]. They represent geographic latitude ϕ of a point on the earth's surface as a function of $\Delta\lambda = |\lambda - \lambda_0|$, the difference between the longitude of that point and the longitude of the stationary satellite for some fixed elevation angle γ.

2. *Determination of the area in which a satellite creates the necessary power flux density.* For this, a calculation (see Chapter 4) of the satellite-earth link to the different points on the earth's surface is performed. This calculation can be explained in simple terms as follows. If the necessary flux density on the earth's surface on the axis of the radiation pattern of the transmitting antenna of the satellite is achieved with some margin Δ dB, then to draw the area, it is necessary, on the basis of the radiation pattern, to determine α, the angle of the deviation of the beam from the axis for which the decrement of the gain of the antenna is Δ dB, and to find the intersection of the cone with vertex at the point where the satellite is positioned and vertex angle 2α, and the earth's surface (Figure 1.10).

When the point to which the axis of the radiation pattern is aimed at the edge of the earth's disk that is visible from the satellite, the beam obviously becomes even more inclined toward the earth's surface and the size of the service area increases slightly (Figure 1.11) [5].

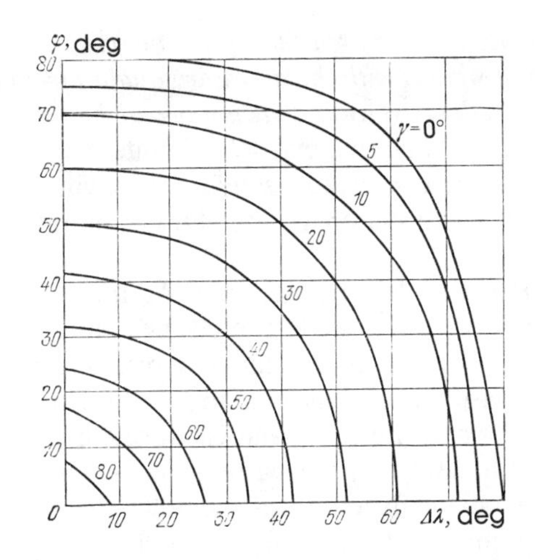

FIGURE 1.9. Lines of equal elevation angles for geostationary satellite.

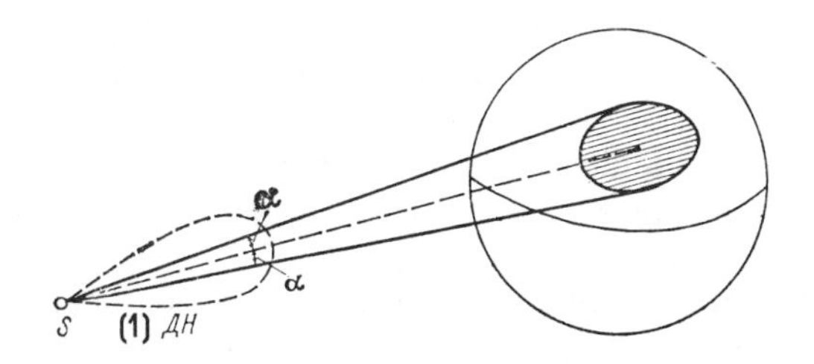

Key: (1) Radiation pattern

FIGURE 1.10. How to determine the coverage area.

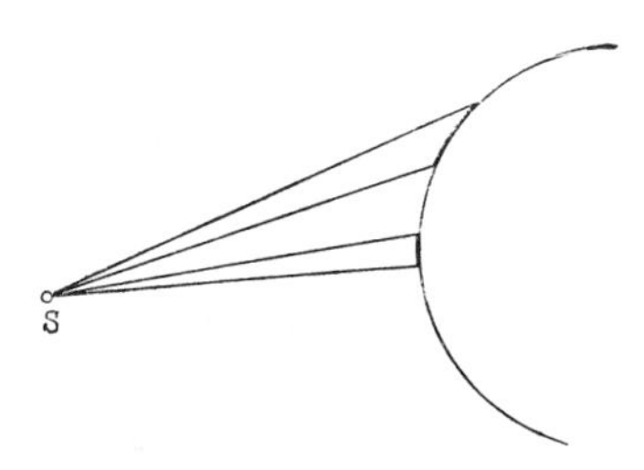

FIGURE 1.11. How to determine the effect of expansion of the service area.

3. *Determination of the area that meets the requirement of reception by a space station of signals with the necessary quality from an earth station in the area with a radiated power that is normal in a given system.* This problem is similar to the one that was solved in item 2, the only difference being that the parameters of the satellite receiving and transmitting characteristics of the earth station are used for the energy calculation of the earth-satellite link.

4. *Determination of the coverage area.* All three areas are drawn on the map and their inside envelope is plotted, i.e., the coverage area is the territory belonging to each of the three depicted areas, in which the conditions of radio visibility and the conditions of the necessary communications quality in the satellite-earth and earth-satellite links are satisfied. In some systems (in satellite broadcasting systems as a rule), different coverage areas are created for up-down and down-up links. For instance, a broadcasting program can be sent to a satellite from the capital of a country and transmitted from the satellite to another part of that country. Then it is necessary to introduce separate concepts of coverage area for down and up links. The coverage area for a down link is defined as the union of areas, determined in accordance with items 1 and 2 given above, and the coverage area for an up link is defined in accordance with items 1 and 3.

Now that the main idea of the construction of the coverage area has been explained, it should be stated that the simplified, purely geometric method, described in item 2, of drawing the area with the necessary flux density is inexact. The calculation by this method merely yields a claimed antenna characteristic, announced by the applicant country in the first step of the registration of a new satellite communications system that is being developed in the International Telecommunication Union (ITU). The claimed antenna characteristic can be defined [6, 3, 1.2] as a set of contours, drawn on a geographic map, corresponding to a transmitting antenna of a satellite repeater with a constant gain (more accurately with a constant decrement of gain of −2 to −4 dB, *et cetera,* relative to the peak of the radiation pattern).

In the design of a real system it is necessary to determine the coverage area in consideration of a multitude of additional factors. For instance, the distances between a satellite and different points on the earth's surface, called the slant range, are different, and consequently the attenuation of the radio signals will be somewhat different (see Chapter 4); the shortest distance is to the point under the satellite (i.e., the point lying on a straight line between the satellite and the center of the earth).

More important and difficult to consider is the fact that additional attenuation occurs in the earth's atmosphere, which is attributed basically to moisture (see Chapter 4), which depends on the angle of incidence of the beam (the angle of elevation of the earth station antennas) and increases appreciably at small angles. This, against geometric considerations presented earlier (see Figure 1.11), renders operation useless at small elevation angles, particularly on frequencies above 10 GHz, and it makes the service area smaller in comparison with the claimed characteristic, plotted purely geometrically.

The positional instability of a satellite in orbit and the instability of the orientation of its antennas exert a significant influence, even with the present-day high accuracy of the insertion and holding of the satellite in orbit. In this connection, in order to accurately calculate the coverage area, it is necessary to determine that part of the earth's surface where a given communications quality is guaranteed for even the worst combinations of parameters, characterizing the instability of a satellite. In other words, it is necessary to find the inside envelope of the coverage areas, determined for different combinations of variables, which determine the position of the satellite and the direction of its antennas. This calculation can be done only with the aid of an electronic computer [3, 5].

The concept of the area of a guaranteed signal level is introduced in [1.5] in order to emphasize the influence of the instability of the beam of the transmitting antenna of a satellite; this concept is close to the definition of the coverage area, takes into consideration both constraints imposed by radio visibility and the power flux density that is created on the earth's surface (in consideration of instability); the only difference is that it ignores the energy relations in the down-up leg (item 3 of the determination of the coverage area).

Moreover, because the performance characteristics of a communications channel usually are statistically given, i.e., a given characteristic must be satisfied for a certain fraction of time, the coverage area should be calculated probabilistically in consideration of the lifetime of various combinations of instability parameters of a satellite. Unfortunately, the necessary calculation procedures have not yet been developed and, furthermore, there are no data on the probability characteristics of the instability characteristics of a satellite, necessary for the calculation. Consequently, the guaranteed coverage area, calculated rigorously in accordance with the principles set forth above, occasionally turns out to be smaller than the coverage area that is actually guaranteed by a communications system.

Finally, we introduce the concept of service area—the part of the earth's surface on which earth stations of a given network are or can be located, i.e., of the area in which it is actually necessary to guarantee that the earth stations will operate normally [6, 1]. Not only all the conditions that determine the coverage area must be met in that territory, but the necessary safety margins must be observed relative to interference from other radio systems, including other satellite communications systems. All calculations of crosstalk interference, carried in coordination with a satellite communications system during the process of registration in the International Frequency Registration Board, must be done for any point in the service area. This coverage area obviously always embraces the service area and is larger. The radio communications regulation, in the interest of saving the frequency spectrum, recommends that the coverage area be as close as possible to the service area.

The representation of the earth's surface may be used for the approximate plotting of the coverage area, and even for the approximate solution of the inverse problem—the selection of the necessary characteristics of the radiation pattern of the space antennas of a geostationary satellite, because the earth is visible from a geostationary satellite [5]. To do this, it is necessary to use a spherical coordinate system with origin coinciding with point S where the satellite is located (Figure 1.12); P is the North Pole of the earth; O is the center of the earth. The position of some point N on the earth's surface in this coordinate system* is determined only by angles α and β (see Figure 1.12), where α is the angle between planes SOP and SAN (or between lines SO and SA, lying in the equatorial plane); β is the angle between a line from the satellite, drawn to point N, and the equatorial plane (i.e., the angle between lines SN and SA); NA is perpendicular to the equatorial plane from point N.

Angles α and β can be determined by the following formulas [5]:

$$\begin{aligned} \beta &= (\arctan R) \sin (\phi/l) \\ \alpha &= \arcsin [(R \cos \phi) \sin (\Delta\lambda/l)] \end{aligned} \tag{1.1}$$

where $l = \sqrt{r^2 + R^2 \cos \phi - (Rr \cos \phi) \cos \Delta\lambda}$; ϕ is the latitude of point N; $\Delta\lambda$ is the difference of longitude between point N and the satellite (the projection of the satellite onto the earth's surface); R is the radius of the earth; r is the radius of the geostationary orbit (SO); l is the length of line SA.

---

*The coordinate system introduced above is intended merely for determining the position of points on the surface of the globe.

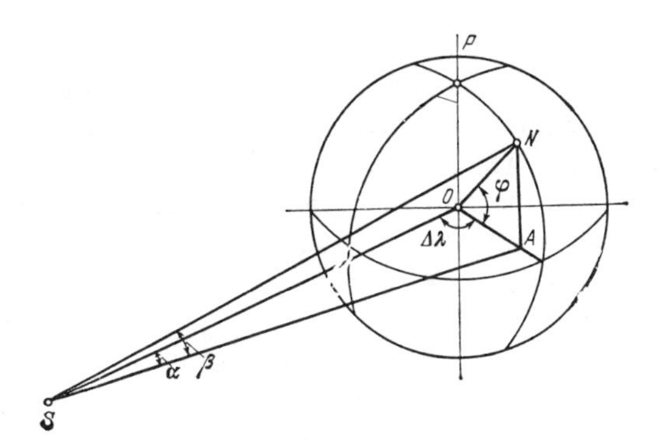

FIGURE 1.12. Spherical coordinate system, connected with a satellite.

According to (1.1) all meridians (*B* in Figure 1.13) [5] and parallels (b) can be drawn in coordinates α, β. Lines *A* (circles) of equal distances *d* and of equal angles of elevation γ can also be drawn on the same coordinate grid (lines g and d for angles β and α).

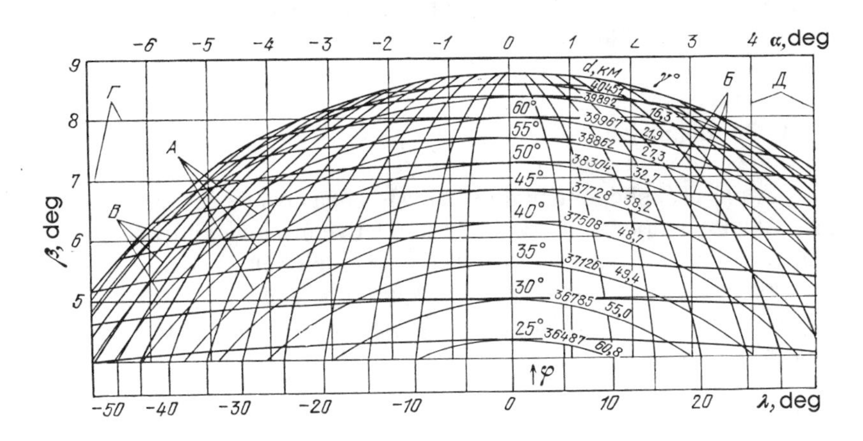

FIGURE 1.13. The globe in a spherical coordinate system, connected with a satellite.

In Figure 1.14 the territory of the USSR, shown in this coordinate system, and visible from a geostationary satellite, is inclined at 100° east longitude. On this drawing it is easy to find the approximate orientation of the antenna beam of a satellite and the character of this beam in terms of angles α and β, necessary for servicing a given territory. It is important to remember that the map that is drawn in Figure 1.14 is suitable only for determining coverage areas of satellites, created from the point 100° east longitude, and a different projection must be drawn for a satellite in any other position. Also when determining the necessary character of the radiation pattern of the antenna beam of a satellite, it is necessary to consider the instability of the orientation of the antenna.

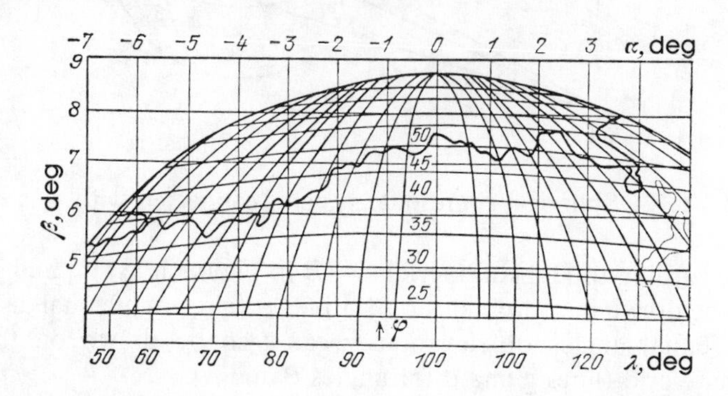

FIGURE 1.14. The territory of the USSR in a spherical coordinate system, connected by satellite.

1.4.3 The Doppler Effect

A physical phenomenon that consists in a change of the frequency of received oscillations, as a result of the relative motion of a transmitter and receiver of these oscillations, is called the Doppler effect. It also occurs as a satellite moves in orbit. If the transmitter is stationary relative to the receiver, the wavelength in meter systems, connected with the receiver or transmitter, is

$$\lambda_0 = c/f_0 \tag{1.2}$$

where c is the speed of light and f_0 is the frequency of the oscillations. But if the transmitter moves in relation to the receiver at velocity v at angle ψ to the communications link (Figure 1.15), then in the meter system, connected with the receiver (the earth station in Figure 1.15), the wavelength changes by an amount equal to the change of the distance in time $T = 1/f_0$ of one period of the radiated oscillation:

$$\Delta\lambda = -(1/f_0)\, v \cos \Psi \tag{1.3}$$

Key: **(1) Satellite**

FIGURE 1.15. Determination of the Doppler effect.

The wavelength* of the oscillation, the frequency, and relative change of the frequency at the receiver are, respectively,

$$\lambda = \frac{c\,(1 - (\nu/c)\cos\Psi)}{f_0} \tag{1.4}$$

$$f = \frac{c}{\lambda} = \frac{f_0}{1 - (\nu/c)\cos\Psi} \tag{1.5}$$

$$\frac{\Delta f}{f_0} = \frac{\nu}{c}\cos\Psi \tag{1.6}$$

---

*The effect of the retardation of time (the effect of relativity theory), which is insignificant when $\nu << c$, is ignored.

The Doppler effect is strongest when the transmitter moves in relation to the receiver along the communications line ($\Psi = 0$ or $\Psi = \pi$):

$$\Delta f = f_0 \, v/c \tag{1.7}$$

i.e., as the transmitter and receiver get closer together the frequency of the radio oscillations increases in proportion to v/c, and as they get farther apart it decreases in the same way.

In a communications link that passes through a strictly geostationary satellite, the Doppler shift does not occur; in actual geostationary satellites it is insignificant, but in elongate elliptical or low-altitude circular orbits the Dopper shift can be significant. It is calculated by calculating $v \cos(\Psi/c)$ for some satellite trajectory, and in this case it is necessary to consider both legs of the communications link (earth-satellite, satellite-earth) and the different positions of the earth station. The results of this analysis for a high-altitude elliptical Molniya orbit are given in Figure 1.16 as two curves—for the maximum (Max) and for the minimum (Min) shifts [17]; also shown there is the corresponding altitude (H) of the satellite. It is important to consider that a satellite in a Molniya orbit usually is turned on for operation only at altitudes above 15,000–20,000 km.

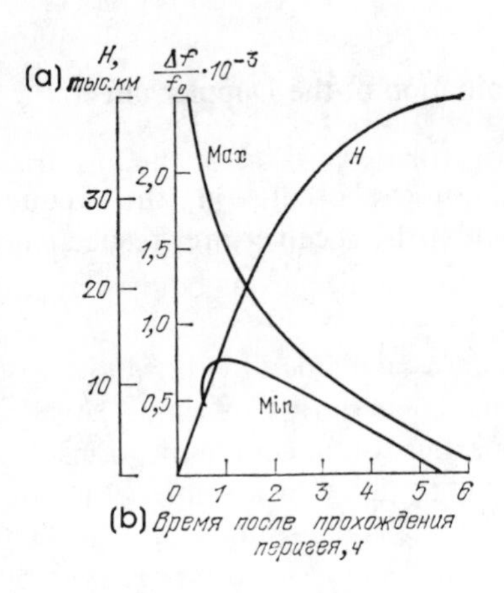

Key: (a) Thousands of km
(b) Time after passing perigee, hours

FIGURE 1.16. The relative Doppler frequency shift and altitude of a satellite as functions of its position in a Molniya orbit.

For circular orbits, the maximum Dopper frequency shift (for one section of a link) can be approximated by the relation:

$$\Delta f/f_0 \approx \pm 1.5 \cdot 10^{-6} s \tag{1.8}$$

where s is the number of revolutions of the satellite around the earth per day ($s > 1$). The greatest summary Doppler shift occurs in a communications link between earth stations that are close together when the shift in both legs (earth-satellite and satellite-earth) are approximately the same, and therefore in the entire link it is doubled.

What is the influence of the Doppler shift on the operation of a communications link? First, the Doppler shift is manifested as frequency instability of the carrier of the oscillations that are retransmitted by a satellite, added to the equipment frequency instability, which occurs in the equipment of a satellite repeater and earth station. This instability can greatly complicate the reception of signals, particularly of narrow-band signals and reduces the noise immunity of reception. Second, the frequency of the modulating oscillations changes a little. Actually, if the carrier frequency f_0 is shifted by Δf [see (1.7)], then the frequency of the upper side component $(f_0 + F)$, attributed to the component F of the modulating process, will be $(f_0 + F)(1 + v/c) = f_0 + f_0 v/c + F + Fv/c$, and of the lower side component, $f_0 + f_0 v/c - F - Fv/c$. Thus, the difference of the side and carrier frequencies, equal to the frequency of the oscillation that is generated after demodulation, is $|F(1 + v/c)|$.

This compression (or expansion) of the spectra of the transmission process is virtually impossible to correct with equipment, so that if the frequency shift exceeds the tolerable limits (for example 2 Hz for certain types of frequency division multiplexing equipment), the circuit turns out to be useless.

The Doppler effect is explained by the velocity of a satellite, i.e., by the derivative of distance in terms of time. However, as was already mentioned, the properties of a communications channel are strongly influenced by the absolute delay of a signal as it propagates through an earth-satellite-earth link. This delay does not cause any distortions of the transmitted message, although for a geostationary satellite or for a satellite at the top of a Molniya orbit, it reaches the appreciable value of 300 ms. In the case of the transmission of one-way messages (TV programs, audio broadcasting, wire photos, and telegraph messages), this delay is not perceived by the consumer, but in the case of two-way communications, a 600-ms delay of response is noticeable. Two-wire subscriber lines and four-wire lines between switching centers are used in two-way telephone communications lines; a certain amount of mismatch always occurs at points of transition from a four-wire circuit to a two-wire circuit, and consequently echoes are formed, which propagate in the opposite direction, reaching the listener's ear after an interval of time equal to twice the time it takes the signal to propagate through the communications link. When this delay is short, echoes are perceived as a kind of backtalk (reverberation), are concealed by the subscriber's voice itself, and have little effect on a conversation. But if the delay is great, they are heard separately as a distinct echo and they create serious interference of the conversation. Therefore special devices, called echo suppressors, which block the return channel to a talking subscriber, must be used on each channel of a satellite communications link (just as in particularly long terrestrial links). However, conversation in a double-hop link becomes difficult, even when echo suppressors are used. Consequently a limit is placed on the use of compound communications links that contain two earth-satellite-earth spans. This limits the use of satellite links in a communications network, particularly in an automated one. Double-hop links may be used in exceptional cases for the sake of saving antennas at an earth station in order to enter into a network of a different satellite through a different station that already has two antennas, or for the sake of preserving the network center structure of a network (see Section 1.6, Figure 1.20).

The delay of signals in a satellite communications link and its change also are of importance in the case of the transmission of certain kinds of information, for example in the transmission of precision time signals or wire photos. In the latter case, because the receiver scanning frequency is maintained constant during an entire communications session (scanning is synchronized at the beginning of a session), and the propagation time changes due to the motion of a satellite, the beginning of each line shifts in time and the received copy acquires a distortion, which is significant when wire photos are transmitted through a satellite in an elliptical orbit. This subject is discussed in greater detail in Section 19.2.

1.5 SOME CONSIDERATIONS ON THE ECONOMIC CHARACTERISTICS OF SATELLITE COMMUNICATIONS AND BROADCASTING SYSTEMS

Not enough economic information is given in this handbook for economic calculations in the design of communications systems or of individual earth stations. This section will merely familiarize the reader with fundamental premises, characterizing the cost effectiveness of satellite communications and broadcasting systems, so that these principles can be taken into consideration during the selection of technical parameters and design principles of systems.

As is generally known, the cost of developing satellite systems is exceedingly high. For instance, according to generalized data from the CCIR [1.5], the cost of building modern multitransponder satellites in 1981 was 15–30 million dollars and more, and the cost of inserting them in a geostationary orbit was 30–60 million dollars; the cost of multifunctional earth stations, depending on the size of the reflector antenna and the number of circuits, ranged from several millions of dollars to several tens of millions of dollars. Nevertheless, thanks to the long service-life of satellites (up to 7 years), the large number of radio frequency transponders on a satellite (up to 8–24), the high carrying capacity of each transponder (up to several hundreds and even up to a thousand and more channels), and the low cost of controlling a satellite in orbit, receiving power from solar batteries, the cost of each telephone channel through a satellite is low. For instance, it costs less than one million dollars a year to lease an Intelsat satellite transponder, which is done from time to time for developing national communications systems, and the rate for an audio frequency channel is only 5.5 cents per minute*, and for a secure circuit it is 12 dollars a day. The rates on Intersputnik channels are about the same (40 francs a minute for a video channel, and 15,000 francs a year for an audio frequency channel). Communications through a satellite is cost effective in many cases at these rates.

A satellite network or a link (the construction and operating costs of which are lower than the costs of a terrestrial network or link with the equivalent carrying capacity and the same number and deployment of connected stations) should be considered cost effective. Costs are best characterized by reduced annual costs R, consisting of capital (startup) costs C (including the cost of developing a satellite, booster, and launch facilities, the cost of building a satellite and of inserting it in orbit, and the cost of building earth stations), and annual overhead O (including the cost of repairing defective satellites, of controlling satellites and of operating earth stations):

$$R = O + (1/6.5)\,C \tag{1.9}$$

---

*In an Intelsat satellite transponder with the SPADE system and a separate TV channel by request (data for 1981–1982).

where the coefficient 1/6.5 corresponds to the standard break-even time of 6.5 years, used in the USSR.

The most important feature of satellite communications systems in the economic sense is the fact that the cost of a satellite link is independent of its length, i.e., of the distance on the earth's surface between connected stations, because this distance has virtually no effect on the length of an earth station-satellite-earth station length, and consequently on the parameters of the equipment of earth and space stations. In contrast, the cost of terrestrial links increases in direct proportion to the length of a line. Hence it follows that for some long enough distance l between connected stations, the cost of a satellite link and the cost of a terrestrial link will be the same. This distance $l = l_{eff}$ is called the cost effective distance [1.6]; it is economical, as a rule, to use satellite communications on a communications link with length $l \geqslant l_{eff}$. There are additional considerations in favor of building a satellite network: the speed of installation, and the local nature of construction of earth stations. The latter is particularly important when a communications line must be laid under difficult geographic conditions (across oceans and mountains, inaccessible or sparsely populated regions). It is also important to consider the other side of the coin; after a satellite link is installed, not all intermediate stations get communications, and it may be necessary in time to build additional terrestrial links or earth stations to provide them with communications. The possibility of building communications links between countries that do not have a common border, without leasing communications lines in third party countries, is an economic advantage of satellite communications systems.

The cost effectiveness of satellite communications systems in a national public communications network increases by virtue of their capability of speedy transfer of channels or of groups of channels to other lines, which makes it possible to establish through satellite systems a flexible reserve for different parts of a network without significant additional costs. Therefore, the cost effectiveness of a satellite communications system can be evaluated most completely by comparing the costs of building a complex communications network with a given reliability and carrying capacity with and without satellite communications. However, this assessment poses a serious scientific problem; in most cases, estimates based on the cost-effective distance l_{eff} are sufficient. For satellite communications systems l_{eff} = 1000–5000 km [1.6], and for satellite TV broadcasting systems l_{eff} = 80–400 km; the distance l_{eff} depends on the number of earth stations in a network, on the existence of a previously constructed terrestrial communications network, the necessary carrying capacity of a communications system, *et cetera.* As technology continues to develop rapidly, l_{eff} tends to decrease: for example, for the new Ehkran and Moskva

TV distribution systems, l_{eff} does not exceed 50 km, i.e., it is cheaper to build earth stations for these systems than to use a radio relay transponder for TV transmission, even for the distance of one radio relay span.

It is clear from the figures given above that satellite TV broadcasting systems and generally distribution systems for transmitting one-way information (TV programs, audio broadcasting, wire photos, *et cetera*) are the most cost effective. The reason is that one common satellite channel, the cost of which is shared by an entire network, is used for an entire network of earth stations; therefore, a satellite can have greater capacity, which makes it possible to build simpler and cheaper receiving earth stations of a distribution network. This principle was formulated and used for the first time in the USSR [1.3].

The question of the choice of the optimum-energy parameters, from the standpoint of economy, of a satellite-earth station leg, and specifically of power P_{sat} of the space transmitter of a satellite and diameter D_a of the antenna of an earth station, is the crucial and most important problem of the economic optimization of the parameters of any satellite communication system. Actually, the cost of an earth station depends to a great extent on the cost K_a of building an antenna, which (the cost) increases rapidly with D_a. For instance, given in [1.6], on the basis of a generalization of numerous data, is the formula:

$$K_a = a_1 + a_2 D^2_{\mathrm{a}} \tag{1.10}$$

which states that K_a increases to the square as D_a increases (a_1 and a_2 are arbitrary constants). By increasing the diameter of the antenna of an earth station, it is possible to reduce the power of the transmitter of a satellite, and consequently to reduce the cost of the satellite which depends on the power of the transmitter of the satellite as follows [1.6]:

$$C_{sat} = a_3 P_{sat}^{a_4 + a_5 t} n_{tr}^{a_6 + a_7 t} \tag{1.11}$$

where n_{tr} is the number of transponders on a satellite, t is the service life of the satellite, and a_3, a_4, a_5, a_6, a_7 are constants.

The optimum values of P_{sat} and D_a, corresponding to the minimum costs for a communications system, can be found by solving equations (1.10) and (1.11) jointly with the equations that determine the necessary energy balance of a satellite-earth link (see Chapter 4). The result, of course, will depend on the number of earth stations N_{st}; the more of them there are in a system, the greater the contribution of the cost of earth stations to the total cost of a system, and the better it is to reduce D_a at the cost of increasing P_{sat} and the cost of the satellite. It was specifically as a result of such an analysis that a comparatively small earth station antenna with D_a = 12 m [4, 1.6] was used for the first time in history in the USSR for the Orbita TV program transmission and communications system, whereas in the Intelsat international system only antennas with D_a = 25–32 m were being used at that time. Later on, D_a began a tendency to decrease in the national systems of some countries (D_a = 12, 7, 5, 3 m) and in the Intersputnik (D_a = 12 m) and Intelsat (D_a = 32, 11, 14 m) international systems.

The material presented above was intentionally simplified in order to explain the basic idea of optimization more clearly. In practice, the procedure by which the optimum values of D_a and P_{sat} are selected is complicated by a number of factors and interactions. For instance, the sensitivity of an earth station is described completely not only by the diameter of the antenna, but also by the noise temperature T_n of the receiver, for which, as is known, a special characteristic is used—the Q-factor of a station, equal to G/T_n (where G is the gain of the antenna of an earth station, and $T_{n\Sigma}$ is the summary noise temperature of the receiver circuit of an earth station). Thus, the necessary Q-factor $G/T_{n\Sigma}$ for an antenna with a smaller diameter D_a can be achieved by decreasing $T_{n\Sigma}$. It is obviously possible to find a combination of a type of quiet input system of an earth station and D_a that is optimum in the economic sense, and this optimum is connected with the optimum power P_{sat} of the space transmitter.

The size of the antenna of an earth station also is connected with the necessary power of the transmitters of earth stations, which guarantee the necessary energy potential in an earth-satellite link. The power of the transmitters of earth stations can be cut back by increasing D_a, but the cost of transmitters at multitransponder stations with a high carrying capacity plays an appreciable role.

As was mentioned in Section 1.2, the effective isotropic radiated power is the product of power P_{sat} and gain of the transmitting antenna G_{sat} of a satellite:

$$\text{EIRP} = P_{sat}G_{sat} \tag{1.12}$$

Obviously, as the service area becomes smaller G_{sat} can increase, and as a result, P_{sat} can decrease. An increase in G_{sat} (and the need to increase the size of the antenna and the steering precision of the antenna of a satellite, related to this) increases the cost of a satellite. Nevertheless, for the sake of reducing P_{sat} and for a whole host of other reasons (see Chapter

6), it may be assumed desirable in all cases to try to maximize G_{sat}, even to the extent of partitioning the entire service area into several separate areas, and a separate satellite signal is generated for each.

When selecting the optimum parameters of a satellite communications system, it is necessary to consider the existing constraints on the power flux density created by a satellite at the earth's surface (see Chapter 3), constraints connected with the satellite antenna and earth station antenna steering precision and others. It is important to remember that the optimum parameters are different for satellite communications systems, operating in different frequency bands.

In the actual design of satellite communications systems, it is necessary, of course, to take into consideration different kinds of technological limits (the maximum weight of a satellite inserted in orbit, the power of the electronic output instrument, *et cetera*), as well as parts and components of a system that are available, the standards that are used in a country or in a system, *et cetera.*

The economic characteristics of a satellite communications system are dependent on its carrying capacity. The reduced cost per telephone channel R_1 in a satellite communications system tends to decrease rapidly as the carrying capacity n_{car} (number of channels) increases, so long as the bandwidth of the satellite is not saturated (see [1.8] and Figure 1.17). However, a comparison of these costs with the cost per channel of terrestrial links shows that in terrestrial links the costs also decrease rapidly as the carrying capacity increases (the dashed lines in Figure 1.17). For terrestrial links, it is necessary to consider that in each communications route between two stations the number of channels n_{com} is less than the total number of channels in a system, equal according to the requirement of equivalence to n_{car}. For stations of a ground network with the identical capacity:

$$n_{com}/n_{car} = 2/[N_{st}(N_{st} - 1)] \tag{1.13}$$

Moreoever, terrestrial links, passing on close or coinciding routes, usually are combined (grouped) into one larger transponder. In this connection, it is necessary to consider the grouping coefficient $K_{gr} = n_{lin}/n_{com}$. For instance, if all the stations are situated in a circle and are connected by a circular terrestrial link (Figure 1.18), then

$$K_{gr} = \frac{1}{8}(N^2_{st} - 1)$$

$$n_{lin} = \frac{1}{4}[(N_{st} + 1)/N_{st}]n_{car} \approx \frac{1}{4}n_{car}$$

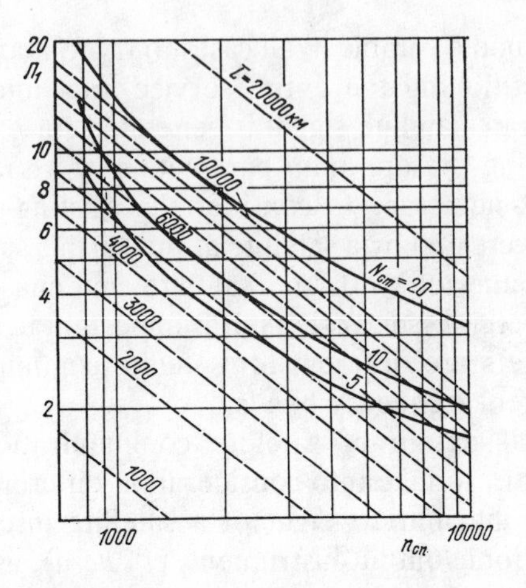

FIGURE 1.17. Reduced costs per channel as function of carrying capacity of satellite and terrestrial (dashed line) communications links.

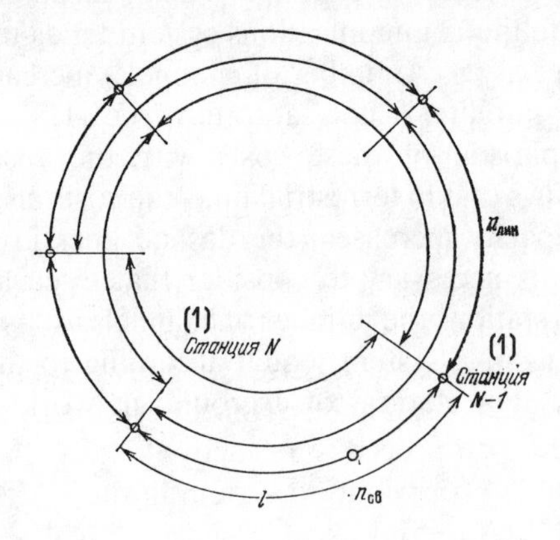

Key: **(1) Station**

FIGURE 1.18. Circular equivalent terrestrial network.

which also determines the conditions for the comparison of satellite and terrestrial communications links.* The fact that distance L, shown in

*This shift is already taken into account in Figure 1.17.

Figure 1.17, is the mean length of a communications link between any pair of stations is also worthy of attention; the mean distance between adjacent stations, even for the simplest system, when the stations are situated and connected in a circle, is

$$\bar{l} = 4\,\bar{L}(N_{st} - 1/N^2_{st} \quad \textbf{(1.14)}$$

Thus, the minimum economically equivalent distances ($\bar{L}_{eq} \approx 10{,}000$ km for $N_{st} = 20$; $\bar{L}_{eq} \approx 6{,}000$ km for $N_{st} = 10$), found from Figure 1.17, should be divided by $N^2_{st}[4(N_{st} - 1)]$, equal, respectively, to 5.25 and 2.78, and then the economically equivalent distance $\bar{l}$ between adjacent stations is only 1.9 and 2.16 thousand km, respectively. Of course, these figures give only an approximate estimate and vary as functions of the time and conditions of the development of a system and of the specific location of the earth stations. The equivalent distance becomes more favorable in a satellite communications system with a service area that is narower than the entire territory of the USSR.

The above discussion on the influence of carrying capacity on the cost effectiveness of a satellite communications systems applies to two-way (principally telephone) communications systems. As was already pointed out, circular (distribution) systems are vastly more cost effective. Therefore, other functions also are usually combined in a system, which increases the efficiency of satellite communications [1.6]. When there are very many earth stations in a distribution network, when it is advantageous to develop for this network a satellite transponder with a high output power, the parameters of distribution and two-way systems become sharply different. In this case it is better, in consideration of reliability and economic optimality, to build these systems with separate satellites [1.9]. It was considered better to proceed in exactly this way during the development of the Ehkran TV distribution system in the USSR.

1.6 FEATURES OF THE USE OF SATELLITE TELEPHONE CHANNELS IN COMMUNICATIONS NETWORKS

1.6.1 The Position of Satellite Channels in Communications Networks

All the communications networks in the USSR, including the telephone network, are parts of the Unified Automated National Communications Network (UANCN) [1.10]. The development of the UANCN is dictated by the modern trend of the development of communications systems, the merging of different networks into a single network, built on the basis of even more powerful transmission and switching systems. This trend is determined by efforts to increase the cost effectiveness and performance reliability of networks.

The construction of a unified network that satisfies the needs of all possible consumers necessitated the development and use of a number of unified requirements, on the basis of which the UANCN is developed. One of the most important requirements is the adoption of some typical nomenclature of channels and circuits, which could be capable of transmitting all kinds of information. An audio frequency channel with a 300–3400 Hz band and groups of channels: primary (12 audio frequency channels), secondary (60 audio frequency channels, i.e., five primary groups), ternary (300 audio frequency channels), *et cetera,* are used as these typical channels. The nomenclature of digital channels was also introduced, on which is based a digital 64 kb/s channel, equivalent to an audio frequency channel. A subprimary 0.512 Mb/s group (8 × 64), a 2.048 Mb/s primary group (32 × 64), an 8.448 Mb/s secondary group, *et cetera,* were also introduced. Audio broadcast and TV program sound accompaniment transmission channels, TV channels and other wide-band channels are also being created in the UANCN.

A primary network is the foundation for the construction of communications networks. A primary network is a complex of typical channels, group circuits and centers, on the basis of which secondary networks are organized. A primary network combines a national primary network, area primary networks, and local primary networks, as is shown in Figure 1.19.

A national primary network connects centers and exchanges, being the centers of areas, which coincide, as a rule, with major administrative regions (oblasts, krays, and autonomous republics). An area primary network includes channels and circuits that connect local networks with the area center and with each other. Local primary networks are the networks of individual cities and rural regions.

A secondary network is intended for the transmission of certain kinds of information or for servicing some group of customers. In the UANCN there are the following main secondary networks: automated telephone communications, telegraph communications, data transmission, TV program distribution, audio broadcasting, facsimile, *et cetera.* A secondary network is a combination of terminals, switchboards, and channels, acquired from the primary network.

An analysis shows that satellite channels, the acquisition of which requires comparatively sophisticated and costly earth stations and communications satellites, are cost effective and efficient primarily in national communications links, between remote centers and exchanges, and also in area networks in remote and inaccessible regions. The physical properties

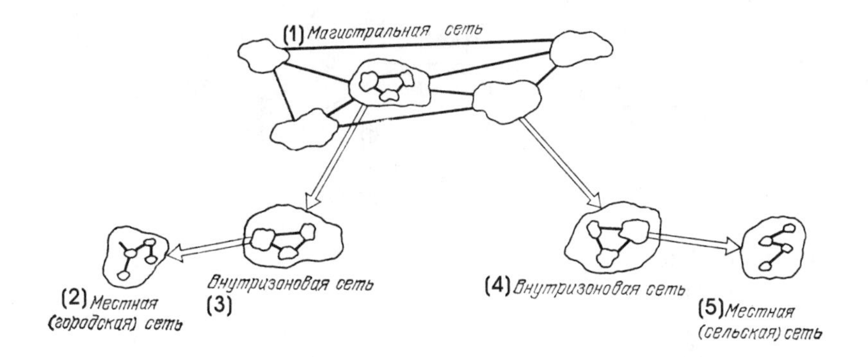

Key: **(1) National network**
(2) Local (city) network
(3) Area network
(4) Area network
(5) Local (rural) network

FIGURE 1.19. Primary network.

of satellite channels determine some of the features of their use in secondary networks, such as the automatic telephone communications, automatic telegraph communications, data transmission, and other networks.

It is important to mention that the quality performance of satellite channels are exactly the same as those of terrestrial links. The features of satellite channels that determine the specific features of their use in communications networks are the following: the long time it takes signals to propagate between two earth stations in an earth-satellite-earth link, reaching 300 ms, and the capability of satellite communications systems to operate in the multiple access mode and with unsecured channels or group circuits. The former effect becomes significant when two (or more) satellite sections (two hops) are connected in the telephone channel (Figure 1.20). By using unsecured channels and group circuits it is possible to maneuver considerably more easily in networks, their versatility increases, and the channels are better loaded, more cost effective, and have greater performance reliability.

1.6.2 Ways of Suppressing Echo Signals

The echo effect appears in communications channels because differentiators are used in them. The fact is that subscriber communications

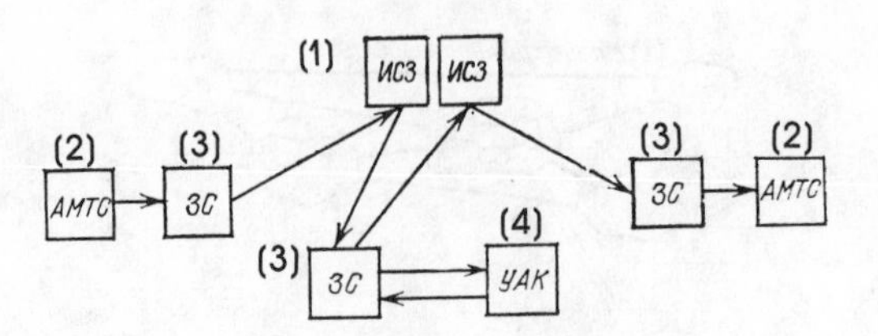

Key: **(1) Satellite**
(2) Automatic long-distance telephone exchange
(3) Earth station
(4) Automatic switching center

FIGURE 1.20. A double-hop satellite channel.

networks in cities and rural regions are built as two-wire networks for economy, whereas long-distance channels are four-wire channels. Special crossovers, called differentiators, are installed at points where four-wire channels are changed to two-wire channels. Signals are reflected in long-distance channels because the systems are imperfectly balanced. An echo of a talking subscriber's own voice, reflected on the receiving end of the channel, returns to him. In cases when the signal propagation time on a channel is short, not longer than 20–30 ms, an echo virtually coincides with a talking subscriber's voice and is masked by it. But in cases when an echo arrives after a long delay, the subscriber hears it as interference. The longer the delay of the signals, the more noticeable an echo becomes and the more interference it creates. This effect is manifested in all very long long-distance channels, but it is particularly strong in satellite channels, in which the delay is longer than in terrestrial channels.

Special networks, called echo suppressors, which must be installed in all satellite channels, are used for combatting the echo effect. The design principle and the connection of an echo suppressor in a channel are illustrated in Figure 1.21. As a rule, echo suppressors are installed at automatic long-distance telephone exchanges. An echo suppressor half-set, consisting of a control circuit and a key, is installed on each end of a channel. When subscriber A talks, the control circuit on the receiving end of the channel, at subscriber B, perceives the speech signals and locks the key in the channel, which B transmits to A. This closes the path to the echo signal, formed in differentiator B. When subscriber A stops talking, the control circuit on end B opens a key, and now subscriber B can answer.

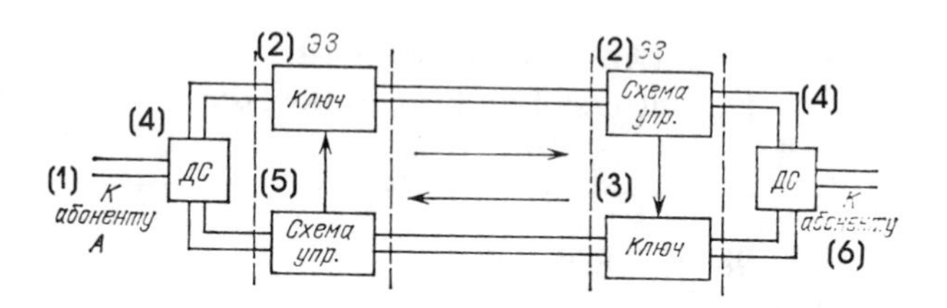

Key: (1) To subscriber A
(2) Echo suppresor
(3) Key
(4) Differentiator
(5) Control circuit
(6) To subscriber B

FIGURE 1.21. Connection diagram of an echo suppressor in a communications channel.

TABLE 1.2
Parameters of an Echo Suppressor

Delay time in a circuit with an echo suppressor, ms	50–400 one-way
Kind of information that can be transmitted in a circuit with an echo suppressor	Two-way conversations, line, control and acoustic signals of a signal system, data at a certain rate
Mode	Idle, locked, failure, neutral
Insertion attenuation in circuit in idle and neutral modes, dB	0–0.5
Input and output impedances, ohms	600
Activation threshold, dB:	
on 300 Hz	−(30–22)
on 3000 Hz	−(24–16)
Time of transition from idle mode to locked mode when information signals enter circuit, ms	4
Attenuation in circuit in locked mode, dB	55
Time of transition from locked mode to idle mode after signal ceases, ms	45±8
Time of transition from locked mode to	

TABLE 1.2 (Continued)
Parameters of an Echo Suppressor

failure mode when signal appears in transmission circuit, 7 dB stronger than signal in receiving circuit, ms	10
Insertion attenuation in failure mode in receiving circuit, dB	6 ± 0.5
Echo suppressor control signal for switching to neutral mode when remultiplex signals must be transmitted	2100 Hz sine wave
Neutralization signal level, dB	−12 ± 6
Neutralization activation time, ms	400 ± 50
Time of return to original state from neutral mode when remultiplex signals or one-frequency sine wave signals cease, ms	200 ± 50

The existence of an echo suppressor in a circuit hampers conversion a little, but according to experience in the operation of satellite communications channels modern echo suppressors have good performance characteristics and do not interfere with communications. When two and more satellite links appear (see Figure 1.20) in a telephone channel between two subscribers, the signal propagation time increases further. In this case, the quality of telephone communications deteriorates, even when echo suppressors are used, because contact between the parties is hampered by the long time spent awaiting answers (up to 1.2 s and longer). Nevertheless, in some satellite communications systems (for instance in the Intelsat system) two hops are permitted in connections between stations, operating through different satellites, for the sake of economy.

In some systems, a long delay can cause automatic connection systems to break down, both in telephone and telegraph automated networks. The operation of automatic connection equipment is calculated for certain tolerable delays during the exchange of service signals between the transmitting and receiving sets of an automatic long-distance telephone exchange. The delays that can occur in two hops in existing automatic long-distance telephone exchange equipment, built before the appearance and extensive adoption of satellite communications, were not anticipated. This can result in an interruption of communications in some cases.

1.6.3 Methods of Eliminating Double Hops in Communications Networks

Methods based on the placement of certain constraints on the connection of satellite channels in a communications network are used for preventing double hops. Also sometimes employed is the automatic counting of the number of satellite links in the connection-making process. Ways of using these methods are examined because they can be used similarly in other networks.

There are two basic types of telephone networks: unswitched channel networks and automatically switched networks. In unswitched networks the channels and group circuits are organized and secured for customers for long periods of time. In this case, during the planning of a network, provisions can be made ahead of time to make sure that a satellite link will not be connected more than one time in all channels. Automatically switched networks, in which most of the information of communications is transmitted, are used most extensively in practice. A typical diagram of an automatically switched mainline network is shown in Figure 1.22. First-class automatic switching centers (UAK 1), connected by the "each to each" principle, and some second-class centers (UAK 2), which service individual isolated regions, are the heart of this kind of network. A network also contains an extensive network of long-distance telephone exchanges, which are terminal exchanges, through which customers use the long-distance network.

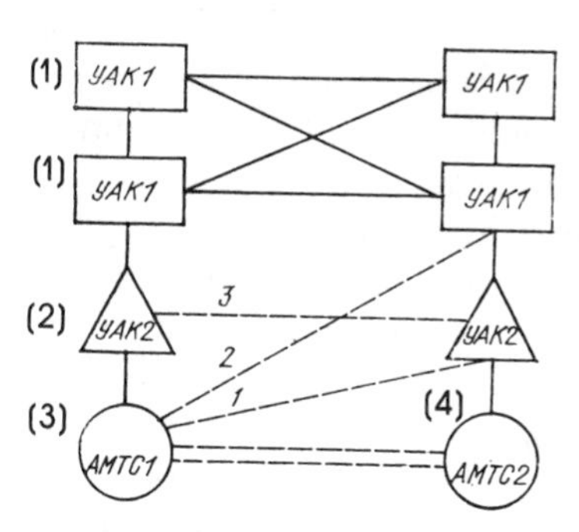

Key: (1) Automatic switching center 1
(2) Automatic switching center 2
(3) Automatic long-distance telephone exchange 1
(4) Automatic long-distance telephone exchange 2

FIGURE 1.22. Diagram of automatically switched telephone network.

An automatically switched network is constructed in such a way that automatic long-distance telephone exchanges with sufficiently heavy mutual traffic are connected by direct channel transponders, bypassing automatic switching centers. A relatively high probability of connection failures due to the unavailability of channels is permitted in these transponders, but this guarantees a heavy load on the channels and reduces cost effectiveness. Calls that cannot be serviced by direct transponder channels (surplus traffic), and also calls between automatic long-distance telephone exchanges without direct channels, pass on switched channels through switching centers.

Several possible channeling paths are examined for making a connection for a call arriving at an automatic long-distance telephone exchange —a direct path, a first alternate path, a second alternate, third alternate, *et cetera*. In Figure 1.22, alternate paths are shown by dashed lines, on which the number of alternate paths is indicated. Alternate paths are also planned but with great losses. If during the process of the successive scanning of alternate paths, an automatic long-distance telephone exchange does not find an open channel for servicing an incoming call, it is sent on a path of last resort, shown in Figure 1.22 by a continuous line. The path of last resort also offers high-quality service, and the failure rate in it does not exceed a given limit for a network. All automatic long-distance telephone exchanges that are serviced by an automatic switching center are connected to it only by a path of last resort, and they are connected to other automatic switching centers by alternate paths.

For eliminating double hops in automatically switched networks, satellite channels should be used only on lines where automatic long-distance telephone exchanges are connected to earth stations, through which they gain access to satellite channels by terrestrial links. As can be seen in Figure 1.22, this applies primarily to direct paths between automatic long-distance telephone exchanges, located in the immediate vicinity of an earth station. Satellite channels also can be used on alternate paths through automatic switching centers if earth stations are located near one of the automatic long-distance telephone exchanges and an automatic switching center that services a second automatic long-distance telephone exchange. Because the second automatic long-distance telephone exchange is connected to its own automatic switching center by the path of last resort, which is always set up on terrestrial links, a double hop does not occur. The first and second alternate paths, shown in Figure 1.22, are examples of these links.

It is not recommended that satellite channels be connected to a path of last resort; i.e., in automatic long-distance telephone exchange to automatic switching center; from second-class automatic switching center to a first-class automatic switching center; and from first-class switching center to second-class automatic switching center sections. The reason is that the channel of a path of last resort consists of several such sections, at least three, and the problem of eliminating double hops is sharply exacerbated. However, satellite channels may be used for connecting two second-class switching centers, because this section of a network is not a path of last resort, but satellite channels are not used in sections between automatic long-distance telephone exchanges and second-class automatic switching centers because they are a part of a path of last resort.

As an exception for remote regions, where it is expensive to build and operate terrestrial links, satellite channels may be used in automatic long-distance telephone exchanges, automatic switching center sections of a path of last resort. Because the path of last resort is built entirely on terrestrial channels, this version does not result in the appearance of double hops in a connection between a given automatic long-distance telephone exchange and automatic long-distance telephone exchanges, located in areas that are serviced by other automatic switching centers. But for a connection between two automatic long-distance telephone exchanges located in the area of one automatic switching center, if they both may use satellite channels, then the direct satellite channels, from one automatic long-distance telephone exchange to the other, must be used.

Area networks, in accordance with the principles adopted by the EASS are built basically on the radial principle. In an area network there is one (sometimes two) automatic long-distance telephone exchanges, which services intra-area communications, connections between city and rural local networks, and also gives local networks access to the national long-distance network.

Connecting satellite channels in area networks by the same principles as terrestrial channels will lead to the appearance of double hops, both between local networks (through an automatic long-distance telephone exchange) and to the long-distance network, if satellite channels are being used in it.

To prevent double hops in area networks, it is necessary to place certain limits on the use of satellite channels in them.

1. The connection between local networks in an area through satellite channels always must be made by direct paths, bypassing automatic long-distance telephone exchanges.
2. To enable local exchanges to gain access to the long-distance network through an area automatic long-distance telephone exchange, which also operates on satellite national channels, either these exchanges must get to the automatic long-distance telephone exchange in their

area through terrestrial links, or a terrestrial link in a national network section must be used. But in cases when it is absolutely impossible to obtain terrestrial links within an area and in a national network section, double-hop work is permitted.

3. In area networks that use satellite channels it is helpful, in addition to automatic long-distance telephone exchanges, to build additional switching centers, which will have a direct access to the long-distance network. This will make it easier to arrange a network of terrestrial links between local exchanges and satellite communications exchange and will improve the reliability of a network.

1.6.4 CONNECTING UNSECURED SATELLITE CHANNELS AND GROUP CIRCUITS IN A NETWORK

Unsecured satellite channels and group circuits may be used in unswitched networks for making alternate paths and for beefing up individual communications links, and also for establishing standby communications links. In automatically switched networks, unsecured channels can be used, if necessary, for increasing the carrying capacity of various communications links, formed by terrestrial or satellite unsecured channels during periods of heavy traffic. Unsecured satellite channels also may be connected in sections where there are not other communications channels, creating temporary communications links. For avoiding the formation of double-hop links, unsecured satellite channels should be connected in the same sections of the national networks as secure channels, i.e., in automatic long-distance telephone exchanges connected to automatic long-distance telephone exchanges; on automatic long-distance telephone exchanges connected to automatic switching center sections that serve a called automatic long-distance telephone exchange; two second-class automatic switching centers connected together; and automatic long-distance telephone exchange sections that serve its automatic switching centers in remote regions. Likewise, in area networks, unsecured satellite channels are connected on the same principles as secure channels.

The use of unsecured satellite channels for backing up broken down terrestrial or secure satellite channels may be used in any section of a network, including a path of last resort. Considering that conserving channels, even with substandard characteristics, is an exceedingly important problem, it is permissible in such situations to connect satellite channels to a network, in spite of the possibility of double hops, in any of its sections, including between automatic switching centers.

1.7 OPERATING CONDITIONS ON SATELLITE CHANNELS IN AUTOMATIC TELEPHONE NETWORKS

In automatically switched telephone communications networks, not only speech signals, but also service signals, exchanged between automatic switching centers and automatic long-distance telephone exchanges during the connection and disconnection processes, are transmitted on the channels. Satellite channels must be capable of transmitting these signals without distortions and without disrupting or altering the operation of automatic long-distance telephone exchanges.

There are different signal systems used in automatic networks. It is important to become more familiar with them and to formulate special requirements on satellite channels, connected with the operation of these signal systems.

Signals of the signal systems are divided into three basic groups:

1. *Line signals,* used for transmitting information about the condition of channels and about the need to perform various operations associated with the connection and disconnection processes.
2. *Control signals,* used for transmitting information about the numbers of called subscribers and certain other commands, exchanged by automatic long-distance telephone exchange registers.
3. *Information sound signals,* tranmitted for the purpose of alerting a calling subscriber that the connection or disconnection process is taking place. These signals are transmitted on audio frequencies and from the standpoint of their transmission on a channel are no different from speech signals.

There are several ways of generating and transmitting line signals and control signals. In the USSR, two basic line signal tranmission systems are used: one-frequency and two-frequency. Signals are transmitted in both systems by audio frequencies in the working band of a channel. In area networks, however, a system is sometimes used, in which line signals are transmitted outside of the telephone channel band, on a specially selected channel, connected to the telephone channel, on a frequency of 3825 or 3800 Hz. The basic characteristics of these systems are given in Table 1.3. Also given in the table are the characteristics of control signals, transmitted in the working band of a channel by audio frequency pulses. The decade tranmission method, whereby every digit is transmitted by a series of pulses, the number of which corresponds to the digit of a number, and coding, when the digits of a number are transmitted in binary code, are utilized.

The future signal system that is being developed in the USSR should be mentioned separately. In this system, line signals and control signals will be transmitted on a separate group (servicing a group of signals) reserved signal channel, formed in one of the telephone channels of a communications line. Signals will be transmitted on this channel in discrete form by binary code at a rate of up to 2400–3000 b/s. One selected signal channel can transmit signals of a signal system that services up to 2000 telephone channels. Message addressing will be used on a channel. The basic characteristics of this system also are given in Table 1.3.

In international communications networks, in which satellite channels are used extensively, several signal systems, specified by International Consultative Committee for Telephone and Telegraph (CCITT), are utilized. The most popular are systems No. 4, No. 5, R-2, and No. 6 [1.11, 1.12, 1.13], data on which are given in Table 1.4.

The characteristic features of the signal systems that have an effect on the performance of satellite channels in automatic telephone communications networks are described.

1. As a rule, line signals, if they are transmitted in the spectrum of a telephone channel, are in the upper part of the spectrum. This places certain requirements on the channels. On FM satellite channels in particular, formed by the "one channel per carrier" method, to improve the signal-to-noise ratio it is desirable to use linear predistortions, thereby elevating the frequency response of a channel in the high-frequency region on the transmitting end and equalizing on the receiving end. Although the use of predistortions produces a good effect, their use of channels operating in automatic networks, in which line signals are transmitted on frequencies above 2000 Hz, is virtually impossible. During the transmission of line signals the circuit becomes overloaded due to the high elevation of the upper frequencies. Significant distortions of the line signals appear, preventing the normal operation of the automatic switching equipment. To keep automatic switching equipment functioning normally, it is necessary either to eliminate completely the predistorting and regenerating circuits in the channels, which detracts from their quality, or to use special devices, which will automatically exclude predistortions on channels only during the transmission of the signals of the signal systems, and reactivate them at the end of these signals.

The operating principle of a predistortion cutoff device is explained in Figure 1.23. Signals on a telephone channel are analyzed continuously by a special channel indicator of signals of the signal system. When a "dial tone" line signal appears on a channel it is detected by this receiver. The channel indicator generates a command, which goes to the channeling equipment and eliminates predistortions. Because the automatic long-

TABLE 1.3

Description of systems	**Parameters of signals**			
	Line			*Control*
	one-frequency system	*two-frequency system*	*signal system using group signal channel*	—
Signal transmission method	Series of pulses with constant duration Signals of different duration Continuous signals until receipt of confirmation from receiving end	Signals with different durations Different frequencies and combinations of frequencies	Discrete signals in binary codes	In frequency systems: Pulse signals, 2 of 6 code, i.e., combinations of pairs of frequencies of 6 base frequencies. Pulse signals, decade method (the number of pulses in a signal corresponds to digit of number)

TABLE 1.3 (Continued)

Description of systems	**Parameters of signals**			
	Line			*Control*
	one-frequency system	*two-frequency system*	*signal system using group signal channel*	—
				In group signal channel signal system, the same as line signals
Signal transmission frequency	2600 Hz—mainline and area networks; 2100 Hz—area networks; 3825 (3800) Hz—area networks on reserve signal channel	1200 and 1600 Hz	Signals transmitted on audio frequency channel at rate of up to 2400–3000 b/s	700, 900, 1100, 1300, 1500 and 1700 Hz 1200 Hz using decade method
Signal levels	−7 to −9 dB at zero relative level point	−7 to −9 dB at zero relative level point	—	−4.3 dB at zero relative level point

TABLE 1.4

Description of systems	Signal parameters							
	Line				*Control*			
	No 4 CCITT	*No. 5 CCITT*	*R-2 CCITT*	*No 6 CCITT*	*No 4 CCITT*	*No 5 CCITT*	*R-2 CCITT*	*No 6 CCITT*
Signal transmission method	Pulses of different duration and on different frequencies	Continuous signals, continuing until receipt of confirmation from receiving end	Continuous signals with acknowledgement; signals are transmitted on special channel	Discrete signals, coded by binary code; transmitted on 2400 b/s telephone channel	Pulses of two frequencies; binary codes with acknowledgement of receipt of each digit	Pulse signals, 2 of 6 code, packet	Pulse signals, 2 of 6 code with acknowledgement of receipt of each digit and receipt of acknowledgements	Discrete signals, coded by binary code, transmitted on 2400 b/s telephone channel
Signal transmission frequency	2040 and 2400 Hz	2400 and 2600 Hz	3825 Hz		2040 and 2400 Hz	700, 900, 1100, 1300, 1500, 1700 Hz	Direct line 1380, 1500, 1620, 1740, 1860, 1930 Hz; Return line 1140, 1020, 960, 780, 660, 540 Hz	
Signal level	−9 dB at zero relative level point of each frequency	−9 dB at zero relative level point	−20 dB at zero relative level point		−9 dB at zero relative level point of each frequency	−7 dB at zero relative level point of each frequency	−11.5 dB at zero relative level point of each frequency	

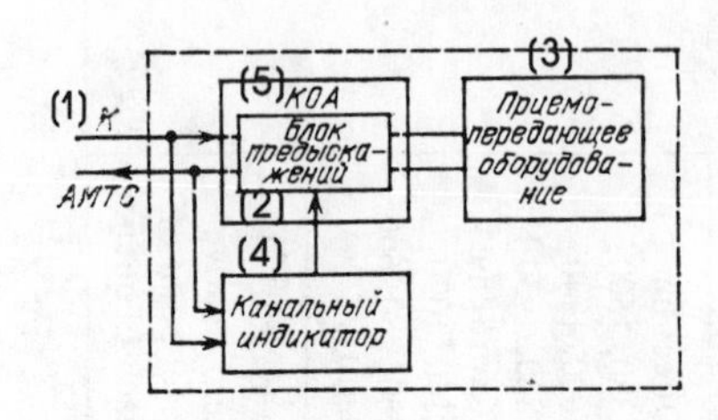

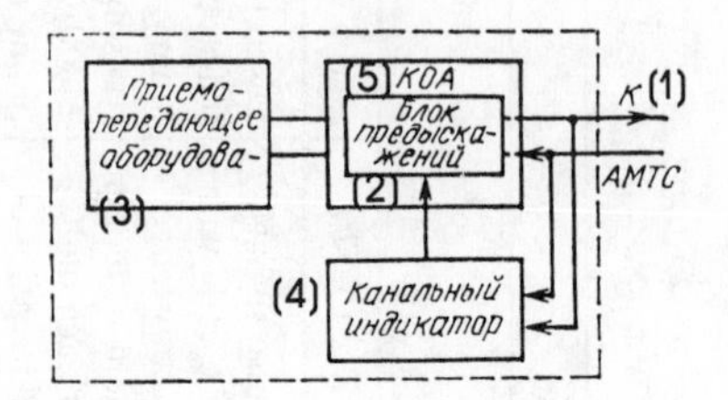

Key: **(1) To automatic long-distance telephone exchange**
(2) Predistortion unit
(3) Receiving-transmitting equipment
(4) Channel indicator
(5) Channeling equipment

FIGURE 1.23. Schematic diagram of device that eliminates predistortions on communications channel.

distance telephone exchange transmits the called subscriber's number after the dial tone, the command to stop predistortion is delayed by the time necessary for transmitting a number (this time varies from one signal system to another), and then predistortions on the channel are activated.

During the connection and disconnection process, automatic long-distance telephone exchanges also exchange other signals. The channel indicator also turns off predistortions when it detects these signals, but in this case only for the time it takes the signal to pass. At the end of the signal, predistortions are immediately reactivated. The above-described automatic predistortion cutoff method was developed for use in the Intersputnik system.

2. When systems are used that transmit line signals on a reserve channel outside of the telephone channel band as, for example, in area communications networks of the USSR or the R-2 system, it is necessary to make provisions for the organization of these channels in satellite systems.

The problem is even more complicated in cases when, as in the R-2 system of the CCITT, a signal is transmitted continuously on a reserve channel, indicating that the telephone channel is unoccupied. This signal stops only after a channel is occupied.

Transmission systems that operate on the "one channel per carrier" principle (see Chapter 18), in which suppression of carriers in pauses of transmission is used for improving the energy characteristics, are used extensively in satellite systems. However, the transmission of an "unoccupied" signal on a reserve signal channel does not permit the use of this effective method of improving the energy relations of a transmission system. To enable systems to operate with carriers suppressed in a pause, it is necessary to convert the R-2 system in a satellite link so that this signal will not have to be transmitted continuously.

3. Because the propagation time in satellite communications systems is long, problems arise in the transmission of control signals in cases when every digit of a subscriber's number is transmitted with confirmation as, for example, in systems No. 4 and R-2 of the CCITT. In the R-2 system, signals have to be transmitted between automatic long-distance telephone exchanges at least four times in order to transmit one digit, and it takes about 1.2 s to do this. It takes at least 17–20 s to transmit a complete international number, containing up to 14 digits, through a satellite channel. This detracts significantly from the quality of service. It is essential in a satellite link to take special measures to speed up the transmission of a number, i.e., to convert the signal system.

4. Satellite communications systems that utilize unsecured satellite channels, made available by request, should be discussed separately. In these systems, each channel can be used alternately by any pair of exchanges (and consequently by any pair of automatic long-distance telephone exchanges) in a communications network. To make a connection in such a network the earth stations must exchange additional signal information, other than that which is exchanged by automatic long-distance telephone exchanges.

5. An analysis of the problems of transmission of signal information on satellite channels shows that, in many cases it is desirable to use a special signal channel in satellite networks, which an entire network shares. Because a time division multiplex system is included in it, this channel is accessible to all the exchanges of a network and enables them to exchange the necessary signal information [1.14].

For instance, when signal systems of the R-2 type of the CCITT are used in terrestrial networks it is possible, using a common signal channel, to transmit line signals between automatic long-distance telephone exchanges without reserving a special signal channel for each telephone channel. The digits of a number can be transmitted quickly on this channel, and this improves the quality of service. When signal systems that utilize a reserve group channel are used in terrestrial networks, it is desirabel to use a common signal as such a channel, which services all the channels of a network simultaneously, significantly increasing the efficiency of the signal channel in a satellite link. A common signal channel is of special importance when unreserved satellite channels are used. In this case, it is necessary to transmit on a common signal channel, not only signals between automatic long-distance telephone exchanges, but also signals which earth stations must exchange in order to establish a satellite channel.

Chapter 2
Performance Characteristics of Satellite Communications Systems

2.0 INTRODUCTION

Satellite communications links, like terrestrial links (cable and radio relay), are compatible with the communications system of the country. Satellite links are standardized on the basis of unified performance standards, which are in effect throughout the USSR. These standards take into account the recommendations of the CCIR and the CCITT of the ITU.

2.1 PERFORMANCE CHARACTERISTICS OF VIDEO CHANNELS

2.1.1 Classification of Satellite TV Channels

National, regional, and local TV networks are provided by satellite transmission systems. National satellite TV channels* provide Orbita-2 and Moskva receiving stations, have performance characteristics compatible with national TV channels on radio relay and cable, and can be used for the international and long-distance exchange of TV programs, including distribution to various republic centers of the USSR. The satellite TV channels for this level are equipped with Ehkran class I receiver installations. Satellite TV channels with Ehkran class II receiver installations are used for collective receiver sets, which are used in the distribution of

---

*The combination of TV picture and sound is defined as a TV channel.

TV programs to population centers with 2000 to 3000 citizens (local TV distribution networks) with the aid of passive TV repeaters or cable TV networks.

2.1.2 Hypothetical Reference Circuits

Performance reference is established by the use of hypothetical reference circuits. These are used to facilitate the normalization and comparison of channels covering different distances; the hypothetical reference circuit of a national video channel is shown in Figure 2.1. It contains one earth-satellite-earth link, and it can contain satellite-satellite links. In a channel at earth stations, there is one modulator-demodulator pair for transferring the modulating spectrum to the carrier, and from the carrier to the modulating frequency. Connecting links between earth stations and switching centers are not included in this hypothetical reference circuit.

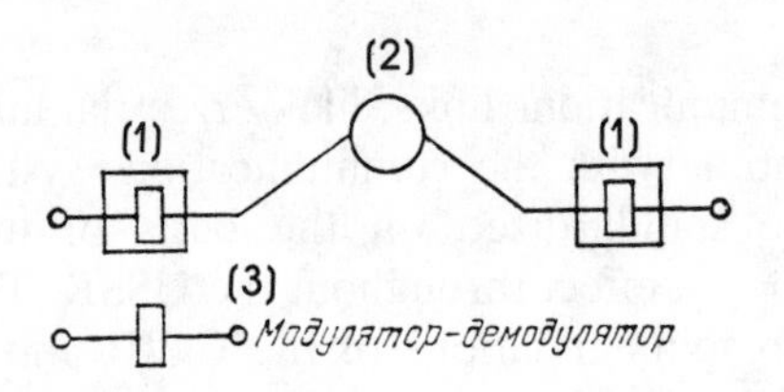

Key: **(1) Earth station**
(2) Satellite
(3) Modulator-demodulator

FIGURE 2.1. Hypothetical reference circuit of video channel of satellite link.

It is assumed, on the basis of the average length and functions of satellite TV channels, that a satellite national video channel is equivalent in terms of the signal-to-weighted-noise ratio to a national channel of a terrestrial circuit consisting of two 2500-km sections. In terms of other characteristics, a satellite video channel is equivalent to a 2500-km national terrestrial channel.

2.1.3 TV Signal Levels

The video channels of satellite links are intended for transmitting black and white and color video signals that meet the requirements of state standards. The sequential-and-memory (SECAM) color television system, in which information about color is transmitted by frequency modulation on subcarriers, is used in the USSR. The levels in a complete color signal of this system are given in Figure 2.2.

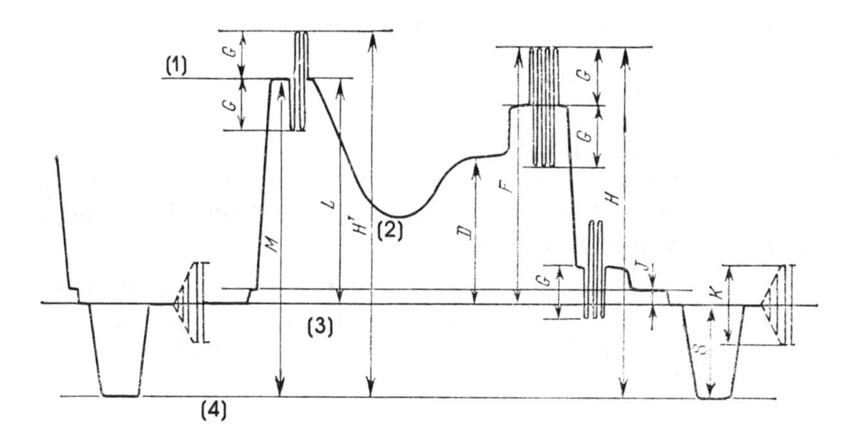

Key: (1) White level
(2) Black level
(3) Blanking level
(4) Sync pulse level

FIGURE 2.2. Levels in complete color signal: *L*- nominal brightness signal amplitude from blanking level to white level is 700 mV; *S*-nominal amplitude of line sync pulse is 300 mV; *M*-nominal amplitude of black and white signal from sync pulse level to white level is 1.0 V; *D*-instantaneous brightness signal level relative to blanking level; *F*-instantaneous amplitude of color signal; *G*-peak amplitude of color signal; *H'*-amplitude of complete color signal of TV signal from sync pulse level to steady state maximum level of color subcarrier at white level is 1.107 V; *H*-instantaneous amplitude of complete color signal of TV signal; *J*-safety margin—distance between black level and blanking (steady state) level 0–50 mV; *K*-amplitude of color subcarrier—214 mV in red and 167 mV in blue lines.

2.1.4 Signal-to-Noise Ratio

The signal-to-noise ratio in a video channel is an important characteristic for the planning of broadcasting networks. The signal-to-weighted-noise ratio on video channels is standardized. It is equal (in decibels) to the ratio of the peak-to-peak amplitude of the video signal (the parameter L in Figure 2.2) to the effective noise level, measured with weighting from 10 kHz to the upper signal frequency of 6 MHz. The amplitude-frequency response with weighting makes it possible to consider the properties of vision in the perception of fluctuation interference in different parts of the video spectrum. Noise is determined by the formula:

$$S/N_w = 20 \log(0.7V/U_{eff.n})$$

where $U_{eff.n}$ is the effective weighted noise voltage.

The standard signal-to-noise ratio must be met for 99 percent of the time of any month. The band in which the weighted noise is measured is formed by series-connected filters:

1. by a weighting filter with the time constant $\tau = 330$ ns (the diagram in Figure 2.3), with an amplitude-frequency response in the form of curve a in Figure 2.4 and with transfer coeffiecient $A = 10 \log(1 + 4\pi^2\tau^2 f^2)$ dB;

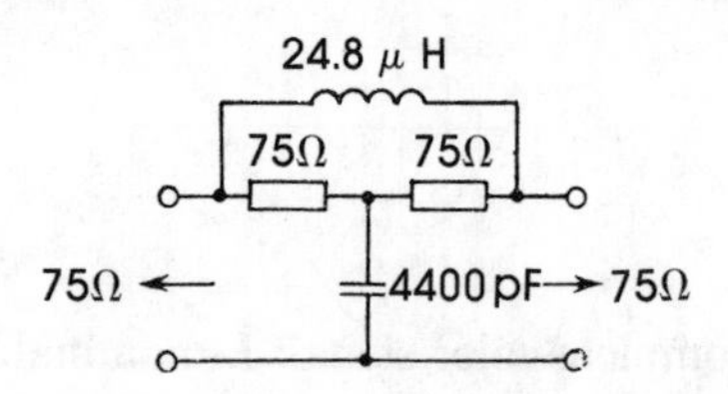

FIGURE 2.3. Diagram of weighing filter according to CCIR Recommendation 421-3.

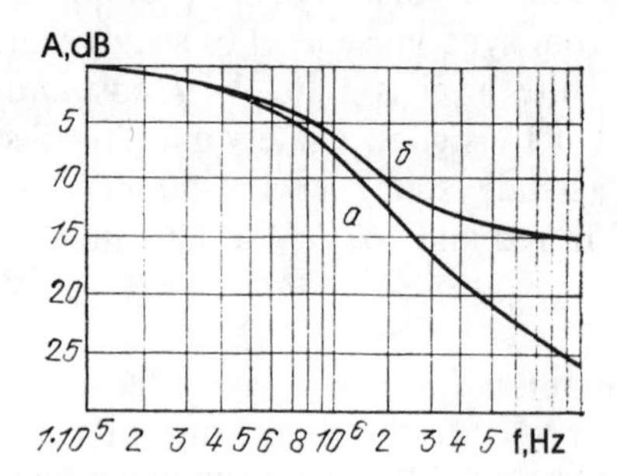

FIGURE 2.4. Amplitude-frequency responses of filters for measuring fluctuation interference: *a*—according to CCIR Recommendation 421-3; *b*—of unified filter.

2. with a low-frequency noise suppression filter (power line and microphone noises) (Figure 2.5(a)) with an amplitude-frequency response in the form of curve b in Figure 2.5(b); the specifications of the elements are given in Table 2.1, and
3. with a low-pass filter, which limits the band in which interference is measured (the diagram of the filter in Figure 2.6(a), the response is given in Figure 2.6(b) and the specifications of the elements are listed in Table 2.1).

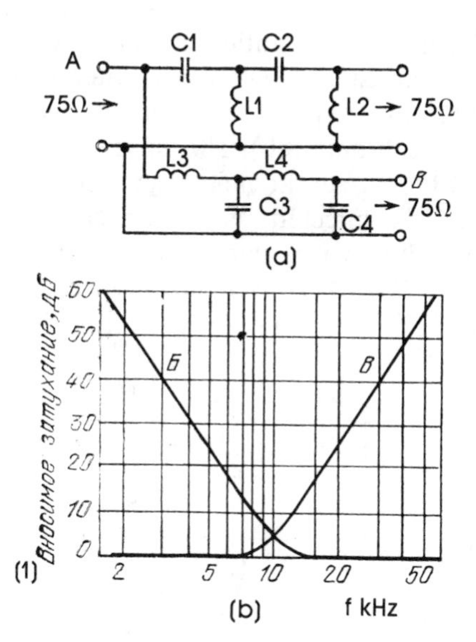

Key: **(1) Insertion loss, dB**

FIGURE 2.5. Filter for separating fluctuation interference and low-frequency periodic interference: (a) diagram; (b) frequency responses (curve b: output of low-frequency interference suppression filter; curve *B*: output of low-frequency interference selection filter).

CCIR Recommendation 567 suggests the use of a filter, unified for different TV standards, with the time constant $\tau = 245$ ns for measuring fluctuation interference in black and white and color TV channels. A diagram of a unified weighting filter is shown in Figure 2.7, its frequency response is shown in Figure 2.4 (curve b), and its transfer coeffiecient (in decibels) is

$$A = 10 \log \frac{1 + [(1 + 1/a)\ \omega\ \tau]^2}{1 + \left[\left(\frac{1}{a}\right) \omega\ \tau\right]^2},$$

and at the asymptote $A_\infty \to 10 \log(1 + a)$, $A_\infty = 14.8$ dB, $a = 4.5$. By using a unified weighting filter, it is possible to convert to the standardization of a signal-to-weighted-noise ratio that is unified for different TV standards. Here the noise for all TV standards is measured in the 5-MHz band, formed by a low-pass filter, the diagram of which is similar to the diagram of a low-pass filter with a 6-MHz band and shown in Figure 2.6(a), but the specifications of the elements are different, and they are given in Table 2.1. The frequency characteristic of a low-pass filter with a 5-MHz band is shown in Figure 2.8. Low-frequency interference also is suppressed by the filter shown in Figure 2.5 (output b). The noise immunity, measured with a unified filter, is 4 dB less than the noise immunity, measured with a filter for black and white TV connected, recommended by CCIR Recommendation 421-3 for noise with a triangular spectrum. The signal-to-noise ratio, found from measurements with a weighing filter in accordance with CCIR Recommendation 421-3 (τ = 330 ns) is given in this book.

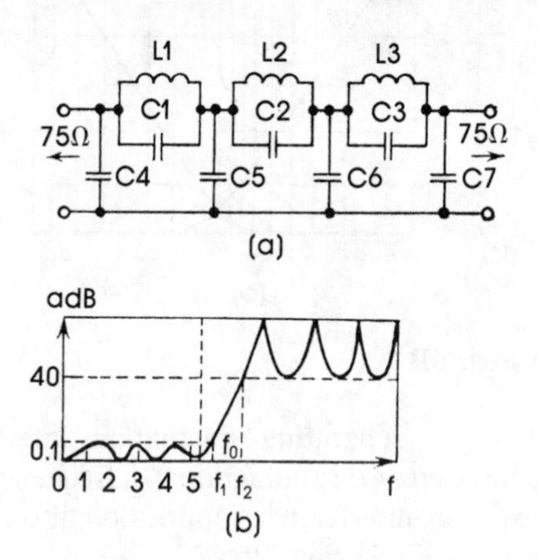

FIGURE 2.6. Low-pass filter: (a) diagram; (b) frequency response (f_w = 6 MHz; f_1 = 5.8842 MHz; f_2 = 6.5382 MHz).

In FM and amplitude modulation (AM) systems, linear frequency predistortions of the video signal on the transmitting end and frequency regeneration on the receiving end are used for increasing the noise immunity of signals. Diagrams of predistorting and regenerating circuits are given in Figure 2.9. The frequency response of the predistorting circuit is given in Figure 2.10, and the transfer coefficient of this circuit is

$$K_{tr.pred} = 10 \log[(1 + 10.21f^2)/(1 + 0.4083f^2)] - 11.0 \text{ (dB)}$$

The factors for converting the influence of regenerating circuits of a predistorted signal and of the weighing circuit, separately and together, to the noise immunity in terms of white noise and in terms of noise with a triangular spectrum, are given in Table 2.2. The output noise of a satellite channel, on which TV transmission is accomplished by frequency modulation, has a triangular spectrum. In the section in which the signal is relayed for consumers, single-sideband AM is used, and the AM signal demodulator output noise has a uniform spectrum.

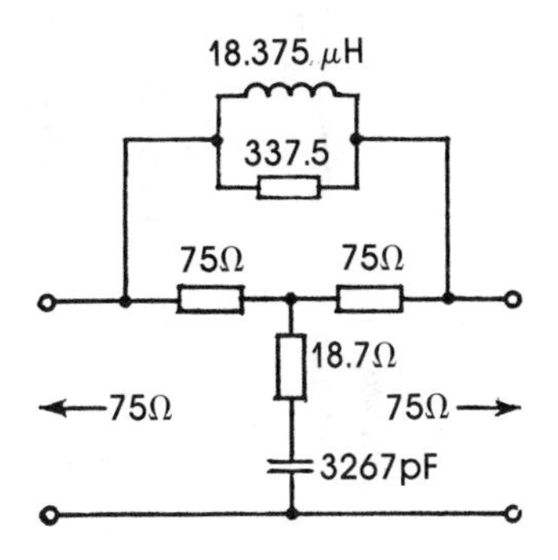

FIGURE 2.7. Diagram of unified weighting filter according to CCIR Recommendation 567.

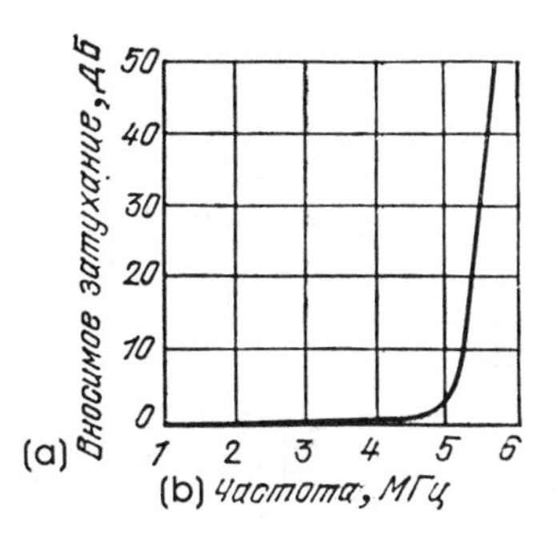

Key: (a) Insertion attenuation, dB
(b) Frequency, MHz

FIGURE 2.8. Amplitude-frequency response of low-pass filter (f_w = 5 MHz).

TABLE 2.1

Symbols in diagram	Specifications of parts			Symbols in diagram	Specifications of parts		
	Dividing network	Low-pass filter $f_w = 6$ MHz	Low-pass filter $f_w = 5$ MHz		Dividing network	Low-pass filter $f_w = 6$ MHz	Low-pass filter $f_w = 5$ MHz
C1, *pF*	139 000	82.93	100	C7, *pF*	—	216.2	259
C2, *pF*	196 000	453.8	545	L1	0.757 mH	2.397 μH	2.88 μH
C3, *pF*	335 000	325.0	390	L2	3.12 mH	1.279 μH	1.54 μH
C4, *pF*	81 000	356.5	428	L3	1.83 mH	1.433 μH	1.72 μH
C5, *pF*	—	469.2	563	L4	1.29 mH	—	—
C6, *pF*	—	385.8	463				

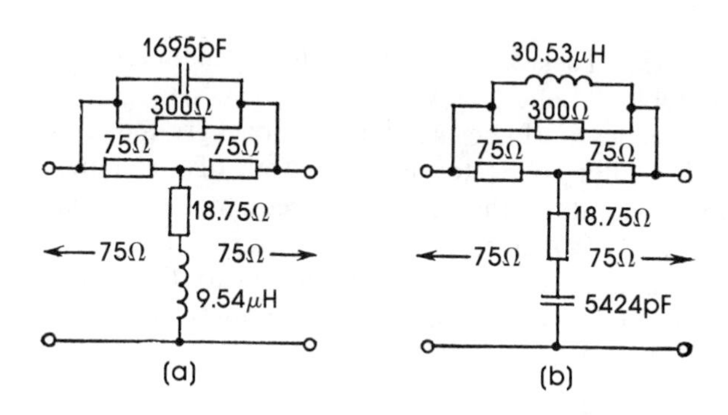

FIGURE 2.9. Diagram of predistorting (a) and regenerating (b) TV signal circuits.

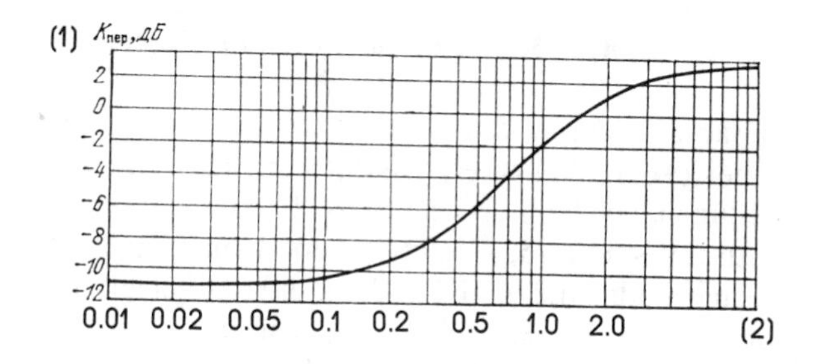

Key: (1) K_{tr}, dB
(2) *f*, MHz

FIGURE 2.10. Amplitude-frequency response of TV signal frequency predistortion circuit.

As can be seen in Table 2.2, the joint effect of regenerating and weighing circuits with a response with $\tau = 330$ ns for triangular noise, yields a gain in noise immunity of 18.1 dB, and for white noise 1.9 dB; with a weighing circuit, with $\tau = 245$ ns, 13.2 and 0.9 dB, respectively. These gains also are considered in calculations of the energy relations of FM and single-sideband AM TV systems. For technical, organizational, and economic considerations, and in accordance with the established link of a channel of a hypothetical reference circuit in national channels, the signal-to-weighted-noise ratio (in consideration of regenerating circuits) is standardized to the edge of the service area at 53–55 dB, on satellite channels using class I sets 53 dB, and on satellite channels using class II sets 48 dB* (see Table 2.3 and [2,1]).

Low-frequency periodic interference (background) is measured on channels with a filter connected to the output of a channel as shown in Figure 2.5(a), output *B*, the amplitude-frequency response of which is shown in Figure 2.5(b), curve *B*. The ratio of the amplitude of the signal from the blanking level to the white level, to the amplitude of the background interference is 35–40 dB on channels (see Table 2.3) and is detemined by the formula:

$$U_s/U_b = 20 \log(0.7V/2.82U_b)$$

where U_b is the background voltage, measured with an effective volt meter, *V*.

2.1.5 Linear and Nonlinear Signal Distortions on a Channel

The tolerable linear and nonlinear distortions of a signal on a national channel are standardized basically in accordance with a state standard. Satellite video channels using class I sets are also standardized in accordance with this standard; satellite video channels using class II sets, tested with a single-sideband AM TV demodulator, meeting the requirements of another state standard, are standardized to less rigid specifications than the previous. Summary data on the characteristics of the video performance channels of satellite TV transmission systems are given in Table 2.3, the last column of which also gives the values of coefficients *p* that determine the formulas for adding distortions in compound satellite channels (see Section 2.4).

2.1.6 Some Information on Measurement Methods

Test signals are used for measuring channel parameters with the carrier not modulated by a video information signal, and test lines are used for measuring channel parameters during the time of transmission (see [20]). Measurements are conducted at the present time using test signals 1, 2, 3a, 3b, described in CCIR Recommendation 421-3 (1974).

---

*A signal-to-weighted-noise ratio lower than the one suggested in CCIR Recommendation 567 (53 dB) is acceptable, for example, in the Intelsat international system: when two carriers are transmitted in a transponder with TV programs, this ratio may be 47.1 dB on video channels, and when one carrier is transmitted it may be 50.1 dB.

TABLE 2.2

Upper modulating frequency, MHz	Color TV system	*Influence of regenerating circuits**, dB*		*Influence of weighing circuits**, dB*				*Influence of weighing and regenerating circuits**, dB*	
		White noise	*Tri-angular noise*	*White noise*	*Tri-angular noise*	*Regenerated white noise*	*Regenerated triangular noise*	*White noise*	*Tri-angular noise*
		a	b	c	d	e	f	g	h
5.0	Unified	−2.1	2.0	7.4	12.2	3.0	11.2	0.9	13.2
6.0	K/SECAM	−1.6	2.3	9.3	17.7	3.5	15.8	1.9	18.1
6.0 5.0**	Unified	−1.6	2.3	8.2	14.6	3.2	13.3	1.6	15.6
6.0		−1.6	2.3	8.0	12.8	3.3	12.0	1.7	14.3

*$a+e=g$, $b+f=h$

**The first number pertains to the band in which unweighted noise is measured, and the second to the band in which regenerated noise, weighted noise, or weighted predistorted noise are measured.

TABLE 2.3

Parameter	Values of parameter for satellite TV system				p *in addition formula*
	Orbita-2	*Moskva*	*Ehkran*		
			Class I	*Class* II	
Amplitude of complete TV signal	1.0 ± 0.1	1.0 ± 0.1	1.0 ± 0.1	1.0 ± 0.1	2
Amplitude of complete color TV signal	1.107 ± 0.11				2
Nonlinear distortions of sync signal, %, not more than	±17	—	—	—	3/2
Amplitude of color subcarrier in red line, mV	214^{+35}_{-50}	214^{+35}_{-50}	214^{+35}_{-50}	214^{+35}_{-60}	3/2
Amplitude of color sync signals in red line, mV	540^{+100}_{-110}	540^{+110}_{-120}	540^{+110}_{-120}	540^{+130}_{-130}	3/2
Brightness to weighted fluctuation interference ratio in brightness signal for 99% of time of any month, dB, not less than	53	53	53 - 55	46 - 48	For power (see Table 2.10)
Brightness signal to background interference ratio, dB, not less than	35	40	40	40	For voltage (see Table 2.10)
Differential gain, amplitude, %, not more than	±18	±15	24	30	3/2

Differential phase, deg, not more than	±8	15	15	20	3/2
Nonlinear distortion coefficient of brightness signal, amplitude, %, not more than	18	15	20	20	3/2
Relative nonuniformity of flat part of rectangular pulses of field frequency, amplitude, %, not more than	10	10	12	12	3/2 for $N > 3$ 1 for $N \leq 3$
Relative nonuniformity of flat part of rectangular pulses of line frequency, amplitude, %, not more than	5	5	8	8	2
Distortions in short time range					
Transient response:					
leading edge duration, ns, not longer than	140	140	140	200	—
first spike, %, not more than	20	20	20	30	—
second spike, %, not more than	10	10	—	—	—
Gain difference of brightness and color signals, %, not more than	±15	±14	±20	±25	2
Time discrepancy of brightness and color signals, ns, not longer than	±100	±150	±150	±200	2
Amplitude-freqency response, dB, not more than:					

TABLE 2.3 (Continued)

Parameter	*Values of parameter for satellite TV system*				*p in addition formula*
	Orbita-2	*Moskva*	*Ehkran*		
			Class I	*Class* II	
up to 1.2 MHz	±1.0	±0.7	±1.0	±1.0	3/2
1.2–4.8 MHz	+1.7 to −1.5	+1.2 to −1.0	+1.8 to −1.6	+1.8 to −3.4	
4.8–6.0 MHz	+2.0 to −3.5	+1.4 to −2.5	+2.0 to −3.5	+2.0 to −5.0	
Group propagation time (GVP) response	See Figure 2.11*	—	—	—	3/2

*The specifications on the group propagation time in accordance with CCIR Recommendation 567 apply to channels of satellite links, utilizing earth stations with G/T = 40:7 dB/K.

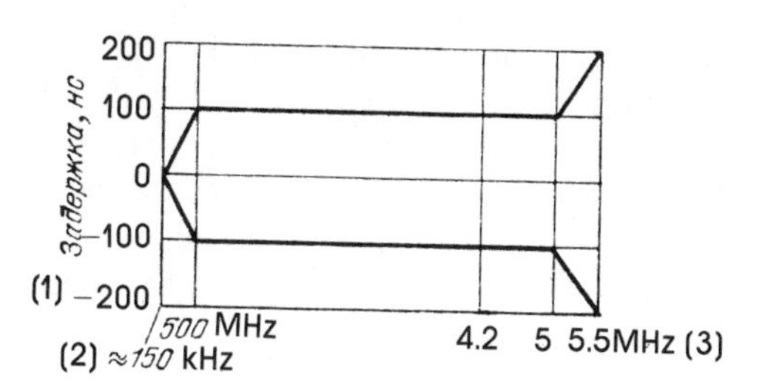

Key: **(1) Delay, ns**
(2) *f*, kHz
(3) *f*, MHz

FIGURE 2.11. Frequency response of group propagation time of picture channel.

Test lines signals are inserted into the lines of frame blanking pulses, are selected from the video signal using line selection units, and are fed to an oscillograph. Signals I–IV, described in GOST 7845-79, are inserted in the test lines for measuring distortions. On a video channel of a studio complex signals, I–IV are inserted at the input of the connecting link that goes to the earth station in lines with the numbers 17, 18, 330, and 331, respectively, and also at the input of the national video channel at the earth station in lines with the numbers 20, 21, 333, and 334, respectively.

Methods of measuring fluctuation and periodic interference were examined above. Methods of measuring linear and nonlinear distortions on a video channel are similar. We will examine two main parameters of a color TV program transmission channel—differential gain and differential phase, compliance with standards on which largely determines both the responses of the HF circuit, and the quality of the picture. Also described are methods of measuring the influence of the color signal on the brightness signal, and nonlinear distortions of the color signal (which are now evaluated with the aid of the new G2 signal), inserted in test lines, and entering in signal IV.

Differential gain (DG) characterizes the amplitude of the color subcarrier as a function of the brightness signal level and is measured on the basis of the change of the level of the sine wave filter on a sawtooth or step brightness signal (using CCIR signals 3a, 3b, or element D2 of signal III), in much the same way the nonlinearity of the video channel is measured. The filter frequency for measuring differential gain is 4.43 MHz, and for measuring nonlinear distortions it is 1.2 MHz.

On the receiving end, nonlinear distortions are determined on the basis of the waveform of the sine wave filter after it is separated from the

HF signal (Figure 2.12). Nonlinear distortions and differential gain are determined by the formula for the differential gain: $(1 - m/M)\cdot 100\%$ (where m is the minimum and M the maximum amplitudes of the sine wave voltage); the differential gain can also be expressed by two values: +DG and −DG, which indicate the difference of the maximum M and minimum m values of the subcarrier and the subcarrier on the blanking level M_0 by the formula:

$$+\text{DG} = 100(M/M_0 - 1) \quad \text{and} \quad -\text{DG} = 100(m/M_0 - 1)$$

Differential phase (DP) characterizes the phase shift of the color carrier as a function of the brightness signal amplitude. This parameter is determined by measuring the maximum ϕ_{max} and minimum ϕ_{min} phase shifts of the 4.43-MHz sine wave filter on a saw-toothed or stepped brightness signal by the formula:

$$\text{DP} = \phi_{max} - \phi_{min} \tag{2.1}$$

Differential phase can also be expressed by two values: +DP and −DP, which represent the greatest positive and negative phase differences of the color signal on different brightness signal levels relative to the phase of the color signal (for example, on element D2) at the blanking level ϕ_0:

$$+\text{DP} = \phi_{max} - \phi_0 \quad \text{and} \quad -\text{DP} = \phi_{min} - \phi_0 \tag{2.2}$$

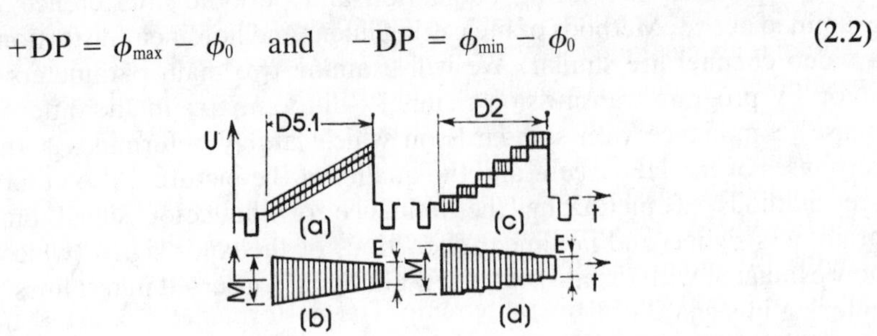

FIGURE 2.12. Test signals for measuring nonlinear distortion (a), (b) and waveform of sine wave filter after being separated from low-frequency signal (c), (d).

The amplitude of the differential phase [see (2.1)] and tolerable deviations toward an increase or decrease from the value of +DP and −DP at the blanking level, are standardized simultaneously in CCIR Recommendation 567 [see (2.2)]. The influence of the color signal on the brightness signal is measured using element G2 after suppressing the received color subcarrier. It is determined by the difference between the amplitude of the brightness signal on the last step of element G2, and the amplitude on the next step U_0 (Figure 2.13), on which there is no subcarrier signal, and it is expressed as a percentage of the amplitude of the rectangular brightness signal pulse by the formula $n_b = (U_{dis}/U_0)/100$, where U_{dis} is the maximum amplitude of the brightness signal distortion, and U_0 is the amplitude of the brightness signal.

Nonlinear distortions of the color signal (static) are determined in percentage points, on the basis of the G2 trilevel color signal [the ratio of the nominal amplitudes of the color subcarrier of the individual steps is $A_{10}:A_{20}:A_{30} = 1:3:5(140:420:700 \text{ mV})$], by the formulas:

$$n'_c = |3A_1/A_2 - 1|\cdot 100, \qquad n'_c = |3A_3/5A_2 - 1|\cdot 100$$

where A_1, A_2, and A_3 are the amplitudes of the subcarrier, respectively, of the smaller, middle, and larger steps of the signal. The largest of the values of n'_c and n''_c obtained are used as distortions.

2.2 PERFORMANCE CHARACTERISTICS OF AUDIO BROADCASTING AND TV SOUND CHANNELS

2.2.1 Hypothetical Reference Circuits

The hypothetical reference circuit shown in Figure 2.1 for a video channel is also valid for a TV sound signal and audio broadcasting channels in satellite systems. An audio broadcasting channel and a sound signal channel* of TV satellite systems each contain one modulator-demodulator pair. And so, in spite of the difference in the structure of TV sound signal and audio broadcasting channels, the same hypothetical reference circuit is used for them. In terms of performance characteristics, depending on length, TV audio broadcasting, and sound signal channels are considered equal to a 2500-km terrestrial circuit. The performance characteristics of TV sound signal and audio broadcasting channels, organized by the same methods, with the hypothetical reference circuit shown in Figure 2.1, correspond to the performance characteristics of an audio broadcasting channel in accordance with the standard in Table 2.4.

---

*TV sound signal channels are set up between TV studios, and audio broadcasting channels are established between a central and local switching-distribution radio communications and radio broadcasting centers.

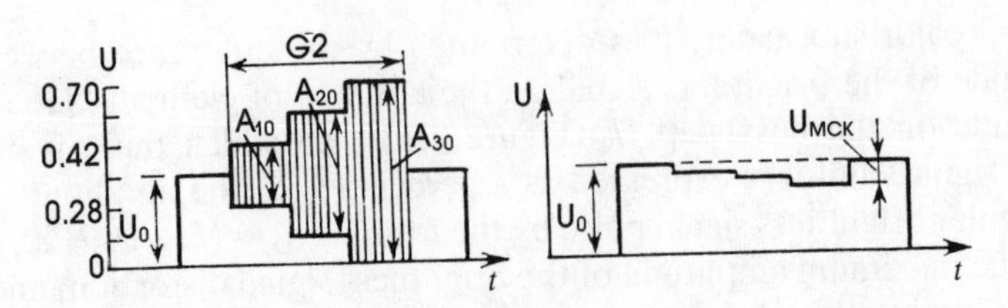

FIGURE 2.13. Determination of nonlinear distortions of color signal and of transitions from color signal to brightness signal on the basis of element G2.

In satellite systems, in which the TV sound signal or audio broadcasting program is supplied from a receiving earth station directly to transmitters or wire broadcasting networks, a somewhat lower signal-to-noise ratio is tolerable. For instance, for a class I TV sound signal channel, in the Ehkran system with class I receivers, the signal-to-psophometric-noise level is 55 dB instead of 57 dB.

On channels of the Orbita-RV system, the standards on distortions, comprising one-third of the distortions permissible on audio broadcasting channels of the corresponding class, are defined in Table 2.4. In higher quality channels, used for transmitting stereophonic broadcast programs, additional parameters also are standardized. According to CCITT Recommendation J.21, these parameters should be the following:

(1) the difference of gains on channels A and B, dB, in the band, Hz	
40–125	1.5
125–10,000	0.8
10,000–14,000	1.5
14,000–15,000	3
(2) the difference of phases on channels A and B, dB, on frequencies, Hz	
40	30°
4000–2000	a straight line segment (degrees on linear scale and frequency on a logarithmic scale)
200–4000	15°
4000–14000	a straight line segment (degrees on a linear scale and frequency on a logarithmic scale)
14,000 . . . 30°	
15,000 . . . 40°	

(3) audible crosstalk conversation between channels A and B
   (a) audible crosstalk interference, measured on a sine wave test signal on frequencies of 40–1500 Hz — 50 dB
   (b) nonlinear crosstalk interference, measured on a sine wave signal on frequencies of 40–1500 Hz — 60 dB

2.2.2 Measurement of Noise

The psophometric noise level is measured with an instrument with the amplitude-frequency response of a weighing filter, corresponding to CCITT Recommendation P.53 (Figure 2.14, curve 1). There is a tendency today to convert to the measurement of psophometric noise with a psophometer with a weighing filter, the response of which is of the form of curve 2 in Figure 2.14 (according to CCIR Recommendation 468-2). Amplitude-frequency response 2 is considered to better take into consideration the properties of the human ear and the frequency responses of playback loudspeakers in the 50–15,000-Hz band. The psophometric noise levels given in Table 2.4, are measured with a psophometer with weighing curve 1 (see Figure 2.14). Weighing white noise with weighing filter 1 in the 15-kHz band gives a reading of 4.3 dB noise immunity, and in the case of triangular noise it is 5.6 dB higher than with weighing filter 2. The output noise power level of filter 1, when its input is supplied, white noise is 4.2 dB, and, at the output of filter 2, 8.5 dB higher than the power level of unweighed noise in the 30–15,000-Hz band (CCIR Report 496-3). Unweighed (integral) interference is measured using a bandpass filter, the amplitude-frequency response of which is shown in Figure 2.15.

According to CCIR Recommendation 468-2, instruments of the quasipulse type, with special dynamic characteristics (Table 2.5), are recommended instead of instruments with quadratic detectors with the corresponding time constant for measuring all kinds of noises and interference. The fluctuation noise immunity, measured with an instrument with a quasipulse detector, is, on the average, 5 dB less than the noise immunity, measured with a mean-square detector and is measured in dBq (q = quasipulse). When harmonic interference is being measured, the readings of the instruments are identical, whereas for certain kinds of pulse interference, the difference of the readings of the instruments exceeds 10 dB (see [2.2]). Companders and predistorting and regenerating circuits

TABLE 2.4

Parameter	Standardized values of parameters of audio broadcasting channels: *Higher class*	*Class* I	*Class* II
Nominal effective transmitted frequency band, Hz	30–15,000	50–10,000	100–6300
Maximum* voltage level input, dB	0	0	0
V	0.775	0.775	0.775
output:			
transmission of program to remote subscriber, dB	+15	+15	+15
V	4.3	4.3	4.3
switching of program directly at reception point, dB	0	0	0
V	0.775	0.775	0.775
Nominal input impedance, Ω:			
on input side	600	600	600
on output side on frequencies:			
from 100 to 15,000 Hz, not more than	20	20	20
below 100 Hz, not more than	30	30	30
Amplitude-frequency response in dB, not more than, in band, kHz	kHz dB	kHz dB	kHz dB
	0.04–0.125 −2.0	0.05–0.1 −4.5	0.1–0.15 −4.5
	0.125–10.0 −0.5	0.1–0.2 −2.6	0.15–0.2 −2.6
	10.0–15.0 −2.0	0.2–6.0 −1.8	0.2–5.0 −1.8
		6.0–8.5 −2.6	5.0–6.0 −2.6
		8.5–10.0 −4.5	6.0–6.3 −4.5

TABLE 2.4 (Continued)

Parameter	Standardized values of parameters of audio broadcasting channels		
	Higher class	*Class* I	*Class* II
Upper limit	+0.5	+1.8	+1.8
Noise immunity relative to psophometric noise, dB, not more than	60	57	51
Noise immunity relative to integral interference, dB, not more than	60	54	48
Nonlinear distortion coefficient, %, not more than, on frequencies, Hz:			
up to 100	1.0	3.0	—
100–200	0.5	2.0	3.0
above 200	0.5	2.0	2.0
Noise immunity relative to audible crosstalk interference, dB	74	70	60

*The level corresponding to 0.707 of the peak signal voltage is called the maximum voltage level on an audio broadcasting channel. The maximum level on an audio broadcasting channel corresponds to the level of 100% modulation of the transmitters.

may be used for achieving the necessary signal-to-noise ratio on audio broadcasting and TV sound signal channels.

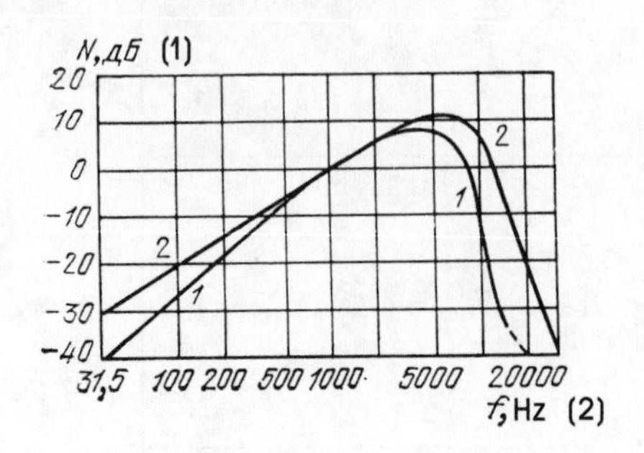

Key: (1) *N*, dB
(2) *f*, Hz

FIGURE 2.14. Amplitude-frequency responses of psophometric filters:
1 according to CCITT Recommendation P.53
2 according to CCIR Recommendation 468-2

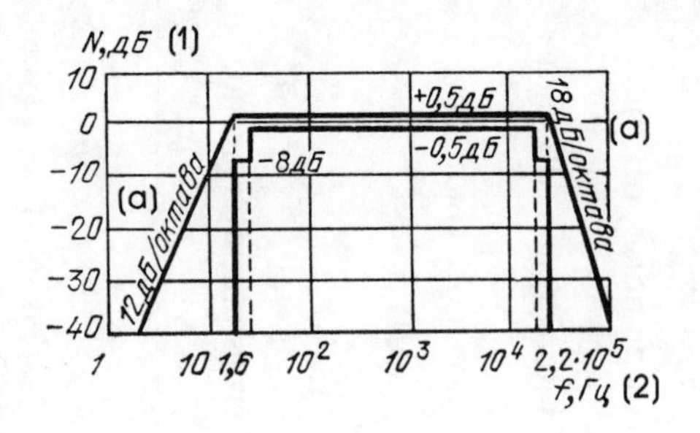

Key: (a) dB/octave
(1) *N*, dB
(2) *f*, Hz

FIGURE 2.15. Tolerances on amplitude-frequency response of filter for measuring unweighed noise by CCIR Recommendation 468-2.

The frequency responses of the predistorting and regenerating circuits with time constants $\tau = 50\ \mu s$ and $\tau = 75\ \mu s$ are given in Figure 2.16. Circuits with $\tau = 75\ \mu s$ are used in satellite systems. Uncontrolled companders are used on the audio broadcasting and TV sound signal channels of the Orbita system. Variable companders, controlled by trans-

TABLE 2.5

Pulse duration*, ms	*1*	*2*	*5*	*10*	*20*	*50*	*100*	*200*
Psophometer reading for single pulse: %	17	26.6	40	48	52	59	68	80 100
dB	−15.5	−11.5	−8	−6.4	−5.7	−4.6	−3.3	−1.9 0

*The test pulse should contain a whole number of periods of a 5-kHz signal and is switched as soon as the voltage is equal to zero. The amplitude of the pulse is 2/3 of the scale of the instrument.

mitting an auxiliary control signal, are used on the audio broadcasting and TV sound signal channels of the Moskva and Ehkran systems.

Variable companders give an additional gain of 0.5*N*, where *N* is the level of the original signal in decibels, relative to the gain produced by an uncontrolled compander in terms of noise immunity, depending on the level of the transmitted signal (see [2.3]). CCITT Recommendation J.17 recommends that the predistorting circuit have the response shown in Figure 2.16e.

The amplitude-frequency responses of the channels are measured by sending a test signal with a 21 dB lower level than the nominal maximum level with the uncontrolled companders blocked. On channels containing frequency predistorters harmonics are measured on 800 Hz. Psophometric noise and integral interference are measured with the uncontrolled companders blocked. Audible crosstalk interference is measured with a spectrum analyzer or selective volt meter with the uncontrolled companders blocked.

2.3. PERFORMANCE CHARACTERISTICS OF AUDIO FREQUENCY CHANNELS AND GROUP CIRCUITS

2.3.1 Hypothetical Reference Circuit of Channels and Group Circuits

Telephone signals are transmitted in satellite systems by analog or digital methods. A channel or circuit of a satellite link, in which a signal is transmitted by analog methods, is a simple channel or circuit, because it does not contain amplifiers in the frequency band of a given channel or group circuit, i.e., it contains a modulator at the transmitting earth station and a demodulator at the receiving earth station. The hypothetical reference circuit for analog channels and circuits of a satellite link is shown in Figure 2.1. The standards on the parameters of an audio frequency channel, depending on the length of a channel, are met by the standards on the parameters of a channel of a 5000-km terrestrial circuit. Shown in Figure 2.17 is the hypothetical reference circuit of digital channels of satellite telephone transmission systems in accordance with CCIR Recommendation 521, which contains an earth-satellite-earth link; but if necessary, also a satellite-satellite link, HF equipment, modems, and the equipment for interfacing with connecting terrestrial links.

2.3.2 Transmission Levels

Absolute level. The level of power, voltage, or current, expressed in terms, respectively, of a power of 1 mW, voltage of 0.775 V or current of

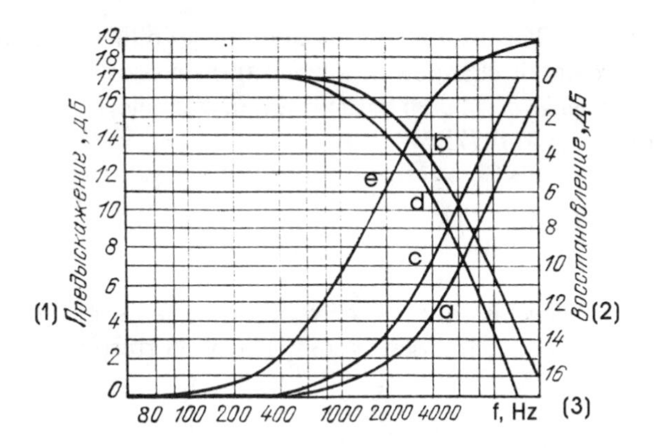

Key: (1) Predistortion, dB
(2) Regeneration, dB
(3) *f*, Hz

FIGURE 2.16. Amplitude-frequency responses of predistorting circuits for $\tau = 50$ μs (a), $\tau = 75$ μs (c), according to Recommendation J.17 (e), and amplitude-frequency responses of regenerating circuits for $\tau = 50$ μs and $\tau = 75$ μs (b and d).

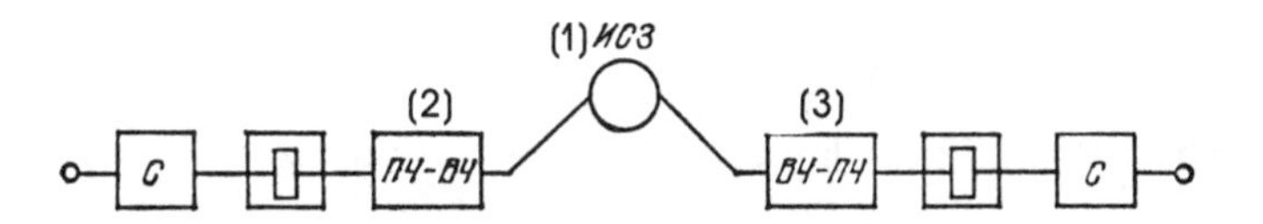

Key: (1) Satellite
(2) IF-HF—HF circuit
(3) HF-IF

FIGURE 2.17. Hypothetical reference circuit of digital channels of satellite transmission systems: *C*-interface with terrestrial links; □-modulator-demodulator.

1.29 mA into a 600-ohm impedance. This level is denoted as p_p, p_v, p_c and is expressed in dBp, dBv, dBc.

Relative level. The level at a given point of a circuit, examined in relation to the level at the beginning of the circuit or at a point conditionally taken as the starting point. The relative level corresponds to the difference of the absolute levels at a given point and at the beginning of a circuit. This level is denoted as $p_{r.p}$, $p_{r.v}$, $p_{r.c}$ and is expressed in dBrp, dBrv, dBrc.

Meter level. The absolute level of power or voltage, recommended for measuring circuits or channels of a given transmission system.

The *voltage level* is equal to the ratio, in decibels, of the voltage (dBv) at a given point x to 0.775 V:

$$p_v = 20 \log(U_x/0.775)$$

The *power level* is equal to the ratio, in decibels, of the power (dBp) at examined point x to the power p_0 = 1 mW, corresponding to the zero level:

$$p_p = 10 \log(p_x/p_0)$$

If at a point where the level is measured the impedance is 600Ω, then the power level is equal to the voltage level. But if the impedance at the level test point is different from 600Ω and is equal to z, then the ratio between the power and voltage levels, expressed, respectively, in decibels, is determined by the formula:

$$p_v = p_p + 10 \log(z/600)$$

Thus, for different impedances z, the ratio of p_v and p_p is different:

$$p_v = p_p - \begin{cases} 9 \text{ dB} & \text{for } z = 75\Omega \\ 6.5 \text{ dB} & \text{for } z = 135\Omega \\ 6 \text{ dB} & \text{for } z = 150\Omega \end{cases}$$

The psophometric noise level in terms of power at the zero relative level point is expressed in decibels dBp0ps.

2.3.3 Performance Characteristics of Audio Frequency Channels

The standards on the performance characteristics of audio frequency channels, established by the same or by digital transmission methods, are given in Table 2.6. Several new parameters, for example the standard on quantization noises, phase jitter, and slip, are added to the standards on digital channels.

TABLE 2.6

Parameter	*Standards on parameters of 2500-km national channel*	*Formulas of addition*	*CCITT documents in which these parameters are examined*
Effective transmitted frequency band of compound audio frequency channel, Hz	300–3400	Does not depend on l and N (l is channel length, N is the number of audio frequency resendings)	G.232
Nominal input impedance of four-wire transit terminal of audio frequency channel, Ω	600	Same	G.232
Asymmetry attenuation of input and output circuits, dB	43	Same	Q.45, K.10
Maximum mean signal power per minute, mW0	1		—
Nominal relative level, dBrp:			
at input	−13	Same	G.232
at output	+4		
Net loss setting error, dB	0.5	Same	M.460, H.22
Difference between mean and nominal net losses of transmission, dB	0.5	Same	M.160
Mean-square deviation of net loss on channel in time from mean on 800 Hz without automatic gain control (AGC), dB	1.5	$\sqrt{N}$	M.160

TABLE 2.6 (Continued)

Parameter	*Standards on parameters of 2500-km national channel*	*Formulas of addition*	*CCITT documents in which these parameters are examined*
Change of frequency of transmittal signal on audio frequency channel, Hz, not more than	0.5	$0.5\sqrt{N+1}$	G.135, G.151, G.225, M.1020, M.910, H.14, H.22
Frequency response of net loss of channel	(See Table 2.7)		G.132, G.141, G.151, M.580
Maximum absolute signal propagation time on simple audio frequency channel between farthest centers of national network of satellite transmission systems, ms including in satellite link, ms	390 300	Proportional to l and to the delay in the circuits Same	G.114 G.114
Deviation of group signal propagation time on channel from propagation time, measured on 1900 Hz	(See Table 2.8)		G.133, G.232C., M.1020, H.12, M.1050
Amplitude response of simple audio frequency channel must be such that its net loss, measured in 300–3400-Hz band, remains constant with an accuracy of, dB:	0.3	$D_N = D_1 N^*$, where $N \leq 7$	G.232

for increase of channel input signal level, dBp0	−18 to +3.5	—	
for increase of channel input signal level, dBp0	9–20	—	
relative attenuation must be increased accordingly by not less than, in dB	1.7 and 8	—	
Mean psophometric noise voltage on simple audio frequency channel	See Section 2.3.4	See Section 2.4	G.123, G.143, G.152, G.153, M.580, G.215
Audible crosstalk interference immunity on simply audio frequency channel, dB, for 90%, not less than, and for 100% of combinations of channels	70 65	58** 52**	
Audible crosstalk inferference immunity between different transmission lines of one simple audio frequency channel, measured in four-wire part of channel on any effectively transmitted frequency, dB, not less than	62	$D_N = D_1 - 10 \log N$	M.1060
Abrupt level changes on simple audio frequency channel, permitted during normal operation of transmission system, dB, not more than	±0.5	—	M.1020, M.910, H.12

TABLE 2.6 (Continued)

Parameter	*Standards on parameters of 2500-km national channel*	*Formulas of addition*	*CCITT documents in which these parameters are examined*
Rate of gradual changes in dB/s	0.5	$D_1\sqrt{l/2500}$	—
Noise immunity of signal from parasitic modulation products, occurring due to pulsations in power circuits at nominal relative transmission level for each product, differing by frequency by ±50 and by ±100 Hz from useful signal, dB, not less than,	57	$D_l = 57 - 10 \log \sqrt{l/2500}$	G.229, G.151, G.229, H.22
in through channel on frequencies up to 400 Hz	45		
Nonlinear distortion coefficient on simple audio frequency channel in %, not more than	1.5	$D_N - D_1 \sqrt{N}$	M.1020
including on third harmonic	1.0		
Phase jitter in 20—300-Hz band	—		M.1020. Instrument Recommendation 0.91

*D_1-distortions on simple channel; D_N-distortions on compound channel with N amplifiers

**Limiting value on national channel.

TABLE 2.7*

Frequency band, kHz	*Deviation of net loss from net loss on 800* Hz *for different numbers* N *of simple connecting audio frequency channels,* dB, *not more than*			
	Excess			
0.3–0.4	1.4	2.3	3.0	3.7
0.4–0.6	0.8	1.2	1.6	1.9
0.6–2.4	0.6	0.9	1.0	1.2
2.4–3.0	0.8	1.2	1.6	1.9
3.0–3.4	1.4	2.3	3.0	3.7
	Decrease			
0.3–3.4	0.6	0.9	1.0	1.2

*The data in the table pertain to 97% of the channels of a network.

2.3.4 Standardization of Noises on Audio Frequency Channel

The following standards on the tolerable noise power in an audio frequency channel at the zero relative level point must be observed in satellite systems in accordance with CCIR Recommendation 353–3:

- the mean psophometric noise power per minute may not exceed 10,000 pW0ps during 20% of the time of any month;
- the mean psophometric noise power per minute may not exceed 50,000 pW0ps during 0.3% of the time of any month;
- the mean unweighed noise power, measured or computed in a time span of 5 ms, may not exceed 1,000,000 pw0 during more than 0.01% of the time of any year.

Noise (1000 pW0ps) due to interference from radio relay lines, operating in jointly utilized frequency bands (CCIR Recommendation 356–4), and noises (2000 pW0ps) due to interference from other satellite systems, also operating in a geostationary orbit (CCIR Recommendation 466–2), and the noises of multiplexing equipment are included in the 10,000 pW0ps standard.

In satellite systems used for communications within a country, for example with Gruppa equipment (Section 18.2) or with time-sharing multiple access message transmission equipment with analog-digital converter equipment for a multichannel signal with frequency division multi-

TABLE 2.8

Frequency, kHz	*0.3*	*0.4*	*0.5*	*0.6*	*0.8*	*1.0*	*1.4*	*1.6*	*2.2*	*2.4*	*2.8*	*3.0*	*3.2*	*3.3*	*3.4*
Deviation of group signal propagation time relative to propagation time on 1900 Hz, ms, *N* = 1	3.5	2.4	1.5	1.1	0.6	0.4	0.15	0.1	0.1	0.15	0.45	0.75	1.35	1.9	3.5
N = 2		60						Not standardized						30	

plexing (FDM), the standard for noise power on an audio frequency channel is 20,000 pW0ps (the mean psophmetric noise power per minute, not exceeded for more than 20% of the time of any month). In satellite systems in which an individual audio frequency channel is subjected to analog-digital conversion by pulse-code modulation for transmission on a line, the following standards on quantization noises are in effect: the mean psophometric quantization noise level per minute when an 800-Hz sine wave signal is fed to the input of an audio frequency channel at the zero relative level point, for the input signal powers, may not exceed

0–12.5 dBp0	−45 dBp0ps (31,600 pW0ps)
−12.5 to −32 dBp0	−50 dBp0ps (10,000 pW0ps)

This standard may not be exceeded for more than 20% of the time of any month.

The psophometric noise level on the very same audio frequency channels, if the input and output of a channel are loaded into nominal impedances, may not exceed −50 dBp0ps (10,000 pW0ps). These standards on noise in audio frequency channels, are organized by individual pulse-code modulation, and must be observed for the following standards on the error probability at the output of the hypothetical reference circuit for digital channels (see Figure 2.17):

- the mean error probability per bit per 10 minutes may not exceed 10^{-6} for more than 20% of the time of any month;
- the mean error probability per bit per minute may not exceed 10^{-4} for more than 0.3% of the time of any month;
- the mean error probability per bit per one second may not exceed 10^{-3} for more than 0.01% of the time of any year.

The standards apply to random errors, but they also may apply to packets of errors, given in consideration of interference and noise due to absorption in the atmosphere and in rain. In telephony, interference from individual components of the interference spectrum is estimated with a psophometer, which takes into consideration the frequency sensitivity function of the ear and telephone. The amplitude-frequency response of a psophmeter is given in Figure 2.18. Psophometric weighing of interference yields gain k_{ps} = 2.5 dB in terms of noises for smooth noise on an audio frequency channel with an effective band of 0.3–3.4 kHz, and up to 4 dB for noises with a triangular spectrum.

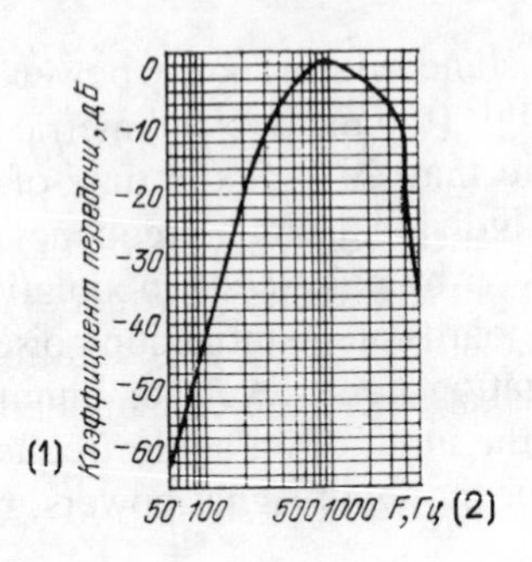

Key: **(1) Transfer coeffecient, dB**
(2) *f*, Hz

FIGURE 2.18. Amplitude-frequency response of psophometer for measuring noise on telephone channel.

In multichannel frequency division systems standard (in accordance with CCIR Recommendation 464-1) linear frequency predistortions, which produce a gain in terms of noises of 4 dB on the upper channel, are used for increasing the noise immunity of the upper channels, where the noises are strongest. The frequency response of the transfer coefficient of a predistorting circuit is given in Figure 2.19, and the transfer coefficient of the circuit is

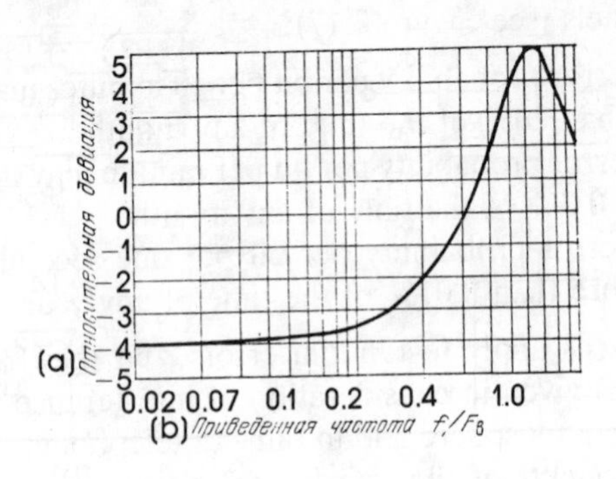

Key: **(a) Relative deviation**
(b) Reduced frequency f/F_u

FIGURE 2.19. Amplitude-frequency response of predistorting circuit for multichannel systems.

$$k_{tr} = 5 - 10 \log\left[1 + \frac{6.9}{1 + 5.25/(f_r/f - f/f_r)^2}\right]$$

(f_r is the nominal resonance frequency of predistorting and regenerating circuits and f_u is the upper modulating frequency).

Diagrams of predistorting and regenerating circuits are given in Figure 2.20. The specifications of the elements are the following:

(a)$R_1=1.81R_0$; $R_2<0.01R_0$; $R_3=R_4=R_0$; $R_5=R_0/1.81$; $R_6>100R_0$;

$$f_r=1.25F_b=\frac{1}{2\pi}\sqrt{\frac{1}{L_1C_1}}=\frac{1}{2\pi}\sqrt{\frac{1}{L_2C_2}};$$

$$\sqrt{\frac{L_1}{C_1}}=0.79R_0; \qquad \sqrt{\frac{L_2}{C_2}}=\frac{R_0}{0.79};$$

(b)$R_1=1.81R_0$; $R_2<0.01R_0$; $R_3=R_4=R_0$; $R_5=\frac{R_0}{1.81}$; $R_6>100R_0$;

$$f_r=1.25F_b=\frac{1}{2\pi}\sqrt{\frac{1}{L_3C_3}}=\frac{1}{2\pi}\sqrt{\frac{1}{L_4C_4}};$$

$$\sqrt{\frac{L_3}{C_2}}=1.47R_0; \qquad \sqrt{\frac{L_4}{C_4}}=\frac{R_0}{1.47}.$$

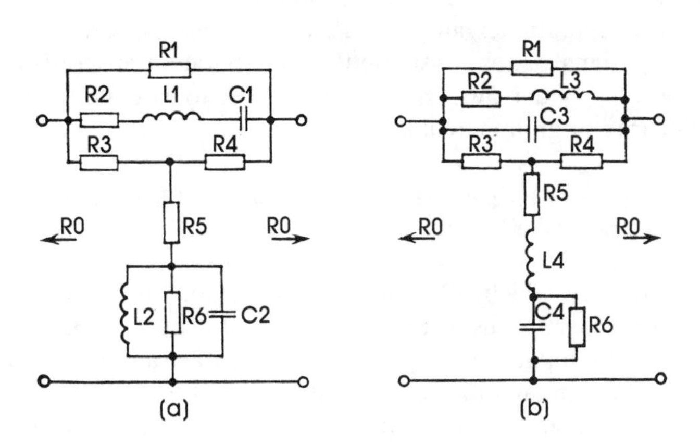

FIGURE 2.20. Diagrams of predistorting (a) and regenerating (b) circuits for multichannel telephony.

Linear frequency predistortions also may be used in single-channel systems for increasing the noise immunity.

2.3.5 Performance Characteristics of Group Circuits

The standards on the performance characteristics of primary and secondary group circuits, in satellite transmission systems, are given in Table 2.9. Virtually all the standards on the group circuits of satellite systems are the same as the corresponding standards on the group circuits of terrestrial transmission systems. The only standard in FDMA-FDM-FM satellite systems, in particular, using the Gruppa equipment (which is different from the standard on terrestrial systems), is the standard on the amplitude response for the limiting increase of the input level of a circuit. The tolerable limiting increase of the level in relation to the nominal relative level at the input of a primary group circuit using Gruppa equipment is 17 dB, while in terrestrial systems it is up to 24 dB (to an accuracy of one dB). In group circuits, formed using timesharing multiple access message transmission equipment and analog-digital converters for a 60-channel FDM signal, this standard is 24–26 dB. In a FDMA-FDM-FM system, an increase in the range of linearity of the amplitude response leads to a decrease of the carrying capacity of a satellite transponder, because it requires an increase of the frequency deviation and, accordingly, of the passband and power of the transponder. The most rigid requirements on the linearity of the amplitude response, embodied in the standards on terrestrial systems, are attributed to an effort to obtain general purpose circuits, intended for the transmission both of telephone signals and of remultiplex and broadcasting signals. In satellite systems, broadcasting signals are transmitted in a separate transponder (in an Orbita-RV transponder), which makes it possible to ease the requirements on the linearity of group circuits.

2.3.6 Performance Characteristics of a Secondary Wideband Channel for Transmitting Facsimile Signals

A secondary wideband channel is set up on the basis of typical secondary group circuits using the channeling equipment of a Gazeta-2 set, connected at the input and output of a circuit, for transmitting facsimile signals in the Orbita and Moskva satellite transmission systems. Listed in Table 2.9 are standards, which apply to a secondary wideband channel for facsimile transmission, starting at the output of a connecting link of the Gazeta-2 set on the transmitting end, and ending at the input of the connecting link of a Gazeta-2 set on the receiving end. These standards apply to up to 12,500-km links, irrespective of the kind of transmission system that is used–cable, radio relay, or satellite. In satellite systems the guaranteed summary noise immunity of a secondary wideband channel is

TABLE 2.9
Standards on Performance Characteristics of Primary and Secondary Group Circuits and Secondary Wideband Channels for Transmitting Facsimile Signal

Parameter	*Primary group circuit*	*Secondary group circuit*	*Secondary wide-band facsimile transmission channel*	*Formulas of addition (Section 2.4)*	CCITT *documents in which parameters are examined*
Working frequency band, kHz	60.6–107.7*	312.3–551.4*	312.3–551.4	Does not depend on l and N	G.211, N is the number of secondary group circuit transits, l is circuit length
Frequency band with group propagation time, kHz			330–530	Same	
Nominal relative levels:					
at transmission circuit input in terms of power, dBrp	−39.0	−36	−36	Same	G.233
in terms of voltage, dBrp	−45.5	−45	−45		
at receiving circuit output in terms of power, dBrp	−5.0	−23	−23		

TABLE 2.9 (Continued)
Standards on Performance Characteristics of Primary and Secondary Group Circuits and Secondary Wideband Channels for Transmitting Facsimile Signal

Parameter	*Primary group circuit*	*Secondary group circuit*	*Secondary wide-band facsimile transmission channel*	*Formulas of addition (Section 2.4)*	CCITT *documents in which parameters are examined*
in terms of voltage, dBrp	−11.5	−32	−32		
Receiving level setting error, dB	±0.3	±0.2	±0.3	Same	—
on frequency, kHz	82	420	420		
Accuracy with which level pattern is maintained on 420 kHz in AGC mode, dB	±1.0	±1.0	±1.0	Same	M.160
Tolerable mean-square deviation of net gain:					
on frequency, kHz	82	420			
dB, not more than	0.75	0.75	—	Same	M.160

Nominal input impedance from input and output, Ω	135	75	75	Same	G.233
Return loss at input and output of circuit in frequency band, percent	15	10	15	Same	—
Tolerable mean-load power in peak-load hour at zero relative level point, not more than, mW0:					
per hour		8	8		G.223
per minute (which corresponds to an excess of level at the nominal relative transmission level point)	4	11	11	Same	
per channel, per dBp0:					
per hour	4.8	9	9	Same	
per minute	6.0	10.4	10.4		

TABLE 2.9 (Continued)
Standards on Performance Characteristics of Primary and Secondary Group Circuits and Secondary Wideband Channels for Transmitting Facsimile Signal

Parameter	*Primary group circuit*	*Secondary group circuit*	*Secondary wide-band facsimile transmission channel*	*Formulas of addition (Section 2.4)*	CCITT *documents in which parameters are examined*
Deviation of transmission level on any frequency relative to level:					
on frequency, kHz	82	420	420	—	G.242, M.450,
not more than, dB	2.0	2.0	2.0	—	H.14, M.910, H.15
in 420–530-kHz band	—	—	1.0	—	
Immunity to audible crosstalk interference between any two simple primary or secondary circuits in working frequency band,					

dB, not less than:					
between circuits of two transmission systems of one satellite transponder	60	—	—	Section 2.4	G.242, G.232, G.233
between direct and return lines of given circuit of one system, dB	60	55	—	Same	
between circuits of one system	—	74	—	Same	
Immunity to selective interference at output of receiving circuit, db, not less than:					
to any interference in frequency band, kHz	60.6–107.7 35	312.3–551.4 35	312.3–551.4 35	35–20 log N ($N \leq 4$)	M.910, H.14, H.15
on carrier fre-	26	26	—	—	N is the number

TABLE 2.9 (continued)
Standards on Performance Characteristics of Primary and Secondary Group Circuits and Secondary Wideband Channels for Transmitting Facsimile Signal

Parameter	*Primary group circuit*	*Secondary group circuit*	*Secondary wide-band facsimile transmission channel*	*Formulas of addition (Section 2.4)*	CCITT *documents in which param-eters are examined*
quency of given group					of transit sections
Change of fre-quency of trans-mitted signal, Hz, not more than	0.5	0.5	—	$0.5\sqrt{N+1}$	G.241, M.910
Immunity to summary in-terference (un-weighted, fluctu-ation and selective):					
at output of circuit in band, kHz with other	60.6–107.7	312.3–551.4	312.3–551.4	—	—
groups of sys-tem in opera-tion, dB, not less than	30	26	22	—	—
Amplitude re-	82	420	—	—	G.223

sponse should be rectilinear for increasing the level of frequency, kHz					
at output of transmission circuit in relation to nominal, dB	24**	26	—	—	
to accuracy, dB, not less than	1.0	1.0	—	$\sqrt{N+1}$	
Resulting nonuniformity of group propagation time response on channel, μs, not more than	On simple channel in 65–103-kHz band (excluding the 82–86-kHz band) 60 μs from minimum value	On simple channel in 330–530-kHz band (excluding the 405–419-kHz band) 30 μs from minimum value	Up to 12,500 km (Figure 2.21)	For primary group circuit $60(N + 1) + 100N$ where 100 μs is the delay in the transit filter For secondary group circuit $30(N + 1) + 20N$, where 20 μs is the delay in the transit filter	M.910, H.14, H.15

* In satellite systems a signal is tranmitted, respectively, in the frequency bands of 12.6–59.7 and 12.3–251.4 kHz.

** The standard on the amplitude response is 17 dB in a primary group circuit, organized with the aid of Gruppa equipment by the FDMA-FDM-FD method.

30–35 dB, compared with the 23-dB standard on this parameter in a through-circuit. Rigid requirements are imposed on the amplitude-frequency response and phase-frequency response for facsimile transmission by single-sideband AM, generated by a Gazeta-2 complex.

Amplitude-frequency distortions of a signal reduce the contrast of an image, while phase-frequency distortions reduce the clarity of a picture. A channel is corrected in order to achieve the necessary amplitude-frequency response and group propagation time. The group propagation time response of a secondary wideband channel for facsimile transmission, from the input to the output of a connecting link of a Gazeta-2 complex, is given in Figure 2.21.

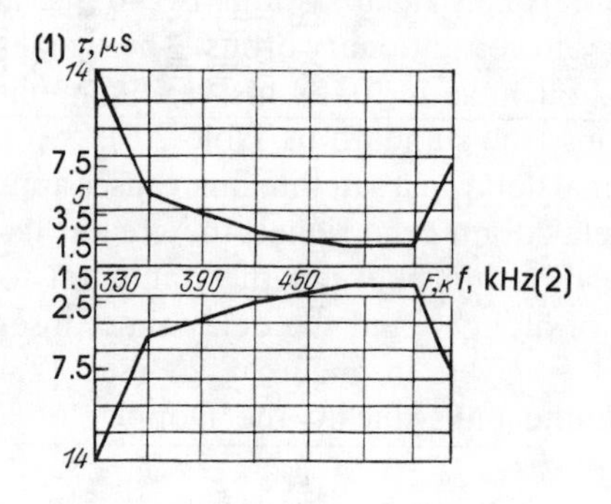

Key: (1) μs
(2) kHz

FIGURE 2.21. Frequency response of group propagation time of secondary wideband channel for facsimile transmission.

A facsimile signal is transmitted in the Orbita-RV system in discrete form on a primary digital channel at the rate R = 2048 kb/s, with an error probability not exceeding 10^{-6}. A facsimile signal reaches an earth station either in analog form, or in digital form, on a primary digital channel of terrestrial links at the rate R = 2048 kb/s (the error probability is 10^{-6}). A digital facsimile signal is recoded at an earth station for synchronous digital signal input into a common communications channel.

2.3.7 Features of the Standardization of Audio Frequency Channels and Group Circuits of Satellite Systems

Numerous parameters of audio frequency channels (listed in Table 2.3) and of group circuits (listed in Table 2.9) are determined, not so much by the satellite link circuit, as by the audio frequency channeling and group circuit equipment. For instance, the input impedance, return loss, mismatch, and asymmetry of the impedances are determined by the input and output circuits of the multiplexing equipment. The amplitude-frequency and group propagation time responses also are determined by the selective circuits of the multiplexing equipment. The stability of the net gain is determined by the multiplexing equipment and channeling equipment of a satellite transmission system.

At the same time, such parameters as the signal-to-noise ratio, the change of frequency of a transmitted signal, and its delay in the line depend primarily on the satellite circuit. For instance, in connection with the great length of an earth-satellite-earth station link, the minimum absolute signal propagation time between earth stations is 240 ms, and the maximum is 280 ms for satellites in geostationary orbits. The signal propagation time on connecting terrestrial links is 10–50 ms. The summary delay of a signal reaches 250–290 ms (the standard in Table 2.6).

The long signal delay in a satellite link causes appreciable echoes, and the greater the delay of an echo signal, the greater its suppression $a_{e\ \min}$ is required. The relationship between the minimum tolerable attenuation $a_{e\ \min}$ in the path of an echo and the delay time, inserted by the channel for satellite links is given in Figure 2.22 (see [2.4]). The change of frequency on satellite links due to the Doppler effect was examined in Chapter 1.

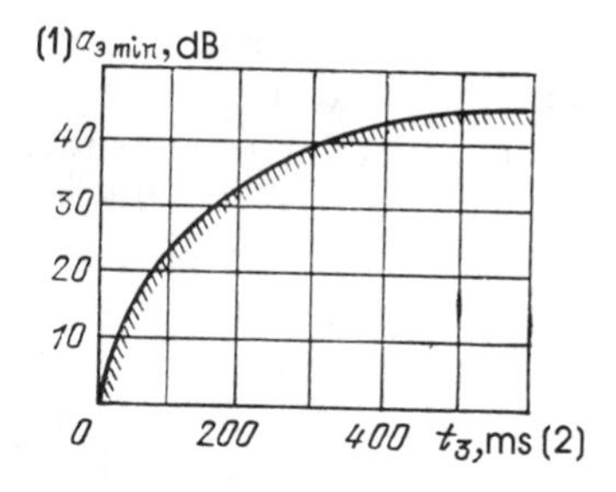

Key: (1) dB
(2) ms

FIGURE 2.22. Minimum tolerable electrical echo signal attenuation as function of delay time for satellite links.

2.3.8 Measurement of the Parameters of Audio Frequency Channels and Group Circuits

The parameters of the channels and group circuits of satellite transmission systems are measured by standard methods (see [2.5]).

However, the mentioned methods pertain to analog transmission systems. We will examine ways of measuring some of the parameters of digital transmission systems and procedures for measuring the group propagation time on audio frequency channels and wideband channels.

Quantization noise during the transmission of audio frequencies by individual pulse code modulation (PCM) are measured by supplying a sine wave signal on 800 Hz to the input of an audio frequency channel. A quantization noise meter (or if there is no such instrument, a linear distortion meter of the S6-5 type), which makes it possible to separate a sinusoidal signal from noise, is connected to the output of the channel. The ratio of the power of noise to the power of psophometric quantization noise (in dB) is

$$P_s P_{n \cdot q \cdot ps} = 10 \log[P_s / P_{n\Sigma} - P_h) + 4]$$

where $P_{n\Sigma}$ is the summary power of an interference in the band of audio frequency channel with a trap on 800 Hz; P_h is the summary power of the second and third harmonics; 4 dB is the weighing coefficient of the psophometer; and P_s is the power of the input signal of the audio frequency channel.

Audible crosstalk interference between audio frequency channels in group circuits, generated by analog-digital conversion of the group FDM signal, should be measured with several unoccupied channels loaded, because, when the output signal of the analog-digital converter of the group signal is weak, quantization noises are strong. Under actual conditions, there is no signal at the input of the analog-digital converter. According to experience in measurements of audible crosstalk interference between channels, its level decreases up to 10 dB when a load appears in other unoccupied audio frequency channels.

The summary psophometric noises at the output of the circuits of analog and digital FDM transmission systems in the frequency band of a telephone channel are measured with an instrument for crosstalk interference. A white noise signal with a level corresponding to the load of a given circuit (see Table 2.9) is supplied to the input of a circuit in the band of the group circuit from the transmitting complex of the instrument for measuring crosstalk interference. In the transmitting complex of the instrument for measuring crosstalk interference, on channels where tests are being conducted, traps are installed, which eliminate the noise. The receiving part of the instrument is connected to the output of a circuit for measuring noises. This method of measuring circuits, using white noise signals, makes it possible to find the complete characteristics of circuits from the standpoint of crosstalk interference, which appears for different reasons in a circuit. The central frequencies of the channels of satellite systems, on which measurements are conducted, the load levels, and white noise bands are regulated on the international level by CCIR Recommendation 482-1.

Phase jitter (or phase fluctuation) results in a shift of the time positions of regenerated pulses for various reasons. The phase jitter effect can occur in the transmission of data both on audio frequency channels and in digital line circuits. The main harmonic components of phase jitter, which influence the transmission of data on audio frequency channels, are transitions of call current, variable current of a network and their harmonics: from the second to the fifth. These components cause phase modulation with a small index of up to 25°, which appears in a band of ±300 Hz on both ends of an audio frequency channel, acting as a carrier. An instrument, the characteristics of which are described in CCITT Recommendation 0.91, and the operation of which is based on the principle of the detection of phase modulation, is used for measuring jitter. Distortions occur in the line circuits of digital satellite transmission systems, connected with the conditions under which the clock frequency signal is generated in regenerators, and caused basically by the line circuits of satellite links. An instrument in which the clock frequency selector is a narrow-pass filter, tuned to that frequency, is used for measuring phase jitter in digital transmission systems. The selected clock frequency voltage is applied to a two-way amplitude limiter. The phase fluctuation of the oscillation thus obtained is selected by a frequency detector (see [2.6]).

The relative group propagation time of a signal on audio frequency channels is determined by the difference of the group propagation times, measured on 1900 Hz, where the group propagation time is the longest, and on the test frequency. Instruments are used for measuring the group propagation time, which utilize the following property of an AM oscillation: if the frequency F of the modulating oscillation is much lower than the frequency of the oscillation being modulated, then phase shift ϕ_F of the envelope of the AM oscillation is proportional to the group propagation time of the modulated carrier oscillation, i.e., when $F << f$, $\tau_{gr} = \phi_F/2\pi F$.

The group propagation time on wideband channels, in particular for facsimile, is measured with panorama instruments—instruments that measure the characteristics of the group propagation time (F4-4, F4-10). Measurements can also be conducted using the points of the test instrument on the transmitting end and a special converter on the receiving end on 330–530 kHz. The nonuniformity of the group propagation time is measured relative to the group propagation time on 420 kHz.

2.4 ADDITION OF DISTORTIONS ON COMPOUND SATELLITE CHANNELS

2.4.1 Addition of Distortions on Compound Video Channels

A compound satellite channel consists of three separate sections: the channel of a satellite link, and the channels of two connecting terrestrial links (between earth stations and the source and recipient of information). Formulas for finding distortions on compound satellite video channels, if the distortions in the individual sections are known, are given in Table 2.10. These formulas can also be used for solving the problem of the distribution of requirements among individual sections of a channel, if the standard on the entire channel is given. The formulas in Table 2.10 were found on the basis of the formulas for determining distortions on a terrestrial channel, contained in a state standard and CCIR Recommendation 567.

2.4.2 Addition of Distortions on Compound Audio Broadcasting Channels

The formulas given in Table 2.10 for determining the parameters of a compound video channel are empirical and yield approximate results. Statistical methods of adding distortions are more promising than addition by empirical formulas, because they make it possible not to establish unnecessarily rigid standards, but to prescribe them with a certain probability of excess. Amplitude-frequency distortions in compound audio broadcasting circuits and on compound audio frequency channels actually are added statistically.

The deviations of transfer coefficient K_{tr} from its value on 1000 Hz are standardized for each circuit and its individual links; the field of tolerances is established for parts of the frequency band. The distribution of K_{tr} on an individual frequency is assumed to be normal, and tolerances are added up as follows: the mean frequency response of several series-connected links or circuits is determined as the sum of the mean frequency responses a_{1i} (in decibels); the mean-square deviations of the amplitude-frequency response a_{2i} are added up quadratically (in decibels). Thus, the amplitude-frequency response of a compound circuit is calculated by the formula:

$$\sum_{i=1}^{N} a_{1i} \pm \left(\sum_{i=1}^{N} a^2_{2i} \right)^{1/2} \tag{2.3}$$

TABLE 2.10

How distortions are combined	*Formula*	*Unit of measurment*	*p*
Arithmetically	$D_N = \sum_{i=1}^{N} D_i$	%	1
Quadratically	$D^2_N = \sum_{i=1}^{N} D^2_i$	dB, ns, %, deg	2
By 3/2	$D_N^{3/2} = \sum_{i=1}^{N} D_i^{3/2}$	dB, ns, %, deg	3/2
By power	$D_N = -10 \log \sum_{i=i}^{N} 10^{-0.1D_i}$		–
By voltage	$D_N = -20 \log \sum_{i=i}^{N} 10^{-0.05D_i}$		–

The middle line of the tolerance field, drawn half-way between the boundaries of the tolerable frequency deviations, may be used as the average amplitude-frequency response. During the calculations it is assumed that the mean-square deviations are equal to the tolerances on the amplitude-frequency response, because the tolerances on the deviation from the amplitude-frequency response in different sections must be given for the identical percentage of time. It is usually assumed that K_{tr} of a section will fall within given tolerances with a probability of 98 percent.

The harmonic coefficient is assumed to be added quadratically (see Table 2.10). Energy (in terms of power) addition of noises is used (see Table 2.10). The standard on audible crosstalk interference is identical, both on a compound circuit, and on its individual links. The least noise immunity of individual sections is taken as this standard.

2.4.3 Addition of Distortions on a Compound Audio Frequency Channel

For determining the performance characteristics of a compound channel, the performance characteristics are established as functions of the lengths of the individual sections and of the number of amplifiers. Such parameters as the mean-square and maximum deviations of the net loss and the harmonic coefficient are added quadratically (see Table 2.10).

The nonuniformity of the amplitude-frequency response of an audio frequency channel, depending on the number of amplifiers, is given in Table 2.6. This table is valid for the addition of distortions of several sections with the identical characteristics. The mentioned standards may be violated in three percent of the channels. For adding distortions that occur in sections with different characteristics, a method is used for adding distortions, which is described in the rules of the addition of distortions in audio broadcasting circuits [see (2.3)].

Energy summation is used for adding noises on a compound audio frequency channel (see Table 2.10). Here, as is known, the noise level in a satellite link does not depend on the distance between the earth stations, and in the terrestrial sections it is determined, depending on length, by the formula:

$$P_{nl} = P_{n.r}\sqrt{l/2500}$$

where P_{nl} is the noise power on a terrestrial channel of legnth l in km; $P_{n.r}$ is the noise power on the reference channel; 2500 is the length of the reference channel in km. Then the noise power on a compound channel is $P_{n\Sigma} = P_{n.sat} + P_{nl}$, where $P_{n.sat}$ is the noise power on a satellite channel. The standard on the noise immunity relative to audible crosstalk interference on a compound audio frequency channel is established in the same way as on a compound audio broadcasting channel.

Given in Tables 2.6 and 2.9 are individual kinds of distortions as functions of the length and number of amplifier sections.

Transponder problems include the tolerable inperfection of the amplitude-frequency response, group propagation time, AM-PM, the selectivity of filters, *et cetera* (during the transmission of carriers, frequency-modulated by analog telephone messages or by video signals to a tranponder using multiple access). These problems are examined quite completely in the literature [2, 16, 17].

In digital systems the above-mentioned imperfection of the characteristics of a circuit, and also the imperfection of the performance of the synchronization circuits of a coherent modem (of the circuits that regenerate the carrier and clock frequency) result in energy losses; the reliability of transmission decreases, and phase jitter and desynchronization occur in the regenerated signal. The problem, most urgent for satellite communications systems, namely determining the requirements on the parameters of HF circuits, carrying wideband PSK signals, is examined in Sections 7.8 and 7.9. A systematic presentation of the principles of the statistical theory of the demodulation of discrete signals is given in [2.8].

Chapter 3

The Frequency Bands Reserved for Satellite Communications and Broadcasting Systems

3.0 INTRODUCTION

At its radio conferences, the ITU, a specialized organization of the United Nations, develops appropriate regulatory rules and procedures based on investigations conducted by its member countries and presented to the CCIR, and assigns frequency bands among different radio services. The fundamental international document that regulates the use of frequencies is the CCIR Radio Regulations [1]. It contains a table of the distribution of frequency bands among services, individual technical limits imposed on the joint utilization of frequencies by different services, procedures for coordinating systems, and also rules for registering frequency assignments in the International Frequency Registration Board (IFRB).

3.1 THE DISTRIBUTION OF FREQUENCY BANDS AMONG SERVICES

The frequency distribution table contains allocations of frequency bands for use by certain radio services between 9 kHz and 275 GHz. The table consists of three columns, each of which corresponds to one of three regions, into which the globe is divided in relation to the distribution of frequency bands. Region 1 includes Europe, Africa, the USSR, and the Mongolian People's Republic; Region 2 includes North America and South America; and Region 3 includes Asia, Oceania, and Australia. (More precise definitions and boundaries of the regions are given in the regulations in the book [1]). This table contains bands, distributed for a whole host of space radio communications services, including fixed, intersatellite, space operation, land, sea, or airborne mobile, radio navigation, radio broadcasting, radio determinations, earth exploration, and numerous other services.

Given below are the frequency bands, distributed in accordance with the CCIR Radio Regulations for fixed (Table 3.1), radio broadcasting (Table 3.2), and mobile (Table 3.3) satellite services. Frequency bands, assigned to the intersatellite service, GHz:

22.5–23.55; 32–33; 54.25–58.2;
59–64; 116–134; 170–182; 184–190.

3.2 INTERNATIONAL REGULATION OF THE UTILIZATION OF FREQUENCIES IN BANDS ASSIGNED FOR THE SATELLITE SERVICES

In addition to the distribution of frequency bands among the services, described in Section 3.1, there are special rules agreed upon on an international level for the use of these bands by satellite services. These rules are set forth in numerous Articles of the CCIR Radio Regulations [7] and in certain other international documents [6]. There are two fundamentally different approaches to the use of frequency bands:

1. on the basis of a gradual buildup of the load on the frequency bands and a geostationary orbit in accordance with international coordination of new systems, and
2. on the basis of international plans, adopted for the utilization of individual frequency bands.

According to the CCIR Radio Regulations any frequency assignment to an earth or space station, both for reception and for transmission, must be applied for and registered at the IFRB:

if the use of a given frequency can cause harmful interference to any service of another administration, or
if a frequency must be used for international communications, or
if formal international recognition of the use of that frequency is desired.

An analysis of the above-listed conditions shows that virtually *all satellite systems must be registered with the International Frequency Registration Board.* The registration process has several stages, including the following: preliminary publication, determination of the need for coordination with other satellite systems or with terrestral services of other administrations, coordination and application for registration.

Preliminary Publication and Determination of the Need for Coordination. An administration that intends to build a satellite system must not earlier than five years, and preferably not later than two years before the planned date of activation of a satellite network, send to the IFRB information, listed in Appendix 4 to the CCIR Radio Regulations [7]. This information

TABLE 3.1

Frequency Bands for Fixed Satellite Service

Space-Earth	*Earth-Space*
2,500–2,690 MHz (Region 2) 2,500–2,535 MHz (Region 3)	2,655–2,690 MHz (Regions 2 and 3)
3,400–4,200 MHz 4,500–4,800 MHz 7,250–7,750 MHz	5,725–7,075 MHz (Region 1) 5,850–7,075 MHz (Regions 2 and 3) 7,900–8,400 MHz 10.7–11.7 GHz (Region 1 only for feed lines to radio broadcasting satellites, operating in a band of about 12 GHz)
10.7–11.7 GHz 11.7–12.3 GHz (Region 2) 12.2–12.5 GHz (Region 3) 12.5–12.75 GHz (Regions 1 and 3)	12.5–12.75 GHz (Region 1) 12.7–12.75 GHz (Region 2) 12.75–13.25 GHz 14.0–14.5 GHz 14.5–14.8 GHz (only for feed lines to radio broadcasting satellites, operating in a band of about 12 GHz for countries outside of Europe) 17.3–18.1 GHz (only for feed lines to radio broadcasting satellites, operating in a band of about 12 GHz)

TABLE 3.1 (Continued)

Frequency Bands for Fixed Satellite Service	
Space-Earth	*Earth-Space*
17.7–21.2 GHz 37.5–40.5 GHz	27–27.5 GHz (Regions 2 and 3) 27.5–31 GHz 42.5–43.5 GHz 49.2–50.2 GHz 47.2–49.2 GHz (only for feed lines to radio broadcasting satellites, operating in the 40.5–42.5 GHz band)
81–84 GHz	50.4–51.4 GHz 71.0–75.5 GHz 92–95 GHz 202–217 GHz 265–275 GHz

Comment. If a region is not indicated in Table 3.1, then the given band is reserved for the fixed satellite service on a global basis. All frequency bands are shared by the fixed satellite service jointly with other services; there are no bands that are assigned on an exclusive basis. In this connection, the conditions for the joint use of frequency bands are set forth in the CCIR Radio Regulations (IRR).

TABLE 3.2

Space-Earth link

Frequency band	*Region*	*Frequency band*	*Region*
620–790 MHz[1]	1, 2, 3	12.3–12.7 GHz[4]	2
2,500–2,690 MHz[2]	1, 2, 3	12.5–12.75 GHz[5]	3
11.7–12.5 GHz[3]	1	22.5–23 GHz[6]	2, 3
11.7–12.2 GHz[3]	3	40.5–42.5 GHz	1, 2, 3
12.1–12.3 GHz[4]	2	84–86 GHz	1, 2, 3

Comments. 1. The 620–790-MHz band may be used for FM satellite TV broadcasting systems if the interested administrations so agree. These systems may not create a power flux density at the earth's surface in other countries greater than -129 dBW/m^2 for angles of incidence smaller than 20°. 2. The use of the 2,500–2,690-MHz band for the satellite broadcasting service is limited to national and regional systems for collective reception by agreement with the interested administrations. 3. The 11.7–12.2-GHz band in Region 3 and the 11.7–12.5-GHz band in Region 1 must be used in accordance with the plan adopted by the World Administrative Radio Conference in 1977 (WARC-1977). 4. The satellite broadcasting service in Region 2 in the 12.1–12.7-GHz band must operate in accordance with the plan which will be worked out at the regional conference in 1982. This same conference must define the lower boundary of the band. 5. The satellite broadcasting service in Region 3, in the 12.5–12.75-GHz band, is limited to collective reception systems with a power flux density at the earth's surface of -111 dBW/m^2. 6. The use of the 22.5–23-GHz band in Regions 2 and 3 for the satellite broadcasting service is permitted if the interested administrations so agree.

must contain data on the orbit of the satellite, information about the service areas, the frequency band, the maximum spectral power density (W/Hz), delivered to the antennas of the earth and space stations, the radiation patterns of the antennas, and also the noise temperature of the receivers, the equivalent noise temperature of the link*, *et cetera*. If after

---

*For using simple transponders with frequency transfer on a space station the minimum equivalent noise temperature (T_{link}) of the link and the transfer coefficient (γ), connected with it, between the outputs of the receiving antennas of the space and earth stations are indicated. In this case $T_{\text{link}} = T_e + \gamma T_{\text{sat}}$, where T_e is the noise temperature of the receiver of the earth station at the output of the receiving antenna of the earth station, and T_{sat} is the noise temperature of the receiver of the space station at the output of the receiving antenna of the satellite.

TABLE 3.3

Frequency band	*Route*[5]	*Kind of service*
235–322 MHz[1]	—	Mobile satellite
335.4–399.9 MHz	—	
406–406.1 MHz[2]	Earth-space	Mobile satellite
1,530–1,544 MHz	Space-earth	Maritime mobile satellite
1,544–1,545 MHz		Mobile satellite
1,545–1,559 MHz		Air mobile satellite
1,626.5–1,645.5 MHz	Earth-space	Maritime mobile satellite
1,645.5–1,646.5 MHz		Mobile satellite
1,646.5–1,660.5 MHz[4]		Air mobile satellite (*R*)
19.7–20.2 GHz[3]	Space-earth	Mobile satellite[6]
20.2–21.2 GHz		
29.5–30 GHz[3]	Earth-space	Mobile satellite
30–31 GHz		
39.5–40.5 GHz	Space-earth	Mobile satellite
43.5–47 GHz	—	
66–71 GHz	—	
71–74 GHz	Earth-space	Mobile satellite
81–84 GHz	Space-earth	Mobile satellite
91–100 GHz	—	
134–142 GHz	—	
190–200 GHz	—	
252–265 GHz	—	

Comments. 1. By prior agreement with the interested administrations. 2. For distress radio beacons, using a power of not more than 5 W. 3. On a secondary basis. 4. The frequencies allocated for the air mobile satellite service (*R*) are reserved for communications between any airship and stations of the mobile service that are intended for maintaining flight safety and regularity on national and international civil aviation lines. 5. A route not indicated in the table means that the frequency band may be used for both lines. 6. All frequency bands for mobile satellite services are shaped on a global basis.

studying this information, published in the weekly circular of the IFRB, any administration considers that its existing or planned services may experience intolerable interference, it should within four months after the date of publication in the weekly circular of the mentioned information, send its comments to the applicant administration. On receiving comments from other administrations about the possibility of unacceptable interference, the applicant administration should exert all efforts to attempt to solve any problems and to satisfy their demands.

At the same time, Resolution 2, relative to the use of the frequency bands by all countries with equal rights for space radio communications services, prescribes that the registration by the ITU of frequency assignments for space radio communications services and their use may not give any permanent priority to any particular country or group of countries and may not impede the development and construction of space communications systems by other countries. Therefore, both parties are obliged to find a mutually acceptable solution during the coordination process. Here coordination must be carried out with those frequency assignments that are in the same frequency band as a planned assignment, correspond to the frequency distribution table, and are listed in the reference register of the IFRB or are already the object of coordination.

For determining the need for coordination with any system, the apparent increase of the equivalent noise temperature of a satellite link caused by interference and created by an examined system, is calculated, and the ratio of this increase to the equivalent noise temperature of the satellite link, expressed in percentage points, is compared with a threshold.

We will examine two possible cases:

1. useful and interfering networks jointly use the same or several frequency bands, and both networks transmit in the same direction;
2. a useful and an interfering network utilize jointly the same or several frequency bands, and the networks conduct transmission in opposite directions (reverse utilization of frequencies).

In the first and most common case (Figure 3.1), when simple transponders with frequency conversion are used in a system, the increment of the equivalent noise temperature of the link may be found from the expression:

$$\Delta T_{\text{link}} = \Delta T_e + \gamma \Delta T_{\text{sat}} \tag{3.1}$$

where ΔT_e is the increment of the noise temperature of the receiver of the earth station at the output of the receiving antenna of the earth station (K); ΔT_{sat} is the increment of the noise temperature of the receiving system of the space station at the output of the receiving antenna of the space station (K); and γ is the transfer coefficient of the satellite link between the output of the receiving antenna of the space station and the output of the receiving antenna of the earth station, usually less than unity and characterizing the extent of the influence of interference, created in the earth-satellite link. Written in greater detail:

$$\Delta T_{\text{sat}} = \frac{P_{\text{e.i}}\, Ge.i\,(\theta_t)\, G_{\text{sat.s}}\,(\delta)}{kL_{\text{u}}} \tag{3.2}$$

$$\Delta T_{\text{e}} = \frac{P_{\text{sat.i}}\, G_{\text{sat.i}}\,(\eta)\, G_{\text{e.s}}\,(\theta_t)}{kL_d} \tag{3.3}$$

where $P_{\text{sat.i}}$ and $P_{\text{e.i}}$ are the maximum power density in a 1-Hz band, averaged in the worst band of 4 kHz for carriers below 15 GHz, and in the 1-MHz band for carriers above 15 GHz, supplied to the antennas of the interfering satellite, and powering the earth station, respectively; $G_{\text{sat.i}}\,(\eta)$ is the gain of the transmitting antenna of the interference; $G_{\text{e.s}}\,(\theta_t)$ is the gain of the receiving antenna of the earth station, subjected to interference, in the direction of the interfering satellite; $G_{\text{e.i}}\,(\theta_t)$ is the gain of the transmitting antenna of the interfering earth station in the direction of the satellite, subjected to interference; $G_{\text{sat.s}}\,(\delta)$ is the gain of the receiving antenna of the satellite, subjected to interference, in the direction of the interfering earth station; k is the Boltzmann constant (1.38×10^{-23} W/Hz·K); L_u and L_d are losses to transmission in open space on the earth-satellite link and satellite-earth link, respectively; and θ_t is the topocentric angular distance between satellites in consideration of tolerances on the longitudinal confinement of a satellite.

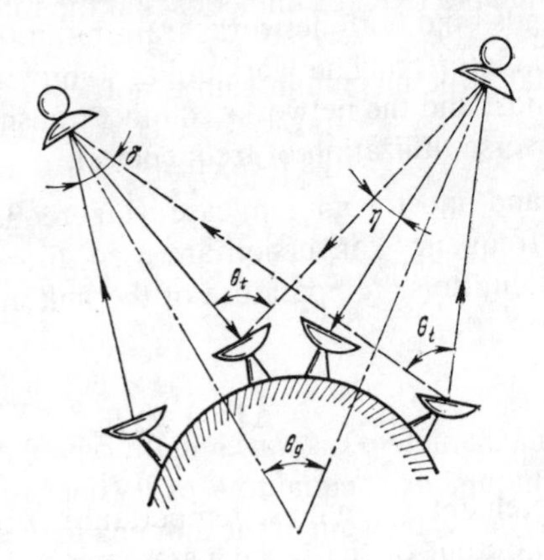

FIGURE 3.1. Calculation of the increment of the equivalent noise temperature.

Losses (in dB) to transmission in free space are

$$L = 20(\log f + \log d + 32.45 \tag{3.4}$$

where f is frequency in MHz and d is distance in km.

The distance between an earth station and geostationary satellite is

$$d = 42644 \sqrt{1 - 0.2954 \cos \psi} \quad (3.5)$$

where $\cos \psi^* = (\cos \xi) \cos \beta$, ξ is the latitude of the earth station, and β is the longitudinal difference between the satellite and earth station.

The topocentric angular distance between two geostationary satellites for a given earth station is

$$\theta_t = \text{arc cos} \left(\frac{d^2_1 + d^2_2 - (84.332 \sin (\theta_g/2)^2}{2d_1 \, d_2} \right) \quad (3.6)$$

where d_1 and d_2 are the distances (in km) from the earth station to both satellites, respectively, computed by formula (3.5) and θ_g is the geocentric angular distance between the satellites.

The gains of the space antennas are determined on the basis of the corresponding application characteristics, submitted by administrations to the IFRB. The gains of the antennas of the earth stations can be determined either on the basis of actually measured characteristics, or on the basis of the appropriate CCIR recommendations. In cases when this information is not available IRR recommends that the following reference radiation patterns be used:

a) for $D/\lambda \geq 100$* (the maximum gain is not less than 48 dB)

$$G(\phi) = G_{\max} - 2.5 \cdot 10^{-3} [(D/\lambda) \phi]^2 \quad \text{for } 0 < \phi < \phi_m$$
$$G(\phi) = G_1 \quad \text{for } \phi_m \leqslant \phi < \phi_r$$
$$G(\phi) = 32 - 25 \log \phi \quad \text{for } \phi_r \leqslant \phi < 48°$$
$$G(\phi) = -10 \quad \text{for } 48° \leqslant \phi \leqslant 180°$$

where D is the antenna diameter; λ is wavelength; ϕ is an angle in degrees, measured from the antenna axis, equal to θ_t or θ_g, depending on the case; and $G_1 = 2 + 15 \log (D/\lambda)$ is the gain of the antenna in the direction of the peak of the first sidelobe of the radiation pattern:

---

*If $\cos \psi < 0.151$, then the satellite is under the plane of the horizon and is not visible from a given earth station.

**If D/λ is not given, then the following formula may be used:

$$20 \log (D/\lambda) \approx G_{\max} - 7.7$$

$$\phi_m = (20\lambda/D)\sqrt{G_{max} - G_1} \quad \text{in deg}$$
$$\phi_r = 15.85\,(D/\lambda)^{-0.6} \quad \text{in deg}$$

b) for $D/\lambda < 100$ (the maximum gain is approximately 48 dB):

$$G(\phi) = G_{max} - 2.5\cdot10^{-3}\,[(D/\lambda)\,\phi]^2 \quad \text{for } 0 < \phi < \phi_m$$

$$G(\phi) = G_1 \quad \text{for } \phi_m \leqslant \phi < 100\,\lambda/D$$

$$G(\phi) = 52 - 10\log(D/\lambda) - 25\log\phi \quad \text{for } 100\,\lambda/D \leqslant \phi < 48°$$

$$G(\phi) = 10 - 10\log(D/\lambda) \quad \text{for } 48° \leqslant \phi \leqslant 180°$$

If modulation is changed on-board the satellite, or if a transmission comes from on-board the satellite, then the increment of the noise temperature is compared with the total noise temperature of a specific examined link (a space or earth station, respectively). In this case the equivalent noise temperature of the entire satellite link and transmission gain (γ) are not used, and ΔT_e and ΔT_{sat} are examined separately. In the latter case, when the systems use the same frequency band for transmitting in opposite directions, only the interference between satellites is analyzed, and the interference between the earth stations must be examined during the coordination process, which is similar to the coordination of earth and terrestrial stations (Section 5.4).

In this case

$$\Delta T_{link} = \gamma \Delta T_{sat} \tag{3.7}$$

or

$$\Delta T_{link} = \gamma P_{sat.i} G_{sat.i}(\eta_s)\, G_{sat.s}(\delta_s)/kL_s \tag{3.8}$$

where $G_{sat.i}(\eta_s)$ is the gain of the transmitting antenna of the interferring satellite in the direction of the satellite that is subjected to interference; $G_{sat.s}(\delta_s)$ is the gain of the receiving antenna of the satellite that is subjected to interference in the direction of the interfering satellite; L_s are losses to transmission in free space in a satellite-satellite link, determined by expression (3.4) as a result of substituting the distance d_s =84,322 sin(θ_g/2) between the satellites, where θ_g is the geocentric angular distance.

If information is available on the polarization that is being used in the systems, then the additional polarization isolation (Y) can be taken into consideration, and then:

$$\Delta T_{\text{link}} = \Delta T_c / Y_d + \gamma \Delta T_{\text{sat}} / Y_u \tag{3.9}$$

or for the second case:

$$\Delta T_{\text{link}} = \gamma \Delta T_{\text{sat}} / Y_{\text{s.s}} \tag{3.10}$$

Polarization isolation in expressions (3.9) and (3.10) is determined in accordance with Table 3.4.

During calculations of ΔT it is necessary for each satellite receiving antenna of a network that is experiencing interference to determine the most unfavorably situated transmitting earth station of an interfering satellite network by superimposing the service areas in the earth-space link of the interfering network on the gain contours of the receiving antenna of the space station, drawn on a map of the earth's surface. The most unfavorably situated transmitting earth station is the one toward which the gain of the receiving antenna of the satellite of the network that experiences interference is the greatest. The most unfavorably situated earth station must be determined in a similar manner for each service area in the space-earth link.

TABLE 3.4

Polarization of Systems		*Polarization isolation coefficient Y*
Useful	*Interfering*	
Left-hand circular	Right-hand circular	4
Left-hand circular	Linear	1.4
Right-hand circular	Linear	1.4
Left-hand circular	Left-hand circular	1
Right-hand circular	Right-hand circular	1
Linear	Linear	1

The computed value of $\Delta T_{\text{link}}/T_{\text{link}}$, expressed in percentage points, must be compared with a threshold of 4% [7]. If the computed ratio does not exceed the threshold, then coordination is not necessary.

When there is interference in just one link, i.e., in an up or down link, the ratio $\Delta T_e/T_e$ and $\Delta T_{sat}/T_{sat}$, expressed in percentage points, must be compared with a 4%-threshold. When there is interference in both up and down links, between which modulation changes onboard the satellite, each of the values of $\Delta T_{sat}/T_{sat}$ and $\Delta T_e/T_e$, expressed in percentage points, is compared with a 4% threshold. If just one of the calculated ratios for one of the interacting systems exceeds the threshold, then coordination between the systems is necessary.

An Example of the Calculation. In the example given here it is suggested that two identical satellite systems be used, each of which has a transponder that utilizes simple frequency transfer and global coverage antennas. All topocentric angles are assumed to be $\theta_t = 5°$. Given this separation for the antenna of an earth station with $D/\lambda > 100$, in accordance with the reference pattern of 32-25 log θ_t, the gain is 14.5 dB in the direction of the satellite of the other network. The starting data that are used for the calculation are given in Table 3.5.

TABLE 3.5

Transmission line	*Conventional designations of the parameter of a line**	*Value*	*Unit of measurement*
Earth-space on 6,175 MHz	$P_{e.i}$	−37	dB (W/Hz)
	$G_{e.i}(\theta_t)$	14.5	dB
	$G_{sat.s}(\delta)$	15.5	dB
	L_u	200	dB
Space-earth on 3.950 MHz	$P_{sat.i}$	−57	dB (W/Hz)
	$G_{sat.i}(\eta)$	15.5	dB
	$G_{e.s}(\theta_t)$	14.5	dB
	L_d	19.6	dB

*$10 \log \gamma = 15$ dB; $T = 105$ k

Because global antennas are assumed to be used in both systems, there is virtually no difference in the levels between the useful and interfering signals, which could be had at the expense of the radiation pattern of the space antenna, and this is the worst case. From (3.2), $\Delta T_{sat} = P_{e.i.} + G_{e.i.}(\theta_t) + G_{sat.s}(\delta) + 228.6 - L_u = -37 + 14.5 + 15.5 + 228.6 - 200 = 21.6$ dBK. Consequently $\Delta T_{sat} = 145$ K. From (3.3), $\Delta T_e = P_{sat.i} + G_{sat.i}(\eta) + G_{e.s.}(\theta_t) + 228.6 - L_d = -57 + 15.5 + 14.5 + 228.6 - 196 = 5.6$ dBK. Consequently $\Delta T_e = 3.6$ K. From (3.1), $\Delta T_{link} = \Delta T_e + \gamma \Delta T_{sat} = 3.6 + 0.032 \cdot 145 = 8.2$ K. Hence $\Delta T/T \cdot 100 = (8.2 \cdot 100)/105 = 7.8\%$. And so, the resulting value of 7.8% exceeds the 4%-threshold. i.e., coordination is necessary between the examined networks.

The coordination procedure can begin six months after the date of publication of the weekly circular, containing information intended for preliminary publication. Here the administration that is applying for a new system sends a request for coordination to the administrations, comments from which were received after preliminary publication. In a request for coordination, the administration must send information about the system, listed in Appendix 3 of the CCIR Radio Regulations [7]. At the same time the mentioned information is forwarded to the IFRB with a list of the administrations to which a request was sent, and it is published in the weekly circular of the IFRB. The CCIR Radio Regulations gives deadlines on the confirmation of the receipt of a request for coordination and on the completion of coordination and permits an applicant to seek help from the IFRB.

In the coordination process, administrations must estimate more exactly the levels of possible interference and take mutual steps to solve the problem. Changes of the positions of the satellites, the conformity of the parameters of the transmitted signals, of the parameters of the antennas and of the powers of the transmitters, *et cetera*, may be examined if necessary. The technical coordination procedure, which consists in a more exact estimation of the possible interference and the working out of the necessary conditions, is not specified at the present time by the CCIR Radio Regulations. In practice, materials of the CCIR on standards on the interference level and the calculation procedure are relied upon for the coordination of systems. Here the calculation is done for noise on a channel and at the receiver output in consideration of the specific kinds of modulation (see Chapter 5).

The Registration of Frequency Assignments. After the successful completion of coordination of a satellite network with all affected administrations the frequency assignments for the space and earth* stations may be registered in the Reference Register of the IFRB. Applications for registration must be sent for this in accordance with Appendix 3 of the CCIR Radio Regulations.

Applications for the registration of frequency assignments, pertaining to a fixed satellite service, must be sent to the IFRB not earlier than three years, and in any case not later than three months before placement

---

*If necessary frequency assignments for earth stations should also be coordinated with the terrestrial services of other administrations (see Chapter 5).

in operation. On receiving an application for registration, the IFRB makes an examination, during which it checks the conformity of this application with the Telecommunications Convention, frequency distribution table and technical limitations, required by the CCIR Radio Regulations (see Section 3.4), and also with requirements on coordination with other systems. If the outcome is favorable, the IFRB registers a requested frequency assignment in the Reference Register of frequencies.

The IFRB writes a special comment in a registration, which explains the preliminary character of this record. Thirty days after placing a system in operation, an administration must inform the IFRB of this, and the above-mentioned comment is deleted. The planned date of placement in operation may, at the request of an administration, be postponed for four months and, in exceptional circumstances, for up to eighteen months. If thirty days after an initially requested or altered date confirmation about the beginning of utilization is not received, the IFRB, after consultations with the administration, may remove the record from the Reference Register.

3.3 THE PLANNED USE OF FREQUENCY BANDS, ASSIGNED TO A SATELLITE BROADCASTING SERVICE

The utilization of frequency bands, assigned for a satellite broadcasting service, differs from the above-described method of using bands by means of international coordination of frequency assignments, which is used for all other space radio communications services. In accordance with Resolution 507, adopted by WARC-1979 ". . . the stations of a radio broadcasting service must be built and operated in accordance with treaties and plans associated with them, adopted by world or regional conferences . . .". Thus, provisions are made for the planned utilization of the frequency bands, listed in Table 3.2. At the same time, in consideration of the possible need to place broadcasting systems in operation before the corresponding frequency plans have been developed, a procedure for coordination is provided for in Resolution 33 of WARC-1979 for broadcasting systems, which is similar to the one described above.

At the present time there are two plans for a radio broadcasting satellite service. One is in the 11.7–12.2-GHz (Region 3) and 11.7–12.5-GHz (Region 1) bands, developed by the WACR in 1977 and 12.2–12.7-GHz (Region 2) developed at RARC '83. This first plan went into effect on January 1, 1979 for a 15-year term and can be re-examined only by a competent conference. According to this plan the 11.7–12.5-GHz band is divided into 40 frequency channels (see Table 3.6), but because they are used over and over again, the total number of channels in all orbital

TABLE 3.6

Number of channels	*Assigned frequency, MHz*	*Number of channels*	*Assigned frequency, MHz*	*Number of channels*	*Assigned frequency, MHz*	*Number of channels*	*Assigned frequency, MHz*
1	11727.48	11	11919.28	21	12111.08	31	12302.88
2	11746.66	12	11938.46	22	12130.26	32	12322.06
3	11765.84	13	11957.64	23	12149.44	33	12341.24
4	11785.02	14	11976.82	24	12168.62	34	12360.42
5	11804.20	15	11996.00	25	12187.80	35	12379.60
6	11823.38	16	12015.18	26	12206.98	36	12398.78
7	11842.56	17	12034.36	27	12226.16	37	12417.96
8	11861.74	18	12053.54	28	12245.34	38	12437.14
9	11880.92	19	12072.72	29	12264.52	39	12456.32
10	11900.10	20	12091.90	30	12283.70	40	12475.50

positions, distributed in the plan, is 984. The plan includes 143 countries (34 in Europe, 52 in Africa, and 57 in Asia), to which frequency channels are assigned in positions, located on an arc of a geostationary orbit from 37° west longitude to 170° east longitude and at 160° west longitude. The distance between adjacent positions is 6° (with the exception of 5° between 29° east longitude and 34° east longitude and 4° between 34° east longitude and 38° east longitude). The necessary bandwidth of a frequency channel in the plan is set equal to 27 MHz. FM color TV programs are supposed to be transmitted on this channel, but other kinds of modulation (for example digital) or other kinds of transmitted signals (for example on several audio broadcasting channels, if the level of interference created in other systems does not exceed the plan) are permitted.

The frequency channels, assigned in this plan for one service area are grouped when possible within 400 MHz. The separation between two frequency channels, fed to a common transmitting antenna, exceeds 40 MHz. The plan uses the carrier-to-noise (P_s/P_n) ratio at the input of a receiver of an earth station as a quality criterion. This ratio is equal to 14 dB for 99% of the time of the worst month, and a decrease of this ratio in an earth-space link must not exceed 0.5 dB for 99% of the time of the worst month. Here the specified ratio of the signal amplitude to unweighed noise, measured in the nominal frequency band of a TV signal at the output of the receiver on the edge of the service area is not less than 33.5 dB for 625-line systems.

By taking into consideration the technical feasibility of implementing the parameters of receivers and antennas and also by conducting an economic analysis, it was possible, for drafting the plan, to use the following values of G/T of an earth station: 6 dB/K for individual reception and 14 dB/K for collective reception. The plan is based on the use of receiving antennas at earth stations, the width of the main lobe of the radiation pattern of which at the 3 dB level (ϕ_0) is 2° for individual reception and 1° for collective reception. Here the following radiation patterns, as functions of the relative angle ϕ/ϕ_0 (Figure 3.2), were adopted:

a) for individual reception with the main polarization in Regions 1 and 3 (curve A):

0	for $0 \leqslant \phi \leqslant 0.25\,\phi_0$
$-12(\phi/\phi_0)^2$	for $0.25\,\phi_0 < \phi \leqslant 0.707\,\phi_0$
$-[9.0+20 \log (\phi/\phi_0)]$	for $0.707\,\phi_0 < \phi \leqslant 1.26\,\phi_0$
$-[8.5+25 \log (\phi/\phi_0)]$	for $1.26\,\phi_0 < \phi \leqslant 9.55\,\phi_0$
-33	for $9.55\,\phi_0 < \phi$

b) for collective reception with the main polarization in Regions 1, 2, and 3 (curve A′):

-0	for $0 \leqslant \phi \leqslant 0.25\ \phi_0$
$-12(\phi/\phi_0)^2$	for $0.25\ \phi_0 < \phi \leqslant 0.86\ \phi_0$
$-[10.5+25 \log (\phi/\phi_0)]$*	for $0.86\ \phi_0 < \phi$

c) for both types of reception with cross polarization in Regions 1 and 3 (curve B):

-25	for $0 \leqslant \phi \leqslant 0.25\ \phi_0$
$-[30+40 \log (\lvert \phi/\phi_0-1 \rvert)]$	for $0.25\ \phi_0 < \phi \leqslant 0.44\ \phi_0$
-20	for $0.44\ \phi_0 < \phi \leqslant 1.4\ \phi_0$
$-[30+20 \log (\lvert \phi/\phi_0-1 \rvert)]$	for $1.4\ \phi_0 < \phi \leqslant 2\ \phi_0$
-30**	

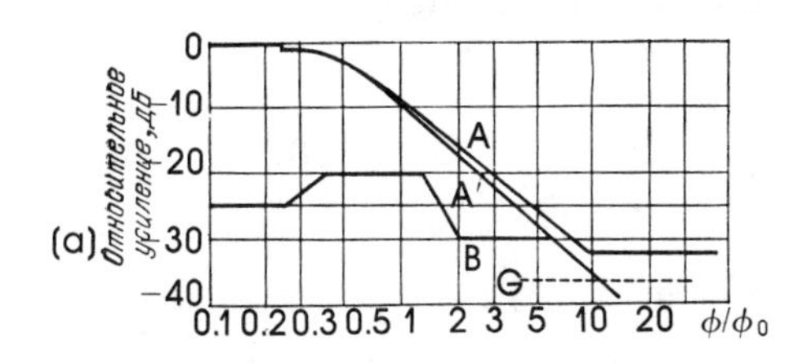

Key: **(a) Relative gain, dB**

FIGURE 3.2. Reference radiation patterns of receiving antennas.

The necessary powers of the space transmitters of satellites, established in the plan, were calculated so as to guarantee, on the boundary of the coverage zone for 99% of the time of the worst month, a power flux density of -103 dBW/m^2. This figure was determined as a result of the optimization of the economic indices of systems in consideration of the requirements of compatibility with terrestrial services. At the same time, it

---

*To the intersection with straight line C, corresponding to an attenuation of the sidelobes, equal to the gain of the antenna, and then along that straight line.

**To the intersection with the curve for the main polarization, and then along that curve.

is specified that for systems intended for collective reception, the recommended power flux density be -111 dBW/m^2. The resolutions of the WARC-1977 permit the construction of systems with intermediate (between -103 and -111 dBW/m^2) power flux densities that are economically optimum for a specific service area [4]. It was assumed for planning purposes that the beam of a transmitting antenna has an elliptical or circular cross section. The minimum beamwidth of a transmitting antenna is assumed to be 0.6°. The necessary gain (G) of a transmitting antenna was determined on the basis of given angular dimensions of the major (a) and minor (b) axes of the ellipse, expressed in degrees, by the approximate formula $G \approx 27{,}843/ab$, or $G \approx 44.4 - 10 \log a - 10 \log b$. During the determination of the parameters of transmitting antennas, an effort was made to make sure that the difference between the effective radiated isotropic powers toward the boundary of the coverage area and on the axis of the beam be 3 dB. (This condition was not met in service areas, the maximum size of which is less than 0.6°.) For the calculations transmitting antenna radiation patterns described by the following expressions for the indicated polarization were used (Figure 3.3):

(a) main:

$-12\,(\phi/\phi_0)^2$ for $0 \leqslant \phi \leqslant 1.58\,\phi_0$

-30 for $1.58\phi_0 < \phi \leqslant 3.16\,\phi_0$

$-[17.5 + 25 \log(\phi/\phi_0)]$* for $\phi > 3.16\,\phi_0$

(b) cross:

$-[40 + 40 \log(|\phi/\phi_0 - 1|)]$ for $0 \leqslant \phi \leqslant 0.33\,\phi_0$

-33 for $0.33\,\phi < \phi \leqslant 1.67\,\phi_0$

$-[40 + 40 \log(|\phi/\phi_0 - 1|)]$* for $\phi > 1.67\,\phi_0$

In the calculations of the necessary transmitter power it was assumed that losses to propagation are made up of losses to propagation in free space and of additional attenuation, which is not exceeded for 99% of the time of the worst month. This additional attenuation is given in Figure 3.4 as a function of the angle of incidence for the five climate zones, shown in Figure 3.5. Here basically angles of incidence, for which the additional attenuation would not exceed 2 dB, were selected in the plan.

---

*To the intersection with the straight line corresponding to a sidelobe attenuation, equal to the gain of the antenna, and from there along that straight line.

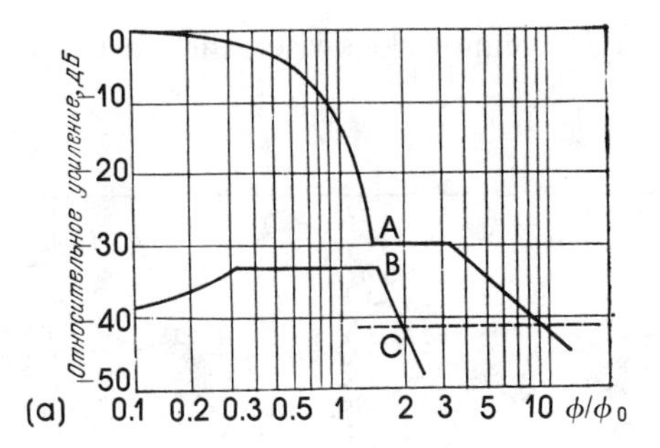

Key: **(a) Relative gain, dB**

FIGURE 3.3. Reference radiation patterns of transmitting antennas.

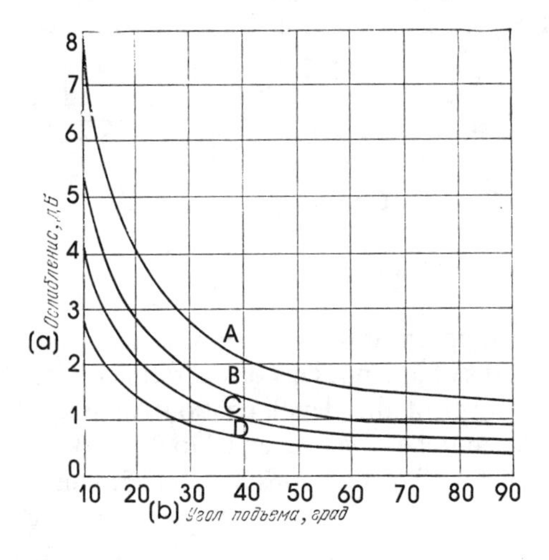

Key: **(a) Attenuation, dB**
(b) Angle of elevation, deg

FIGURE 3.4. Additional attenuation as functon of angle of elevation for different climate regions. A: for climinate region 1, B: for climate region 2, C: for climate regions 3 and 4, D: for climate region 5.

Two orthogonal circular polarizations are used in the plan:

1: right-handed, clockwise: the vector **E** rotates clockwise in the plane perpendicular to wave propagation, if viewed in the direction of propagation;

2: left-handed, counter-clockwise: the vector **E** rotates counter-clockwise.

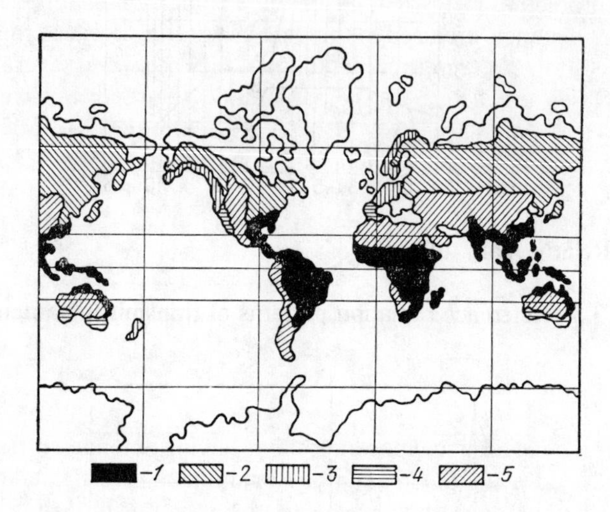

Key: -1 -2 -3 -4 -5

FIGURE 3.5. Climate zones

The additional polarization isolation was determined on the basis of the radiation patterns of antennas in consideration of depolarization in the atmosphere. The level of the depolarization component relative to the main polarization component was set equal to −27 dB for climate regions 1 and 2, and −30 dB for climate regions 3, 4, and 5. During the selection of the optimum positions, limits on the angles of incidence were taken into consideration, and also the need to satisfy the requirement that an interruption of operation when the earth hides the sun when solar batteries are used as power sources will occur during hours of the night or later. This requirement necessitated a shift westward by 15°–20° relative to the corresponding service areas.

The requirement that a space station be held in orbit with an accuracy of better than ±0.1° in north-south and west-east directions was adopted in the plan. The tolerable deviation of the axis of the radiation pattern of a transmitting antenna from the nominal direction may not exceed 0.1° in any direction, and the angular rotation of the beam around the axis may not exceed 2°. To protect terrestrial services operating in the same band, the energy of the carrier is dispersed artificially with a peak-to-peak deviation of 600 kHz, which corresponds to a 22-dB decrease of the spectral power flux density, measured in the 4-kHz band, in comparison with the corresponding power flux density, measured in the entire band [6].

The protection ratios R = 31 dB on a coincident channel and 15 dB on adjacent channels are used in the plan as criteria. Because there are no rigorous solution and experimental data, in the case when several interference processes are active at the same time their powers were added together. Here the analysis was carried out for protection ratio with the equivalent margin M, determined by the expression:

$$M = -10 \log (10^{-0.1M_1} + 10^{-0.1M_2} + 10^{-0.1M_3})$$

where M_1, M_2, and M_3 are the protection margins on a coincident, and on the upper and lower adjacent channels, determined by the formula:

$$M_i = 10 \log (P_s/\Sigma P_{\mathrm{int}.i}) - R_i$$

where P_s is the useful signal power at the receiving antenna output; $\Sigma P_{\mathrm{int.i}}$ is the summary power of the interfering signals on an investigated channel at the receiver antenna output; and R_i is the required protection ratio for single interference on an investigated channel.

In the plan, adopted by the WARC-1977, five positions in the geostationary orbit are assigned for Soviet satellites: 23°, 44°, 74°, 110°, and 140° east longitude. The total number of programs transmitted at the same time is 70, which makes it possible to transmit at least four union-wide programs across the country, which is covered by eight service areas, and one–two programs in each of the union republics, and several autonomous republics and national okrugs.

Shown in Figure 3.6 are eight service areas, calculated on the basis of data, recorded in the plan, which can be used for transmitting four central TV broadcasting programs in five–six broadcasting belts with the corresponding time shift. The characteristics of the frequency assignments, reserved for the USSR, are given in the book [4]. The analysis that was conducted proved the plan to be highly effective and the adopted parameters to be close to optimum [4].

3.4 SPECIAL RULES THAT APPLY TO SPACE RADIO SERVICES*

Several special rules are set forth in the CCIR Radio Regulations

*For the limits on the power flux density at the earth's surface, created by emissions of space stations, see Chapter 5.

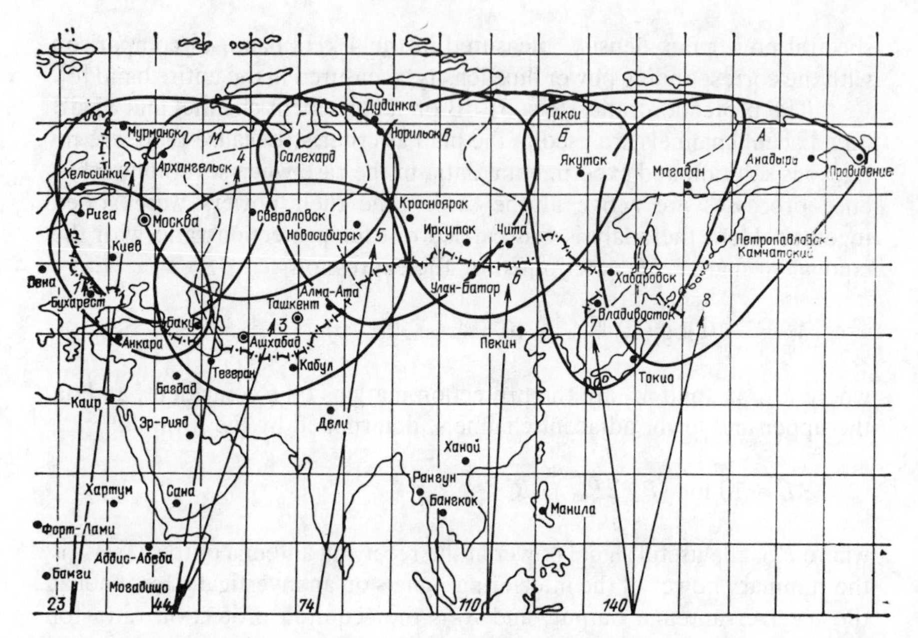

FIGURE 3.6. Service areas, corresponding to planned frequency assignments for the USSR.

that apply to space radio services. We will mention here those which pertain either to all or to fixed and broadcasting satellite services.

General Requirements. According to Item 2612 of the CCIR Radio Regulations, space stations must be capable of ceasing operation immediately in cases when necessary. Nongeostationary satellites and their earth stations must cease transmission or reduce the level of radiation at times when there is not enough angular separation from geostationary satellites and an unnacceptable level of interference is created [7, Item 2613]. All possible technical means of minimizing radiation in other countries, if no preliminary treaty has been made with these countries, must be used when determining the characteristics of a space station of a radio broadcasting satellite service (Item 2674).

The Satellite Station-Keeping Precision. Space stations onboard a geostationary satellite, which utilize any frequency band, assigned for a fixed or radio broadcasting satellite service (with the exception of the 11.7–12.5-GHz band) must be capable of maintaining their position to an accuracy of ±0.1° by longitude and must maintain their position to an accuracy of ±0.1°, with the exception of test satellites, which must have a station-keeping precision by longitude of ±0.5°.

Space stations, used in other satellite services, apart from fixed and radio broadcasting satellite services, must have a station-keeping precision of ±0.5°. Space stations onboard a geostationary satellite, placed in operation prior to January 1, 1987, the preliminary publication on which preceded January 1, 1982, must also be capable of holding their longitudinal position to an accuracy of ±1°, and efforts should be made to achieve an accuracy of ±0.5°. The above-listed requirements on the station-keeping precision need not be observed if a given space station does not create unacceptable interference with any other system.

Antenna Beam-Pointing Precision. Space stations onboard a geostationary satellite must be capable of maintaining an antenna beam-pointing precision such that the maximum deviation of the axis of the main lobe of the radiation pattern from the nominal position will not exceed 10% of the width of the main lobe at the −3-dB level, or 0.3°. (The larger of the two figures is taken.) The requirement applies to directional antennas that have less than global coverage. The transmitting antennas of satellite broadcasting space stations, operating in the 11.7–12.5-GHz band, must meet the requirements of the plan of the WARC-1977.

Chapter 4
Power (EIRP) Calculation of Satellite Links

4.1 ENERGY FEATURES OF SATELLITE LINKS. OBJECTIVES AND PROBLEMS OF THE CALCULATION

Satellite communications links consist of two sections: earth-satellite and satellite-earth. In terms of power, both sections are under great stress: the former because the power of the transmitters tends to decrease and earth stations tend to get simpler, and the latter because of constraints on the weight, size, and energy consumption of a satellite transponder, which limit its power.

The main feature of satellite links is the existence of great losses of the signal due to the attenuation of its EIRP in paths of great physical length. For instance, when the orbital altitude of a satellite is 36,000 km, the attenuation of the signal may reach 200 dB. In addition to this main attenuation in space, the signal in satellite communications links is subjected to the influence of many other factors. These include absorption in the atmosphere, Faraday rotation of the polarization plane, refraction, depolarization, *et cetera.* On the other hand, the receivers of satellite and earth stations, apart from their own time-varying noise, are acted upon by different kinds of interference in the form of radiation from space, from the sun, and from planets. Under these conditions, the correct and exact consideration of the influence of all factors will make it possible to plan the optimum system and to guarantee that it performs faithfully under the most difficult conditions, and at the same time will make it possible to eliminate superfluous energy reserves, which make the earth and space equipment unnecessarily complicated.

As stated in Chapter 2, the standards on some of the performance characteristics of satellite channels (for instance the signal-to-noise ratio) are of a statistical nature. This means that disturbances also have to be evaluated statistically, i.e., not only a quantitative measure of the effect of a given factor, but also the probability (frequency) of its appearance must enter into the calculations. The nature and the number of the signals that are transmitted, and the character of their processing in a satellite transponder must be considered. In the simplest case, for example the transmission of TV programs, a transponder operates in the single-signal mode, which is typical of terrestrial radio relay links and it merely amplifies the signal to be retransmitted. In the case of the transmission of telephone signals with multiple access, several signals separated by frequency, time, or waveform, and having a mutual influence on each other, must be taken into consideration in the power calculation of satellite links. Depending on the type and purpose of a system, different kinds of signal processing may be used on a satellite, all the way to the complete regeneration of the signal. This precludes the accumulation of noise and distortion in each section of the link.

A procedure for calculating satellite links, both in the single-signal, and in the multisignal modes, is presented in this chapter. For the single-signal mode, the calculation should be done section by section (Section 4.3), and in the multisignal mode, either section by section or with the aid of a generalized equation (Section 4.4), which takes into consideration all the effects and factors that accompany the passage of several signals through a common satellite transponder path.

4.2 COUPLING EQUATIONS FOR TWO SECTIONS

A schematic diagram of one section of a communications link and the signal level pattern are given in Figure 4.1. The EIRP of the transmitting station is

$$\mathrm{EIRP} = P_{tr}\eta_{tr}G_{tr} \tag{4.1}$$

where P_{tr} is the effective power of the transmitter output signal; η_{tr} is the transfer coefficient (in terms of power) of the waveguide path (the efficiency of the path) between the transmitter and antenna; and G_{tr} is the gain of the transmitting antenna relative to the isotropic radiator. The attenuation in free space, determined by the decay of the power flux density with distance from the transmitter is

$$L_0 = 16\pi^2 d^2/\lambda^2 \tag{4.2}$$

where λ is the wavelength and d is the slant range (the distance between the transmitting and receiving antennas). In addition to these main losses, as mentioned in Section 4.1, there are additional losses L_{add} along the path; the total loss along the path is $L_\Sigma = L_0 L_{add}$.

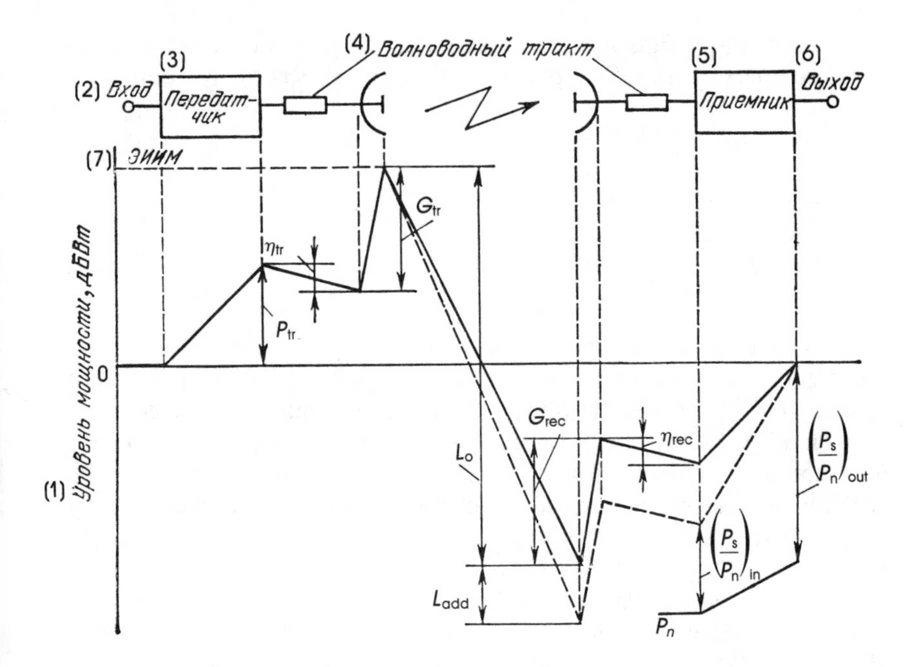

Key: **(1) Power level, dBW**
(2) Input
(3) Transmitter
(4) Waveguide path
(5) Receiver
(6) Output
(7) EIRP

FIGURE 4.1. Schematic diagram and level pattern of one section of a communications link.

At the point of reception, an antenna is installed with gain G_{rec}, which is connected to the receiver by a waveguide signal path with transfer coefficient η_{rec}. When the wave impedances of the antenna, of parts of the path and of the receiver are matched the signal power at the receiver input is

$$P_{rec} = \frac{\text{EIRP}}{L_0 L_{add}} G_{rec}\eta_{rec} = \frac{\lambda^2 G_{tr} G_{rec} \eta_{tr} \eta_{rec}}{16\pi^2 d^2 L_{add}}$$

In this form the above expression is the familiar formula that is used for calculating line of sight radio links [4.1].

When the parameters of the antenna are given in the form of the effective aperture area S_{rec}, connected with the gain by the relation:

$$G_{rec} = 4\pi S_{rec}/\lambda^2$$

the previous expression may be written as

$$P_{tr} = P_{rec} \frac{4\pi d^2 L_{add}}{G_{tr} S_{rec} \eta_{tr} \eta_{rec}} \qquad \textbf{(4.3)}$$

With (4.3) it is possible to determine the necessary transmitter power for a given receiver input signal power. Notice that wavelength λ does not appear in it. Consequently, when the transmitting antenna has a constant gain on all frequencies and the receiving antenna has a constant effective aperture area (it can operate efficiently as frequency increases), the receiver input signal power will not depend in the first approximation on frequency (actually it does depend on frequency to some extent, since L_{add} is determined largely by the frequency band).

During the calculation of a link, the receiver input signal-to-noise ratio $(P_s/P_n)_{in}$ often is given not the signal power at the receiver input. Then $P_{rec} = P_n(P_s/P_n)_{in}$ may be substituted into (4.3), where P_n is the total noise power at the receiver input.

Because, in the frequency bands in which satellite systems operate, noise from different sources are additive in nature, their cumulative power is expressed by the formula:

$$P_n = kT_\Sigma \Delta f_n \qquad \textbf{(4.4)}$$

where $k = 1.38 \cdot 10^{-23}$ W/Hz·deg is the Boltzmann constant; T_Σ is the equivalent noise temperature of an entire receiver system in consideration of set and external noises; and Δf_n is the equivalent (energy) noise band of the receiver.

A schematic diagram and power level pattern of a satellite link, consisting of two sections, is given in Figure 4.2. Using (4.1) and (4.4), the power relations are:

for the earth-satellite section*:

---

*Here and in the ensuing, the subscript e is attached to all the indices that pertain to earth equipment, and the subscript sat is attached to the indices that pertain to space equipment; values that pertain to the earth-satellite section have the subscript 1, and those that pertain to the satellite-earth section have the subscript 2.

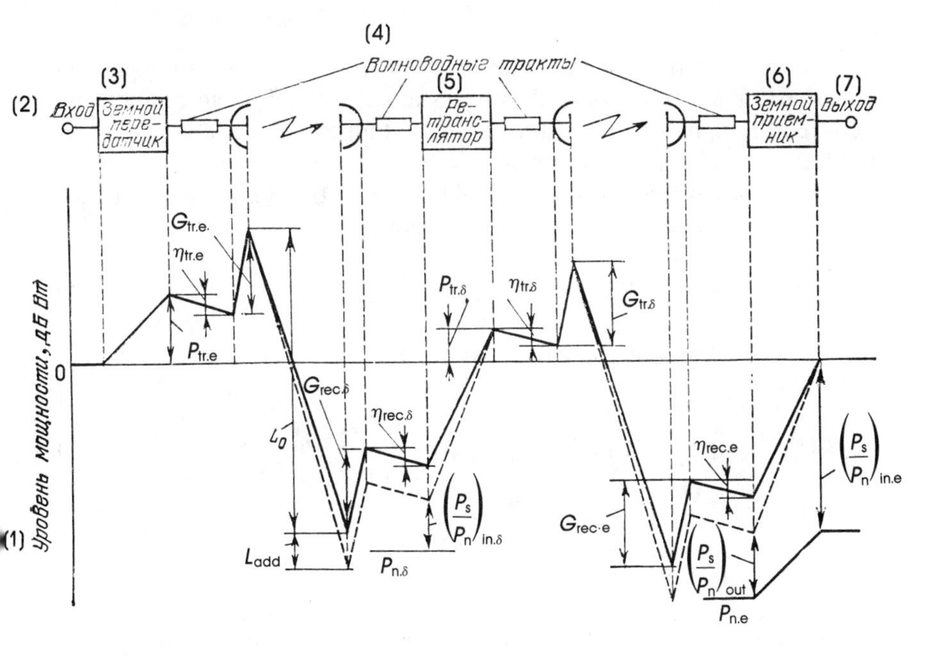

Key: (1) Power level, dBW
(2) Input
(3) Earth transmitter
(4) Waveguide paths
(5) Transponder
(6) Earth receiver
(7) Output

FIGURE 4.2. Schematic diagram and level pattern of communications link consisting of two sections.

$$P_{\mathrm{tr.e}} = \frac{16\pi^2 d^2_1 L_{1\,\mathrm{add}} P_{\mathrm{n.e}}}{\lambda^2_1 G_{\mathrm{tr.e}} G_{\mathrm{rec.sat}} \eta_{\mathrm{tr.e}} \eta_{\mathrm{rec.sat}}} \left(\frac{P_{\mathrm{s}}}{P_{\mathrm{n}}}\right)_{\mathrm{in.sat}}$$

where $P_{\mathrm{n.sat}} = kT_{\Sigma\mathrm{sat}} \Delta f_{\mathrm{n.sat}}$;
for the satellite-earth section:

$$P_{\mathrm{tr.sat}} = \frac{16\pi^2 d^2_2 L_{2\,\mathrm{add}} P_{\mathrm{n.e.}}}{\lambda^2_2 G_{\mathrm{tr.sat}} G_{\mathrm{rec.e}} \eta_{\mathrm{tr.sat}} \eta_{\mathrm{rec.e}}} \left(\frac{P_{\mathrm{s}}}{P_{\mathrm{n}}}\right)_{\mathrm{in.e}}$$

where $P_{\mathrm{n.e.}} = kT_{\Sigma\mathrm{e}} \Delta f_{\mathrm{n.e}}$.

To convert from the equations for individual sections to the general equation for an entire link, it is necessary to establish the relationship between the signal-to-noise ratios at the output of the link and in each of the sections. As stated in Section 4.1, when signal processing in space is not used, the noise of each section is combined; here the cumulative signal-to-noise ratio at the end of the communications link is

$$(P_n/P_s)_\Sigma = (P_n/P_s)_{in.sat} + (P_n/P_s)_{in.e} \tag{4.5}$$

The signal-to-noise ratio in each section obviously must be greater than at the end of the link:

$$(P_s/P_n)_{in.sat} = a(P_s/P_n)_\Sigma \qquad (P_s/P_n)_{in.e} = b(P_s/P_n)_\Sigma \tag{4.6}$$

It follows from (4.5) and (4.6) that

$$a = b/(b - 1) \tag{4.7}$$

Expression (4.7) makes it possible to distribute a given ratio $(P_s/P_n)_\Sigma$ between the two sections of a communications link. For instance, by setting the excess of the signal-to-noise ratio in the satellite-earth section equal to 1 dB (b = 1.26), we find that the necessary excess in the earth-satellite section should be 7 dB ($a \approx 5$). The above distribution of margins a and b presumes that the noise bands of the satellite transponder and earth receiver are equal; if $\Delta f_{n.e.} < \Delta f_{n.sat}$, then the noise power at the space receiver input should be computed in the band $\Delta f_{n.e.}$.

In consideration of the above, the equations for a satellite link, consisting of two sections, finally acquires the form:

for the earth-satellite section:

$$P_{tr.e} = \frac{16\pi^2 d^2{}_1 L_{1\ add} k T_{\Sigma sat} \Sigma f_{n.e}}{\lambda^2{}_1 G_{tr.e} G_{rec.sat} \eta_{tr.e} \eta_{rec.sat}} a \left(\frac{P_s}{P_n}\right)_\Sigma \tag{4.8}$$

for the satellite-earth section:

$$P_{tr.sat} = \frac{16\pi^2 d_2^2 L_{2\ add} k T_{\Sigma e} \Delta f_{n.e}}{\lambda^2{}_2 G_{tr.sat} G_{rec.e} \eta_{tr.sat} \eta_{rec.e}} b \left(\frac{P_s}{P_n}\right)_\Sigma \tag{4.9}$$

4.3 GENERALIZED LINK EQUATION OF A COMMUNICATIONS SYSTEM

For solving problems associated with the optimization of the parameters of satellite systems and of the sound distribution of attenuations and noises between the sections of a communications link, it is desirable to use a generalized power equation, which connects the parameters of both sections. In application to the single-signal mode of operation of a transponder, this equation can be derived on the basis of the following relations. The signal power of a transmitting earth station at the input of a space receiver is

$$P_{s.in.sat} = \frac{\mathrm{EIRP}}{L_{\Sigma 1}} G_{rec.sat} \eta_{rec.sat} \quad \textbf{(4.10)}$$

Because the transponder output power in the single-signal mode is used for transmitting just a signal and noise, the fraction of the space EIRP that is used for radiating a signal is

$$\mathrm{EIRP}_{sat.s} = \frac{\mathrm{EIRP}_{sat} P_{s.in.sat}}{P_{s.in.sat} + P_{n.sat}} \approx \mathrm{EIRP}_{sat}$$

because $P_{s.in.sat} >> P_{n.sat}$, and the fraction of the EIRP that goes to noise is

$$\mathrm{EIRP}_{sat.n} = \mathrm{EIRP}_{sat} P_{n.sat} / (P_{s.in.sat} + P_{n.sat}) \approx \mathrm{EIRP}_{sat} P_{n.sat} / P_{s.in.sat} \quad \textbf{(4.11)}$$

The signal power of a satellite at the receiver input of the earth station is

$$P_{s.in.sat} = \mathrm{EIRP}_{sat.s} \eta_{rec.e} G_{rec.e} / L_{\Sigma 2} \quad \textbf{(4.12)}$$

The noise power from the satellite (the noise of the earth-satellite section) at the receiver input of the earth station is

$$P_{n.e.sat} = \mathrm{EIRP}_{n.sat} G_{rec.e} \eta_{rec.e} / L_{\Sigma 2} \quad \textbf{(4.13)}$$

The power of the receiver set noise $P_{n.e}$, and the resulting signal-to-noise ratio at the receiver input are

$$(P_s/P_n)_{\Sigma} = P_{s.in.e} / (P_{n.e.sat} + P_{n.e}) \quad \textbf{(4.14)}$$

After substituting the values from (4.10)–(4.13) into (4.14), and making conversions we obtain

$$\left(\frac{P_s}{P_n}\right)_{\Sigma} = \frac{\mathrm{EIRP}_e \mathrm{EIRP}_{sat} G_{rec.e} G_{rec.sat} \eta_{rec.e} \eta_{rec.sat}}{\mathrm{EIRP}_e G_{rec.sat} \eta_{rec.sat} P_{n.e} L_{\Sigma 2} + \mathrm{EIRP}_{sat} G_{rec.e} n_{rec.e} P_{n.sat} L_{\Sigma 1}} \quad \textbf{(4.15)}$$

A generalized power equation for a satellite communications system, consisting of n earth stations, operating through a common satellite transponder, can be derived in like manner. A schematic diagram of such a system is shown in Figure 4.3. The transmitter of the earth station radiates a signal toward the satellite; the power of this signal at the input of the satellite transponder is determined by (4.10). Along with this signal, the input of the transponder is acted upon by signals of the other $n - 1$ earth stations of the system, which for the first signal are interference, and also by time varying noise of the earth-satellite section. The total power of the interference at the transponder input is

$$P_{\text{t.sat}} = \sum_{i=1}^{n-1} sp\ P_{si} P_{\text{n.sat}} \tag{4.16}$$

where s is a coefficient that takes into consideration the relative difference of the power of the interfering signals, and p is a coefficient that takes into consideration the lifetime of the interfering signals.

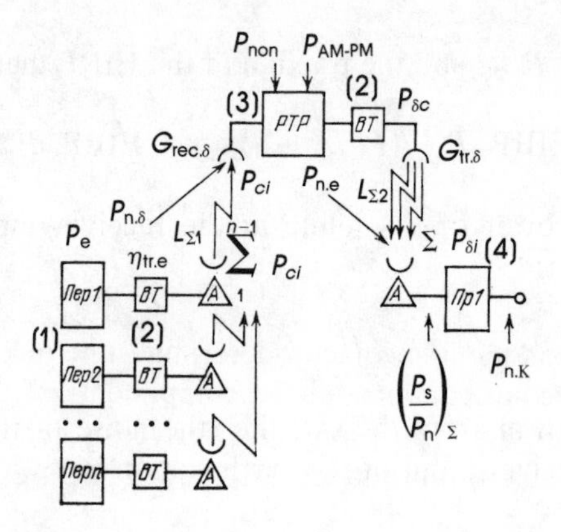

Key: **(1) Transmitter**
(2) Earth-satellite section
(3) Transponder
(4) Receiver

FIGURE 4.3. Schematic diagram of multistation communications system.

The additive mixture of the signal, interference, and noise is amplified in the common transponder channel and is radiated toward the earth stations of the system. However, because there are elements in the transponder channel that have nonlinear amplitude and phase responses, and alsoAM-PM converter elements (see later in Chapter 7), the signal in the general case is suppressed and the corresponding interferences P_{non} and $P_{\text{AM-PM}}$ appear, which also are radiated by the transponder.

The transponder output power that goes into the signal of one station is

$$P_{\text{sat}i} = \frac{P_{\text{sat}} c P_{\text{s.in.sat}} K_{\text{su}}}{P_{\text{s.in.sat}} + P_{\text{su.sat}} + P_{\text{non}}\, p + P_{\text{AM-FM}}} = \frac{c P_{\text{sat}} K^{p}{}_{\text{su}} \text{EIRP}_{\text{e}} G_{\text{rec.sat}}}{L_{\Sigma 1} \Sigma P_{\text{su}}} \tag{4.17}$$

where

$$\Sigma p_{\text{su}} = \sum_{i=1}^{n} sp\, P_{si} + P_{\text{n.sat}} + P_{\text{non}}\, p + P_{\text{AM-FM}}$$

c is the transponder output power utilization coefficient in the multisignal mode; and K_{su} is the coefficient of suppression of the signal by interference. The remainder of the transponder power is spent on the radiation of other signals and interference. The generalized equation that connects the power parameters of a multistation satellite communications system is

$$\left(\frac{P_{\text{s}}}{P_{\text{su}}}\right)_{\Sigma} \approx \frac{\text{EIRP}_{\text{e}} \text{EIRP}_{\text{sat}} G_{\text{rec.e}} G_{\text{rec.sat}} \eta_{\text{rec.e}} \eta_{\text{rec.sat}}}{L_{\Sigma 1} L_{\Sigma 2}}$$
$$\times \frac{c K_{\text{su}}^{p} l}{\text{EIRP}_{\text{sat}} G_{\text{rec.e}} \eta_{\text{rec.e}} / L_{\Sigma 2} + P_{\text{n.e}}}$$
$$\times \left(\sum_{i=1}^{n} sp\, \frac{\text{EIRP}_{\text{sat}} G_{\text{rec.sat}} \eta_{\text{rec.sat}}}{L_{\Sigma 1}} + P_{\text{n.sat}} + \text{P}_{\text{non}} \text{P} + P_{\text{AM-FM}} \right)^{-1} \tag{4.18}$$

In this equation the first cofactor determines the receiver input signal power, and the second contains all the components of interference that occur in multiple access. The expression derived is general for all signal separation methods in a transponder; the specific features of each method are taken into consideration in (4.18) by the corresponding coefficients. For instance, in the case of time division, the transponder output power is used completely, every signal occupies the entire frequency band of the receiver and interference in the form of adjacent signals does not coincide in time with the main signal; consequently, $c = 1$, $l \approx 1$, and $p = 0$; for frequency division $c < 1$, $p = 1$, and $l \approx 1/n$.

4.4 THE ABSORPTION OF SIGNAL POWER IN THE ATMOSPHERE

Before the equations derived in Sections 4.2 and 4.3 can be used for calculation, the additional losses of signal energy and of noises of external (atmospheric) and internal (set) origin must be determined quantitatively. In the frequency bands reserved for satellite systems, the influence of the atmosphere is manifested as the attenuation (absorption) of radio waves in the troposphere and ionosphere, distortion of the beam trajectory as the result of refraction, a change of waveform, rotation of the polarization plane of radio waves, and the appearance of interference due to the thermal emission of the atmosphere and absorption noises.

A quantitative estimate of the first of the listed factors, the absorption of radio waves in the atmosphere, is given by the coefficient L_a. It has been established that in bands above 500 MHz the troposphere, more accurately gases of the troposphere (oxygen and water) vapors, and also rain and other hydrometeors (the ionosphere and other gases of the troposphere, for example carbon dioxide and nitrogen, play a minor role) are mainly responsible for absorption.

For quantitative estimation, it is convenient to use the following formula:

$$L_a = L'_{O_2} l_1 + L'_{H_2O} l_2 \tag{4.19}$$

where L'_{O_2} and L'_{H_2O} are the coefficients of linear absorption (dB/km) in oxygen and water vapors, and l_1, l_2 are the respective equivalent signal paths in these media. The absorption coefficients for the standard atmosphere are determined in the works [4.2]–[4.4] and are given in Figure 4.4, from which it follows that absorption is decidedly a frequency-dependent phenomenon; resonance peaks occur on frequencies of 22 and 165 GHz (for water vapors), and also on 60 and 120 GHz (for oxygen). The equivalent signal path in the standard atmosphere obviously depends not only on the equivalent thickness of the atmosphere, but also on the elevation angle β of the antenna of an earth station and elevation h_e of the earth station above sea level:

$$l_1 = (h'_{O_2} - h_3) \csc \beta \qquad l_2 = (h'_{H_2O} - h_e) \operatorname{cosc} \beta \tag{4.20}$$

where h'_{O_2} and h'_{H_2O} are the equivalent thicknesses of the atmosphere for oxygen and water vapors [4.3].

The results of computations by (4.19) and (4.20) in consideration of data presented in the report [4.4] are given in Figure 4.5; they determine absorption in a calm (undisturbed) atmosphere without hydrometeors, which represents a kind of constant component of losses that occur 100% of the time. The estimation of the attenuation of a signal in hydrometeors:

$$L_r = L'_r l_e \tag{4.21}$$

turns out to be a more difficult problem than in a calm atmosphere, because absorption in this case depends on the kind of hydrometeors (rain, snow, fog), the rate of precipitation, the size of the area of precipitation and the distribution of rates through the area and the distribution of the sizes of the particles of hydrometeors. These factors influence both the coefficient of linear absorption L'_r and the equivalent signal path l_e in (4.21). Liquid hydrometeors (rain, fog, and wet snow) are responsible for most of the attenuation of a signal; attenuation in solid structures (hail and dry snow) is considerably weaker. Suspended particles (aerosols) have virtually no effect on signal absorption and need not be considered under ordinary conditions.

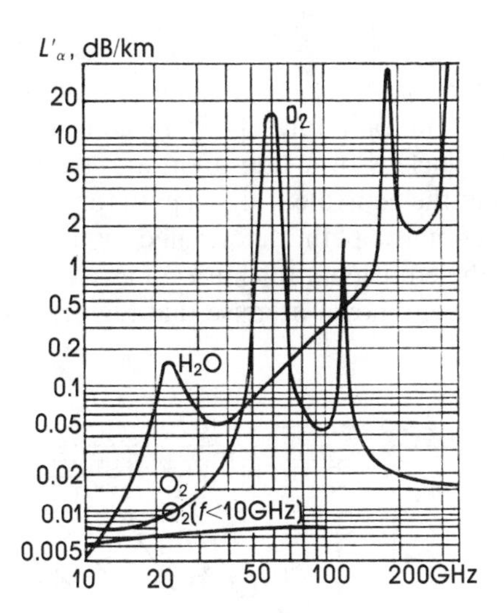

FIGURE 4.4. Absorption coefficient for oxygen O_2 and for water vapors H_2O as function of frequency.

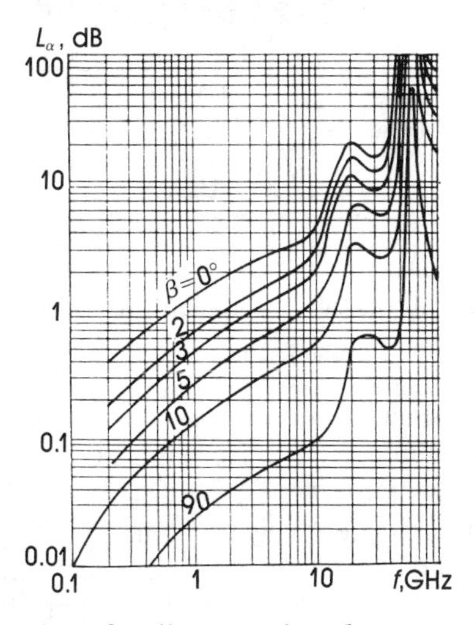

FIGURE 4.5. Absorption of radio waves in calm atmosphere (without rain) as function of frequency for different elevation angles.

The linear absorption coefficients in rain at different rates, averaged on the basis of numerous measurements and recommended by the CCIR, are given in Figure 4.6. These functions in the range of rates of rain $0.1 \leq \epsilon \leq 40$ mm/hr and in the frequency range $4 \leq f \leq 30$ GHz can be approximated, with an error of not more than 20%, by the expression:

$$L'_r(\epsilon, f) = A[1 + (f/f_0)^5]\exp(\epsilon/\epsilon_0) \quad \textbf{(4.22)}$$

where A is a constant; $f_0 = 12.5$ GHz; and $\epsilon_0 = 17.5$ mm/hr. Values of l_e, computed for different rain rates [4.5] in consideration of the spatial localization of the nucleus of a rain area, are given in Figure 4.7, from which it follows that the equivalent signal path for high rates is significantly shorter than the geometric path, determined by the formula $l_e = (h_r - h_e) \csc \beta$, where h_r is the equivalent thickness of a rain area. This quantitative result confirms the well-known premise that high-rate rains are, as a rule, strongly localized.

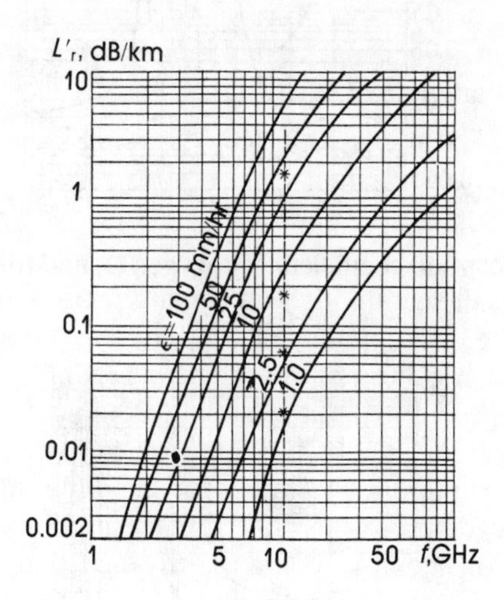

FIGURE 4.6. Signal absorption coefficient in rain of different rates as function of frequency.

With the aid of Figures 4.6 and 4.7, it is possible to calculate the absorption in rains of different rate. Here it is necessary to solve one more important problem, on statistics, i.e., the distribution of the probability of precipitations of different rates. This problem does not submit to theoretical solution and is based entirely on experimental weather data. However, when using these data it should be remembered that they have acceptable reliability for processing results for a period of not shorter than seven to ten years, pertain to a certain point on the earth and, strictly speaking, cannot be extrapolated for an entire satellite service area, which can cover several climate areas; they usually characterize the mean annual or mean monthly amount of precipitations (i.e., they give distributions that are averaged over a year or month), whereas the existing CCIR and EASS standards require a knowledge of the mean hourly and mean minute distributions.

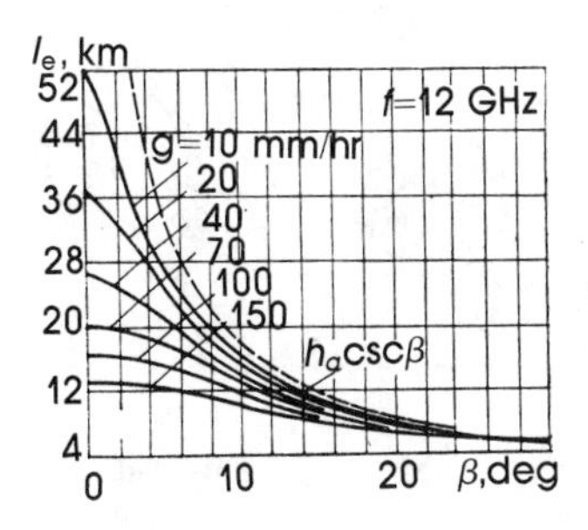

FIGURE 4.7. Equivalent signal path in rain of different rates as function of elevation angles of antenna of earth station.

The most complete data on the statistics of rain in the USSR are given in the book [19]. A map of the USSR, zoned by the rates of precipitations, is shown in Figure 4.8. The statistical distributions of the mean minute rates of rain in different climate regions are given in Figures 4.9 and 4.10; here the numbers of the curves in Figure 4.9 and 4.10 correspond to the numbers of the climate regions in Figure 4.8. It is noteworthy that these distributions apply to the worst month of the year and the expected frequency of the appearance of precipitations, with a distribution close to the one indicated, for the different climate regions is two to four times every five years [19]. The distribution of the mean minute rate of precipitation, computed for the average year, obviously will be different from these data. The approximate factors for converting from the worst month to an average year are given in Table 4.1 [19].

The relationships among the mean daily, mean hourly, and mean minute rates of precipitation are also mentioned. They can be helpful in the absence of local data on the mean minute rate. An analysis of numerous Soviet and non-Soviet results of weather observations shows that for climate regions with a moderate rate of precipitation (Europe, the European part of the USSR, Siberia) in the range of probabilities of 0.001–1%, it may be assumed, with an error of not more than 20%, that

$$\epsilon(\tau = 1 \text{ min}) \approx 5\epsilon(\tau = 1 \text{ day}), \epsilon(\tau = 1 \text{ min}) \approx (1.3\text{–}1.5)\epsilon(\tau = 1 \text{ hr})$$

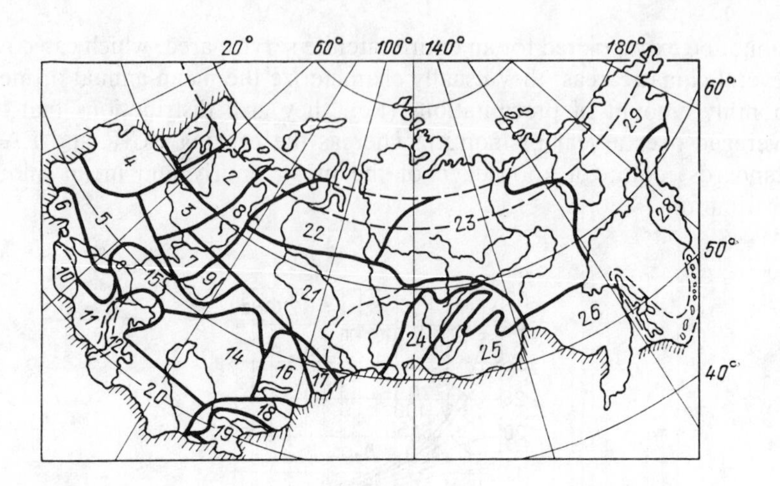

FIGURE 4.8. Map of the USSR, zoned by rates of precipitation.

The distribution function of signal losses in rain obviously is determined by the distribution of the rate of precipitations and by the functional relationship between the coefficients of linear absorption and the rate of precipitations, given by (4.22).

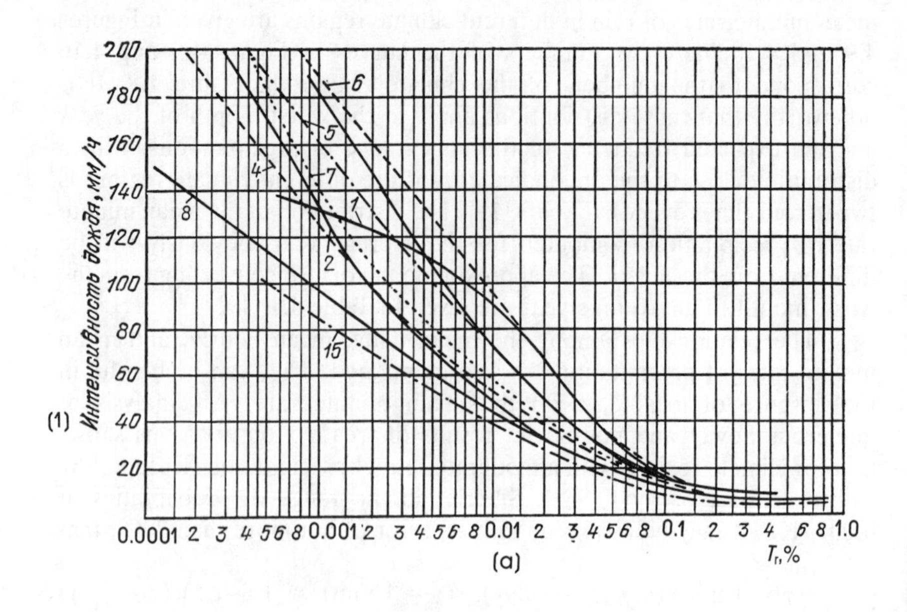

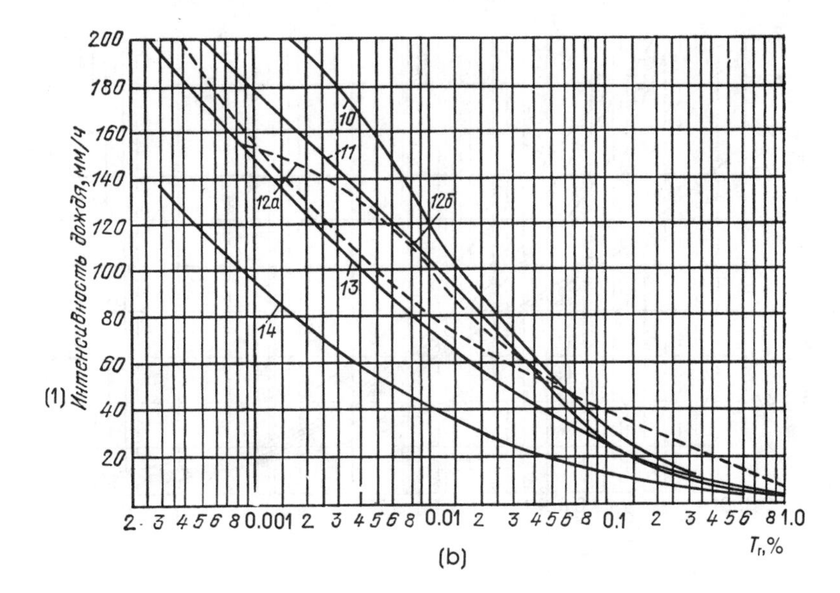

Key: (1) Rate of rain, mm/hr

FIGURE 4.9. Statistical distributions of mean minute rates of rain: (a) European territory; (b) Caucasus.

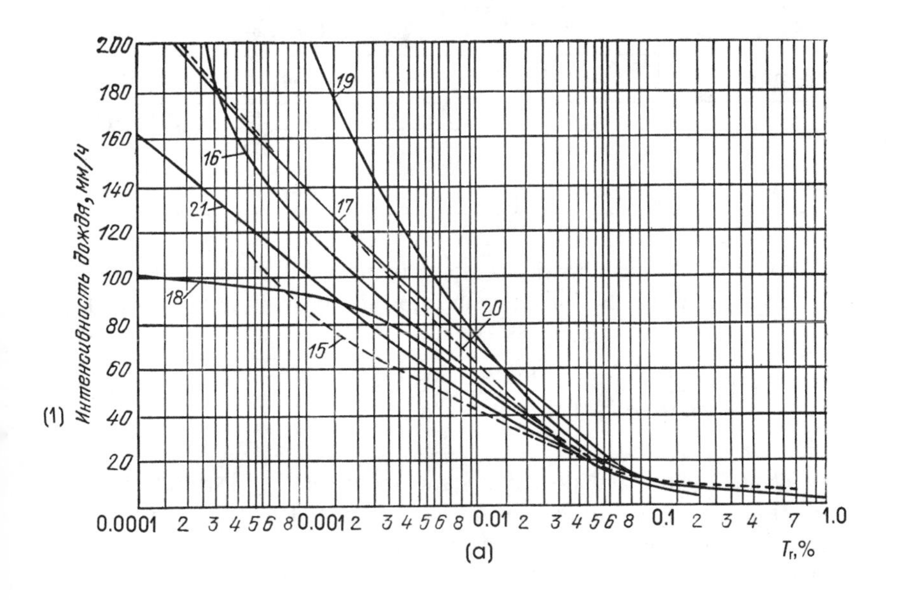

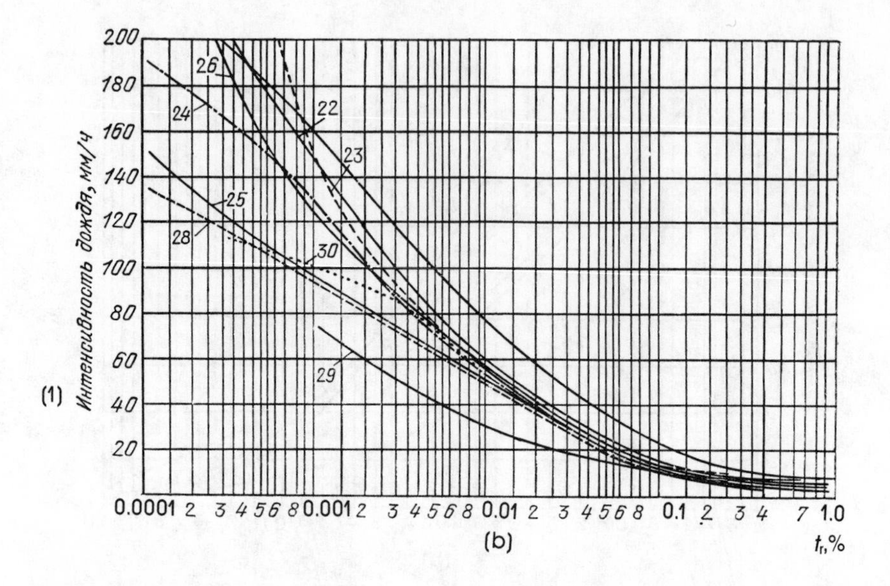

Key: **(1) Rate of rain, mm/hr**

FIGURE 4.10. Statistical distributions of mean minute rates of rain: (a) Central Asia and Kazakhstan; (b) Siberia and Far East.

TABLE 4.1

	$T_r(\epsilon)_{mon}/T_r(\epsilon)_{year}$ for regions			
ϵ, mm/hr	1	3	4	10
20	5	4.5	4–4.5	2–3
50	7.5	6.3	6.5	4.5
100	9	8	8–8.5	5–7

Processing of the statistical data in Figures 4.9 and 4.10 shows that the mean minute distributions of rain rates are logarithmically normal:

$$w(\epsilon) = \frac{1}{\epsilon\sigma_\epsilon\sqrt{2\pi}} \exp\left[-\frac{(\log \epsilon - \log m)^2}{2\sigma_\epsilon^2}\right] \quad (4.23)$$

where σ_ϵ^2 and m are the dispersion and mean of the distribution. The attenuation of radio waves in rain which is not exceeded for a given fraction of the time of year or of the worst month can be determined on the basis of the data presented. The results of the calculation in application to the

European part of the USSR (on the basis of the average distribution for the first, second, third, fourth, and fifth climate regions), for the probabilities $T_r = 1$ and 0.1% that are most often used, are presented in Figure 4.11 as the attenuation in rain L_r as a function of frequency f and of the elevation angle β of the earth station antenna.

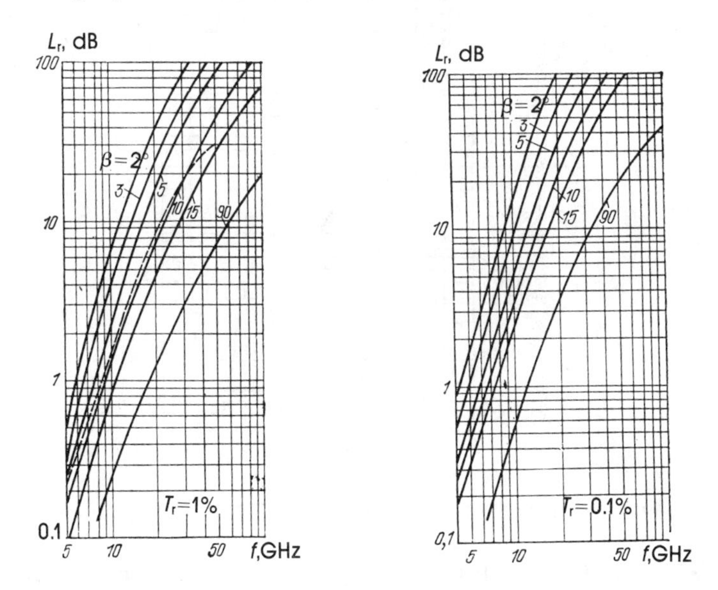

FIGURE 4.11. Dependence of statistical signal absorption in rain for different T_r.

The next most significant absorber of the power of radio waves is fog. The intensity of fog is measured by the limiting optical visibility S (in meters), and its absorptivity L'_{fog} (in dB/km) is determined by the absolute humidity ρ (in g/m^3). The relationship between these parameters can be represented as empirical formulas [4.6]: $\rho = 3S^{-4.3}$; $L'_{fog} = 0.483\rho/\lambda^2$. The mean vertical extent of fog is small and usually does not exceed 0.5–0.6 km, but then the horizontal extent can reach 100 km, and the lifetime of this region can be significantly longer than that for rain. Curves for determining the statistical absorption in fog on 4 GHz are given in Figure 4.12 for climate regions of Europe, similar to the first and second climate regions of the USSR; this much absorption will occur considerably less frequently in the rest of the USSR.

In some climate regions of the USSR, snow (especially wet snow) and also hail can exert a considerable influence on the level of received signals. The linear absorption coefficient in dry snow and in hail $L'_{dr.sn}$ is considerably smaller than in rain of the same rate; the ratios of $L'_{dr.sn}$ and L'_r for a precipitation rate of 10 mm/hr are given in Table 4.2 for different frequency bands [19].

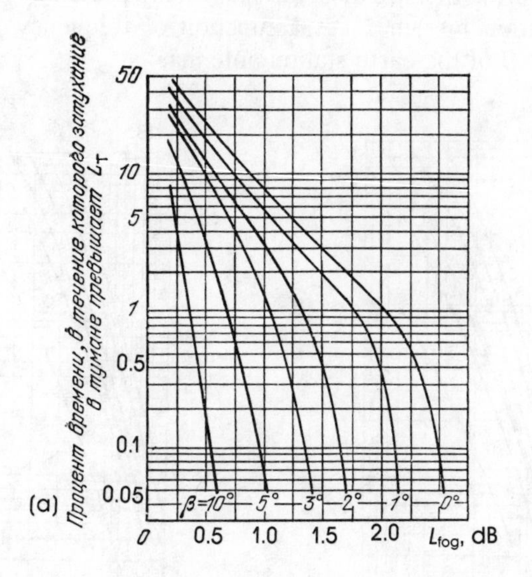

Key: **(a) Fraction of time during which attenuation in fog exceeds L_{fog}**

FIGURE 4.12. Integral distribution of losses in fog for different elevation angles.

TABLE 4.2

f, GHz	8	11	18	25	35
$L'_{dr.sn}$, dB/km	0.0067	0.0107	0.0312	0.0362	0.281
L'_{r}, dB/km	0.085	0.24	0.78	1.5	2.6

Absorption caused by wet snow is about the same as in rain of equal rate, but in individual cases (when big flakes of wet snow fall) it can be four to six times greater than in rain. Experiments show that this phenomenon is fairly rare, and for calculations for the worst month only absorption in rain needs to be taken into consideration for practical purposes. The ionosphere also has an effect on wave propagation conditions, but absorption in it on frequencies above 1 GHz is exceedingly small and in accordance with the report [4.6], $L_{ion} \approx 2500/f^2$ does not exceed $2.5 \cdot 10^{-3}$ dB, even when the antenna is at low elevation angles.

4.5 LOSSES DUE TO REFRACTION AND IMPRECISE POINTING OF THE ANTENNA AT THE SATELLITE

Next, losses of signal power due to refraction, i.e., due to the curvature of the trajectory of a signal as it passes through the atmosphere (ionosphere and troposphere), are treated. Ionospheric refraction (in degrees) can be determined by the formula (see [4.7]):

$$\delta_{\text{ion}} \approx -\frac{57.3 \cos \beta}{f^2 \sin^3 \beta}$$

from which it follows that it is inversely proportional to the square of frequency and becomes negligible at $f > 5$ GHz. Tropospheric refraction is not a function of frequency. For the standard model of the atmosphere at low elevation angles, the constant (regular) component of tropospheric refraction in degrees is $\delta_{\text{tr}} \approx (n - 1)\text{ctn}\, \beta$. The total refraction $\delta = \delta_{\text{ion}} + \delta_{\text{tr}}$ is given in Figure 4.13 [4.7] (tropospheric refraction is shown by the continuous lines, and ionospheric refraction by the dashed line). The influence of refraction on the performance of satellite systems is manifested as an increase of the satellite search range by the earth station antenna in the heavens; the satellite can be "lost" during programmed steering of the antenna in the range of small elevation angles; signal losses occur in fixed receiving antennas, pointed at the true position of a satellite.

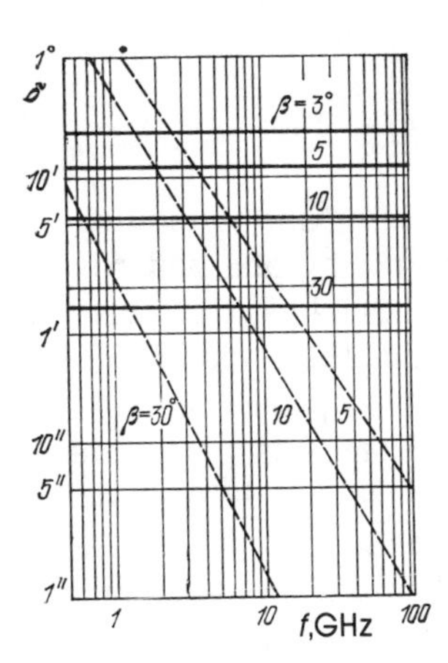

FIGURE 4.13. Atmospheric refraction as function of frequency.

These consequences of refraction can be eliminated or minimized by taking into account beforehand, the regular component of refraction on the basis of the data presented here, because the irregular fluctuations of refraction usually are small and do not exceed 40″ for elevation angles $\beta \geqslant 5°$. In the case of automatic beam-pointing of antennas by the peak of the incoming signal the influence of refraction is virtually eliminated. Here, however, a new component of losses appears, losses due to imprecise pointing of earth station antennas at satellites; it is determined by the angular deviation of the axis of the main lobe of the radiation pattern from the true direction of the satellite, and also by the width and shape of this lobe. Usually, the pattern in the principal part of the main lobe is given by:

$$G(\phi) \approx 2\,\frac{J_1\left[\dfrac{2\pi R}{\lambda}\sin\phi\right]}{\dfrac{2\pi R}{\lambda}\sin\phi} \approx \frac{1}{1 + (\phi/\Delta\phi_{0.5})^2}$$

where J_1 is a first-kind Bessel function of the real argument, and R is the reflector radius of the antenna. The beam pointing losses are

$$L_{bp} = G(\phi = 0)/G(\phi) \approx 1 + (\phi/\Delta\phi_{5.0})^2 \qquad \textbf{(4.24)}$$

In modern beam-pointing systems, the antenna usually is controlled on two axes (for example the azimuthal and elevation axes). Here the angular pointing error on each of the axes can be represented by the sum of three components:

$$\phi_{A,\beta} = \phi_m + \phi_{fl} + \phi_v$$

where ϕ_m is the angular error due to the imperfection of the mechanical part of the system (gear play and reflector deformations); ϕ_{fl} is the fluctuation error due to the influence of noises in tracking channels; ϕ_v is the dynamic (rate) error due to the motion of the antenna during tracking.

The first component depends on the design of the antenna and usually is given in certificate data on an antenna (its statistics are not given); the second is computed on the basis of the expected signal-to-noise ratio on the receiving channels and has a normal distribution with the parameters m_{fl} and σ_{fl}; the third depends on the rate of the relative motion of the satellite and can be determined on the basis of data on the orbital parameters as a result of the solution of the equation:

$$\phi_v(t) = K_t|\bar{v}(t)\bar{r}|\ [1/d(t)]$$

where K_t is the transfer coefficient of the tracking channel; $\bar{v}$ is the velocity of the satellite in space; r is the unit radius vector; and d is the distance to the satellite (slant range).

This equation is solved regularly during the calculation of target indications for earth stations of a system, and for this reason, for determining $\sigma_v(t)$, it is sufficient to carry out statistical processing of these target indications for several earth stations of a system. The results of this processing in the Molniya and Ehkran satellites are given in Figure 4.14, from which it follows that the greatest velocities of Molniya satellites do not exceed 0.2 deg/s, and it is even less for geostationary satellites. The distribution of $F(\omega)$ is nearly normal; accordingly the probability density of the angular beam-pointing error in each plane is

$$\omega(\phi_{A\beta}) = \frac{1}{(\sigma_{\mathrm{fl}} + \sigma_v)\sqrt{2\pi}} \exp\left[-\frac{(\phi + \phi_{\mathrm{m}} - m_{\mathrm{fl}} - m_v)^2}{2(\sigma_{\mathrm{fl}} + \sigma_v)^2}\right] \tag{4.25}$$

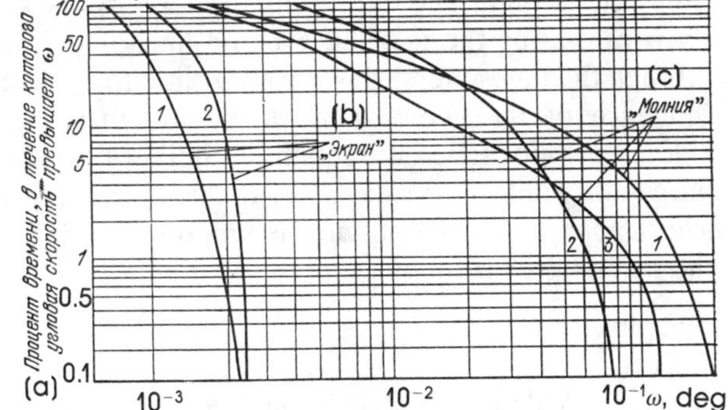

Key: **(a) Fraction of time during which angular velocity exceeds** ω
(b) Ehkran
(c) Molniya

FIGURE 4.14. Integral distribution of angular velocity of satellite relative to terrestrial stations.

Expressions (4.24) and (4.25) can be used for calculating the magnitude and density of the probability of antenna beam-pointing errors on each axis. The cumulative beam-pointing error in the equitorial plane is determined by the familiar rule:

$$\phi_\Sigma = \sqrt{\phi_A^2 + \phi_\beta^2}$$

and the density of the error probability has the generalized Rayleigh distribution:

$$\omega(\phi_\Sigma) = (\phi_\Sigma/\sigma^2_{\phi\Sigma})\exp(-0.5\phi^2_\Sigma/\sigma^2_{\phi\Sigma}) \quad \textbf{(4.26)}$$

Accordingly, the distribution function of losses due to antenna beam-pointing is

$$\omega(L_{bp}) = \omega[1 + (\phi_\Sigma/\Delta\phi_{0.5})^2] \quad \textbf{(4.27)}$$

in consideration of (4.26).

4.6 PHASE EFFECTS IN THE ATMOSPHERE

The Faraday effect and its corollary (the phase dispersion of signals) are connected with the influence of the atmosphere. As is known, the Faraday effect is caused by the fact that when a linearly polarized wave propagates through the ionosphere, this wave is split into two components under the influence of the geomagnetic field, and these components propagate at different velocities. Consequently, there will be a phase shift between them, which will lead to rotation of the polarization plane of the summary wave. Under certain simplifying assumptions [4.7], the angle of rotation of the polarization plane is

$$\Psi \approx 2.37 \cdot 10^{17}/f^2$$

The results of calculations by this formula for a few frequencies and antenna elevation angles are given in Table 4.3, from which it follows that the Faraday effect causes an appreciable change of direction of the polarization vector on frequencies below 5 GHz; on frequencies above 10 GHz this phenomenon is negligible. The influence of this effect is manifested by the fact that, when linearly polarized signals are used for communications, signal losses will occur between colinear antennas (transmitting and receiving):

$$L_\phi = 20 \log \cos \Psi$$

TABLE 4.3

f, GHz		0.1	0.5	1	2	5	10
Ψ deg	β=90°	2400	95	25	6	1	0.3
	β=0°	6000	250	60	18	3	0.1

To avoid this on frequencies below 10 GHz, circular polarization exclusively is used in satellite systems; in higher frequency bands phase effects do not prevent the use of linear polarization. Phase effects in the atmosphere, more accurately their frequency dependence, produce phase dispersion of the components of transmitted signals and consequently distort them on the receiving end. Like Faraday rotation, these effects are inversely proportional to the square of frequency. The total phase shift of a signal is

$$\Psi_0 = \frac{2\pi f}{c} \int n \, dl$$

where n is the refraction index of the atmosphere; c is the speed of light; and the group delay time of the signal is

$$\tau = d\Psi_0/dt$$

The approximate difference of the group delay time $\Delta\tau$ for the outer components of a wideband signal with band Δf should be such that no signal distortions will occur.

For a quantitative estimation of the bandwidth of the amosphere, we will use $\Delta\tau\Delta f = 0.1$, and then $\Delta f \leq \sqrt{3 \cdot 10^{-11} f^3}$. The results of calculation by this formula are given in Table 4.4, from which it follows that the widest signal band that can be transmitted through the atmosphere without phase distortions is approximately 25 MHz in the 1-GHz range and reaches up to 270 MHz in the 4–6-GHz range. These limits should be remembered when designing wideband TV and telephone lines, especially on frequencies below 4 GHz.

TABLE 4.4

f, GHz	0.5	1	5	10
Δf, MHz	10	25	270	750

4.7 LOSSES DUE TO MISMATCH POLARIZATION OF ANTENNAS

According to the book [4.7], the normal transfer coefficient of power between two antennas in the general case, when both antennas have elliptical polarization, is

$$K = \frac{1}{2}\left[1 + \frac{4e_1e_2}{(1 + e_1^2)(1 + e_2^2)} + \frac{(1 - e_1^2)(1 - e_2^2)}{(1 + e_1^2)(1 + e_2^2)} \cos 2\Psi\right] \quad (4.28)$$

where e_1 and e_2 are the coefficients of ellipticity (the ratio of the minor semiaxis of an ellipse to the major semiaxis) of the polarization of the transmitting and receiving antennas, respectively; and Ψ is the angle between the corresponding semiaxes of the polarization ellipses of the transmitting and receiving antennas.

If the relative positions and parameters of the polarization ellipses of the antennas are known, (4.28) can be used for determining signal power losses L_{loss} for a given receiving antenna in comparison with an antenna, the polarization ellipse of which coincides with that of the transmitting antenna. Curves of the maximum losses for $\Psi = 90°$ are shown in Figure 4.15 for different amounts of ellipticity of polarization, and the transfer coefficients are given in Table 4.5 for different combinations of polarization of transmitting and receiving antennas.

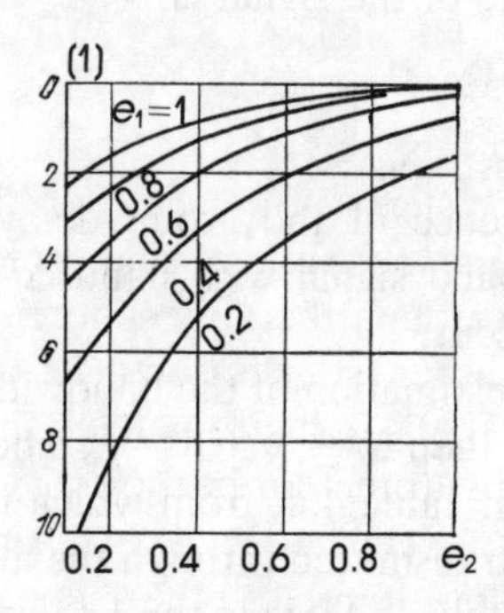

Key: (1) L_{loss}, dB

FIGURE 4.15. Losses due to mismatch polarization of transmitting and receiving antennas as functions of polarization ellipticity.

TABLE 4.5

	Transfer coefficient for different polarizations of transmitting antenna		
Polarization of receiving antenna	*Linear*	*Right-hand circular*	*Left-hand circular*
Linear	$\frac{1}{2}(1 + \cos 2\Psi)$	$\frac{1}{2}$	$\frac{1}{2}$
Right-hand circular	$\frac{1}{2}$	1	0
Left-hand circular	$\frac{1}{2}$	0	1

As follows from Table 4.5, when both antennas have linear polarization, the transfer coefficient is maximum, if both polarization vectors are colinear. In the case of communications through bodies in space, this condition obviously will be violated, both because of a change of the relative position of the target and the earth station, and due to the above-examined phase effects in the atmosphere. Therefore, it is better to use circular polarization in the 1–10-GHz range; on frequencies above 10 GHz, the choice of polarization is determined basically by the effects of depolarization of signals in the atmosphere, which are examined below. Polarization losses cannot be estimated statistically, and therefore, for energy calculations of satellite links, they must be assumed to be constant, i.e., to exist 100% of the time.

4.8 DEPOLARIZATION OF SIGNALS IN THE ATMOSPHERE

In connection with the adoption of frequency bands above 10 GHz and the use in satellite systems of polarization division, it is of practical importance to describe one more effect, connected with the propagation of electromagnetic waves through the atmosphere: the effect of depolarization in hydrometeors. Earlier (see Section 4.5), during examination of the effect of absorption of a signal in hydrometeors, no stipulations were made relative to the shape of hydrometeor particles; more accurately, they were considered to be spherical. This model of hydrometeors does not produce the effect of depolarization. In reality, the shape of natural hydrometeors, especially of raindrops (the main factor of absorption) not only is not spherical, but even asymmetric with respect to the vertical axis when they fall at an angle. This produces various amounts of attenuation and phase shifts for the vertical and horizontal components, and consequently, it is a cause of depolarization of a radio wave and of the appearance of a cross-polarization component on the receiving end. For instance, if the total attenuation is 30–40 dB, the difference of the attenuations of horizontally and vertically polarized waves (called differential attenuation) $l_{\mathrm{dif}} = L_{\mathrm{hor}} - L_{\mathrm{vert}}$ reaches 6–8 dB on frequencies of 20–30 GHz.

The depolarization of a signal produces a cross-polarization component, which is estimated by the coefficient of isolation of cross-polarization signals, which is the ratio of the powers of normally and orthogonally polarized signals. The isolation of cross-polarization signals is plotted in Figure 4.16 as a function of attenuation in rain [4.8], from which follows that, when $L_r \geqslant$ 20–30 dB, the magnitude of the cross-polarization component will reach $-(15)-10$ dB and will be perceptible interference on the receiving end. The frequency response of the isolation of cross-polarization signals in rain at a constant rate (ϵ = 100 mm/hr) is given in

Figure 4.17; it shows that cross-polarization decreases as frequency increases, and also how the isolation of cross-polarization signals is correlated with the rate of precipitation. The existence of this correlation is also mentioned in the article [4.9] with the comment that the statistics should be compiled over a period of many years.

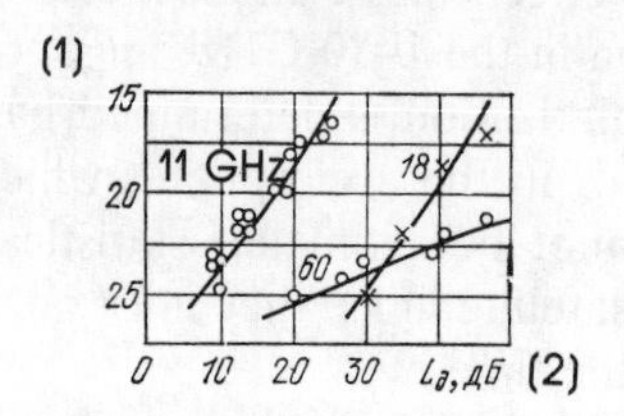

Key: (1) Isolation of cross-polarization signals, dB
(2) L_r, dB

FIGURE 4.16. Cross-polarization coefficient as function of attenuation in rain.

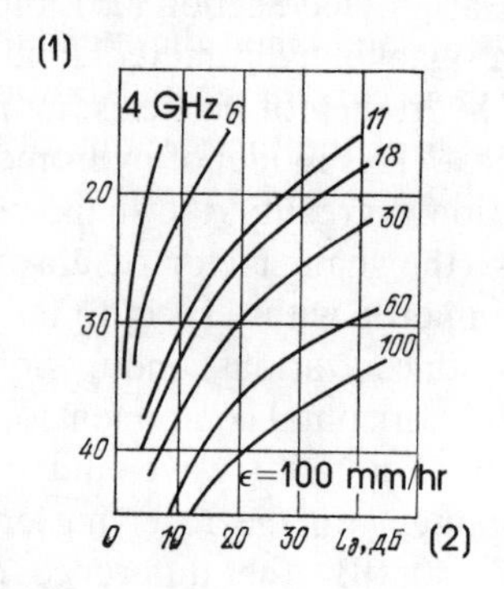

Key: (1) Isolation of cross-polarization signals, dB
(2) L_r, dB

FIGURE 4.17. Cross-polarization coefficient as function of attenuation in rain.

All that was said above applies to circularly polarized radio waves. Linearly polarized waves, strictly speaking, should not produce cross-polarization components. However, this is true for vertically falling rain with symmetrically shaped raindrops. In reality, rain always falls at an angle, and the actual inclination of raindrops does not always correspond to the slope of the shower, because a common shower contains raindrops with positive and negative angles of inclination with an overall disbalance that corresponds to the angle of inclination of the shower. Because of these factors, linearly polarized radio waves also will experience depolarization, especially when the angle of inclination of the polarization vector is different from the angle of inclination of the shower. The maximum depolarization obviously will occur when the relative angle of inclination of the polarization vector is 45°, and theory predicts that the level of cross polarization will be the same in this case as for circular polarization. This is well illustrated in Figure 4.18 (based on materials in the articles [4.8], [4.10], and [4.11]), which shows the cross-polarization component as a function of attenuation in rain on 20 GHz for:

1: circular polarization (experimental, median value);

2: vertical polarization (theoretical) for a shower with a zero angle of inclination;

3: vertical polarization (experimental points for inclinations of 15°–25°);

4: vertical polarization (experimental crosses) for 45°-angles of inclination. As follows from Figure 4.18, linearly (vertically) polarized radio waves theoretically are less vulnerable to depolarization, but in practice this difference balances out, and when the inclination of the polarization vector is 45° (which will occur for earth stations that are located at longitudinally distant points of the service area of the satellite), it will all but vanish.

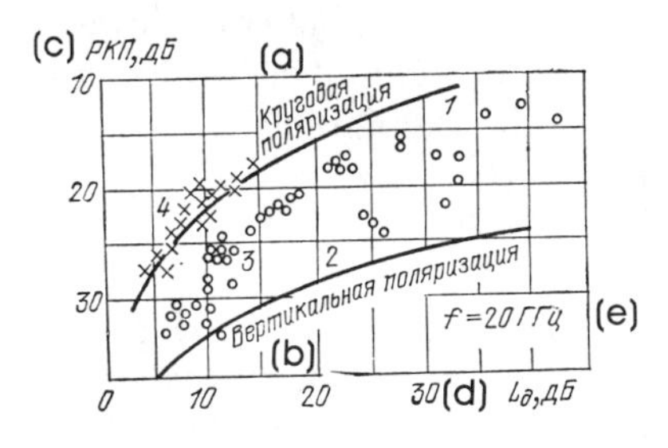

Key: (a) Circular polarization
(b) Vertical polarization
(c) Isolation of cross-polarization signals, dB
(d) L_r, dB
(e) f=20 GHz

FIGURE 4.18. Cross-polarization coefficient of radio waves as function of losses in hydrometeors.

The following important practical conclusions can be drawn from the material presented above.

1. The effect of depolarization of a signal on frequencies above 5 GHz is related to the absorption of radio waves in hydrometeors, and consequently a statistical quantitative estimate of this effect should be correlated with the statistics of rain.

2. For moderate climatic conditions in the USSR, where the rate of rain is low, the examined effect will lead to the appearance of real interference (isolation of cross polarization $\leq$ 25 dB): for TV broadcasting systems [ϵ 1%) $\leq$ 5 mm/hr] on frequencies above 18 GHz; for telephone communications systems [ϵ(0.01%) $\leq$ 12.5 mm/hr], on frequencies above 12 GHz.

3. Linear (vertical) polarization theoretically is preferable to circular, but because of the difficulty of the practical implementation of these advantages in the entire service area of a satellite, the question of the choice of the kind of polarization on frequencies above 10 GHz cannot be answered on the basis of this effect alone and requires an analysis of other factors.

4.9 NOISE OF THE ATMOSPHERE, PLANETS, AND RECEIVER SYSTEMS

It is important for the energy calculation of satellite radio links to determine the total noise power at the input of the receiver of the satellite and of the earth station, produced by different sources. As was demonstrated in Section 4.2, the receiver input noise power can be determined by (4.4); we will give a quantitative estimate of the variables that enter in it. The equivalent noise bandwidth of a receiver is

$$\Delta f_{\mathrm{n}} = \frac{1}{K^2(f_0)} \int_{-\infty}^{\infty} K^2(f)\, df$$

where $K(f)$ is the frequency response of the IF circuit of a receiver, and is usually a little wider than the frequency band $\Delta f_{0.7}$ of the IF circuit. Assuming $\Delta f_{\mathrm{n}} = \gamma \Delta f_{0.7}$, the values of coefficients γ fro n single-loop (γ_1) and two-loop (γ_2) IF amplifier stages are given in Table 4.6.

TABLE 4.6

n	1	2	3	5	10
γ_1	1.57	1.22	1.15	1.11	1.09
γ_2	1.11	1.038	1.022	1.01	1.002

The total equivalent noise temperature of a receiver system, consisting of an antenna, waveguide path, and the receiver itself (converted to the antenna radiator) is

$$T_{\Sigma} = T_A + T_0[(1 - n_B)/n_B] + T_{rec}/n_w \qquad \textbf{(4.29)}$$

where T_A is the equivalent noise temperature of the antenna; T_0 is the absolute ambient temperature (290 K); T_{rec} is the equivalent noise temperature of the receiver due to set noises; and n_w is the transfer coefficient of the waveguide signal path. The purpose of this section is to give a quantitative determination of the components in (4.29) for computing noise power (4.4), entering in the coupling equations presented in Sections 4.3 and 4.4.

The equivalent noise temperature of an antenna can be represented as the components:

$$T_A = T_c + T_a + T_a + T_{a.e} + T_{n.A} + T_{dome} \qquad \textbf{(4.30)}$$

which are attributed to different factors: T_c = the reception of cosmic radio emission; T_a = radiation of the atmosphere in consideration of hydrometeors; T_e = radiation of the earth's surface through the sidelobes of the antenna; $T_{a.e}$ = reception of radiation from the atmosphere, reflected from the earth; $T_{n.A}$ = antenna noises due to losses in its elements; and T_{dome} = the influence of the antenna dome (if there is one). The general procedure for determining these components is based on the fact that an antenna, located in an infinite space of an absorbing medium with an inhomogeneous kinetic temperature, in thermodynamic equilibrium, absorbs and reradiates power, equal to the radiated power. In this case:

$$T_A = \frac{1}{4\pi} \int_{4\pi} T_b(\beta,\Psi)G(\beta\Psi)\, d\Omega$$

where $T_b(\beta, \Psi)$ is the brightness emission temperature in direction β, Ψ in a spherical coordinate system; and $G(\beta, \Psi)$ is the antenna gain (relative to an isotropic radiator) in the same direction.

The concept of "brightness temperature" is introduced for characterizing sources of emission; it is defined as the temperature of a perfectly black body, which, on a given frequency and in a given direction, has the same brightness as the source being examined. The concept of average or effective emission temperature:

$$T_{av} = \frac{1}{\Omega_s} \int_{\Omega s} T_b(\beta, \Psi)\, d\Omega$$

where Ω_s is the solid angle of a source of emission, is used for describing sources of radiation with an unevenly distributed brightness temperature.

If the angular dimensions of the source of radiation are larger than the width of the main lobe of the antenna radiation pattern Ω_A, then $T_{av} = T_b$, and otherwise

$$T_{av} = T_b\Omega_s/\Omega_A \tag{4.31}$$

To simplify the ensuing calculations we will assume that the antenna gain in the main lobe is constant and is equal to G_{main}, and in the rear lobes and sidelobes it is also constant and equal to G_{side}, and then

$$T_A = \frac{G_{main}}{4\pi}\int_{\Omega main} T_b(\beta, \Psi) + \frac{1}{4\pi}\sum_{i=1}^{n}\int_{\Omega side\, i} G_{side\ i}T_b(\beta, \Psi)\, d\Omega$$

By solving this expression for all noise components (4.30) in consideration of (4.31) we obtain:

for an earth antenna:

$$T_{A.e} = T_{b.c}(\beta) + T_{b.a}(\beta) + c(T_{b.e} + T_{b.a-e}) + T_{n.A} + T_{dome}(\beta) \tag{4.32}$$

for a space antenna:

$$T_{Asat} = T_{b.a} + T_{b.e} + 2c\ T_{b.c} + T_{n.A} \tag{4.33}$$

where c is a coefficient that takes into account the integral energy level of the sidelobes [4.3]:

$$C = \frac{1}{2}\frac{\sum_{i=1}^{n}\int_{\Omega side\, i} G_{side\ i}(\beta, \Psi)\, d\Omega}{\int_{\Omega main} G_{main}(\beta, \Psi)\, d\Omega}$$

A quantitative estimate of c for different types of antennas is given in the book [4.12]; depending on how the antenna reflector surface is irradiated c = 0.2–0.4 (Table 4.7). As follows from (4.32), the first component of the noise temperature of an antenna is determined by the brightness temperature of space (isophot lines that provide a quantitative estimate of $T_{b.c.}$ given in the articles [1.7] and [4.7]). Most of it comes from radio emissions of the galaxy and of point radio sources (the sun, moon, planets, and certain stars).

The frequency response of the values of $T_{b.c}$, averaged through the heavens, is shown in Figure 4.19, from which it follows that cosmic radiation is significant on frequencies below 4–6 GHz, and the maximum on a given frequency is 20–30 times different from the minimum due to the great nonuniformity of the radiation of different parts of the sky; the greatest brightness occurs in the center of the galaxy; there are also some local peaks. It is important to note that the emission of the galaxy has a continuous spectrum and is weakly polarized; therefore, when it is received by a polarized antenna (with any kind of polarization), it may be assumed with sufficient accuracy that the received emission will be of one-half intensity (i.e., half of all the radiated power that enters the antenna aperture is received).

TABLE 4.7

Direction of radiation	*Fraction of radiated power with irradiation decreasing toward edges of aperture by*	
	6 dB	**10 dB**
In main lobe	0.59	0.81
In sidelobes of front hemisphere	0.03	0.04
In sidelobes of rear hemisphere	0.38	0.15

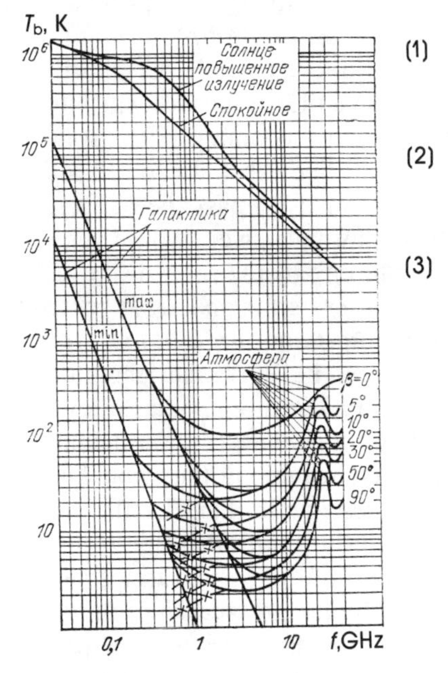

Key: (1) Sun, increased emission
Calm
(2) Galaxy
(3) Atmosphere

FIGURE 4.19. Brightness temperature of galaxy, sun, and atmosphere as function of frequency.

Shown in the same figure is the contribution of solar radiation in a calm state (during years of minimum activity) and in a "disturbed" state, inherent to years of maximum activity. The sun is the strongest source of radio emission and can completely interrupt communications when it enters the main lobe of the radiation pattern of an antenna. However, the probability of such an event is small and, in the first approximation:

$$p \approx D^2/(\pi-\beta)^2 \approx 0.6 \cdot 10^{-4}$$

where $D \approx 1.4°$ are the angular dimensions of the dangerous region of the sun (the angular size of the sun itself is 0°32′). For geostationary orbits it is approximately an order of magnitude higher and is $(2\text{–}5) \cdot 10^{-4}$, depending on the longitude of the satellite. For a geostationary satellite, the maximum time it takes to pass the danger zone is

$$t_{max} = (\Delta\phi + D)/(U_{sun} \pm U_{sat})$$

where $\Delta\phi$ is the width of the main lobe of the antenna radiation pattern; and U_{sun} and U_{sat} are, respectively, the angular velocities of the sun and satellite relative to the earth station.

The + sign in the above formula pertains to the case when a satellite is traveling eastward, and the − sign westward. It is important to note that a satellite passes through the center of the solar disk rather rarely, and usually it crosses it on lines that are displaced relative to the center; accordingly the time it takes to pass the danger zone is shorter than the time calculated by the formula given above. The exact date and time when earth antennas are "lit up" by the solar disk usually are calculated on the basis of orbital data of a satellite and are reported to earth stations along with satellite target indications.

The next strongest radio source, the moon, cannot for practical purposes interrupt communications, because its equivalent temperature is no greater than 220 K. Other sources (the planets and radio stars) make a significantly smaller contribution; the probability that an antenna will encounter these sources is less than for the sun, because their angular dimensions are small.

The radio emission of the earth's atmosphere is thermal in character and is caused entirely by the previously-examined absorption of signals in the atmosphere. In view of thermodynamic equilibrium, the medium (the atmophere) makes the same amount of energy on a given frequency that it absorbs, and accordingly

$$T_{b.a} = T_{a.av}(L_a - 1)/L_a$$

According to calculations, done for the above-described model of the atmophere, the mean thermodynamic temperature of the atmosphere for elevation angles $\beta \geqslant 5°$ in the examined frequency bands is

$$T_{a.av} \approx T_0 - 32\ \text{K} \approx 260\ \text{K}$$

The brightness temperature of a calm atmosphere (without rain) on different frequencies and for different elevation angles can be found by using the values of L_a in Section 4.5 (see Figure 4.5). The effect of precipitations in principle can be taken into consideration by the same procedure, i.e., by determining $T_{b.a}$ through losses L_r in rain, given in Figure 4.11. Although numerous studies show that the direct correlation between rain and the temperature of the sky is minor (i.e., noises of the sky can increase due to rain clouds, when rain itself is not falling), there is nevertheless a correlation with the statistics of rain over a period of many years.

The separate calculation of the temperature of the sky and of the temperature of rain and their subsequent summation produce an error (the result is approximately doubled), and therefore the calculation is done by the formula:

$$T_{b.a.} = T_0(L_a L_r - 1)/L_a L_r \tag{4.34}$$

The results of the calculations are given in Figure 4.20 as the frequency responses of the bright temperature of the atmosphere with rain for different elevation angles for probabilities of 1% and 0.1% of the time; it follows from the figure that the maximum noise temperature of the sky does not exceed 260 K and begins to play a significant role on frequencies above 5 GHz. The estimate of the temperature of the atmosphere, given above, essentially pertains to the troposphere; the radio emission of the ionosphere on frequencies above 1 GHz can be ignored, because, according to Section 4.5, absorption in the ionosphere is inversely proportional to the square of frequency.

The brightness temperature of the earth is determined by its kinetic temperature T_{0e} = 290 K and by the coefficient of reflection of electromagnetic energy from the earth's surface:

$$T_{b.e} = T_{0e}(1 - \phi)^2 \tag{4.35}$$

The combined reflectivity is determined by the familiar Fresnel formulas [4.1]:

for horizontal polarization:

$$\dot{\phi}_h = \frac{\sin\beta - \sqrt{\epsilon + i\,60\,\sigma\lambda - \cos^2\beta}}{\sin\beta + \sqrt{\epsilon + i\,60\,\sigma\lambda - \cos^2\beta}} \tag{4.36}$$

for vertical polarization:

$$\dot{\phi}_v = \frac{(\epsilon + i\,60\,\sigma\lambda)\sin\beta - \sqrt{\epsilon + i\,60\,\sigma\lambda - \cos^2\beta}}{(\epsilon + i\,60\,\sigma\lambda)\sin\beta + \sqrt{\epsilon + i\,60\,\sigma\lambda - \cos^2\beta}}$$

where ϵ is the dielectric constant of the earth and σ is the electrical conductivity of the earth. Values of ϵ and σ are given in Table 4.8 for a few kinds of the earth's surface.

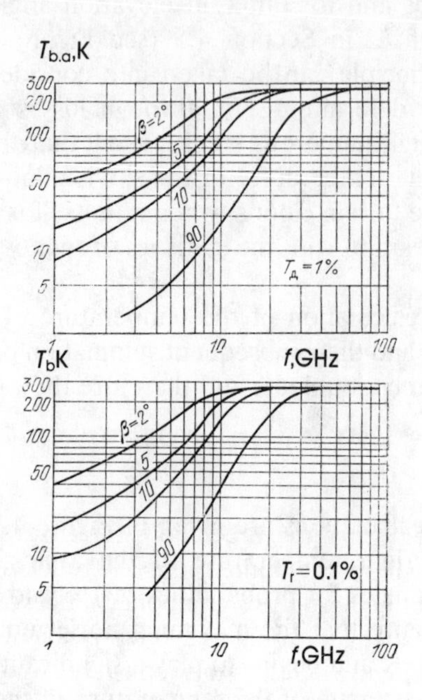

FIGURE 4.20. Brightness temperature of atmosphere (in consideration of rain) as function of frequency.

TABLE 4.8

Kind of earth's surface	ϵ	σ, *mho/m*
Sea water	80	1–6
Fresh water	80	10^{-3}–$5 \cdot 10^{-3}$
Moist soil	5–30	10^{-2}–10^{-3}
Dry soil	2–6	10^{-4}–10^{-5}

The results of calculation by (4.35) in consideration of horizontal and vertical polarizations (4.36) for reflection from parts of the earth's surface, described in Table 4.8, are given in Figure 4.21 (the numbers of the curves in Figure 4.21 correspond to the enumeration in Table 4.8). To determine $T_{b.e}$ for circular polarization, it is necessary in the first approximation to average the values of $T_{b.e}$ for horizontal and vertical polarizations.

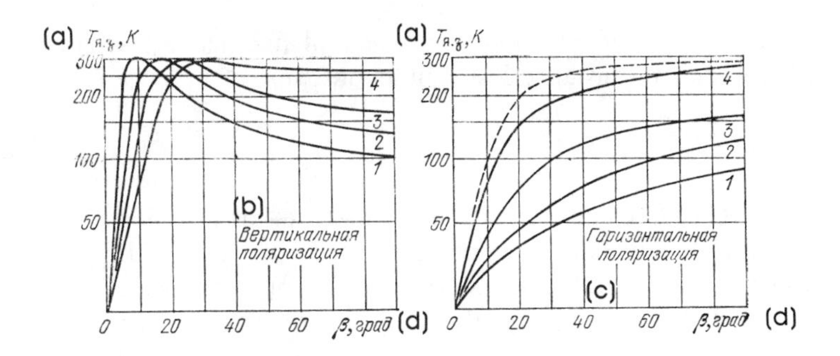

Key: **(a) $T_{b.e}$, K**
(b) Vertical polarization
(c) Horizontal polarization
(d) deg

FIGURE 4.21. Brightness temperature of earth as function of elevation angle.

The estimates given above are valid for mirror reflection, whereas on high frequencies, where the dimensions of reflecting irregularities are related to wavelength, the diffuse component, determined by the kinetic temperature of the earth and equal to 290 K, will be of greater significance. Therefore, given in Figure 4.21 is a dashed curve, which takes into account the diffuse component and the averaging of two linear mirror components (for circular polarization). This curve was plotted on the basis of experimental results and is recommended later for practical use.

The brightness temperature of atmospheric radiation, reflected from the earth, is

$$T_{b.a-e} = T_{b.a}\,\phi^2 \tag{4.37}$$

Because on frequencies above 10 GHz $T_{b.a} \approx T_{0e} = 290$ K, it then follows from (4.34) and (4.37) that

$$T_{b.e} + T_{b.a-e} \approx 290\text{ K} \tag{4.38}$$

i.e., the component of atmospheric noise, reflected from the earth, augments the thermodynamic emission of the earth and their sum produces radiation with a brightness temperature that is close to 290 K. It is noteworthy that for small elevation angles, where the component $T_{b.e}$ is small, this effect is exacerbated, because in this case $\phi \rightarrow 1$, and the contribution of atmospheric reflection predominates. Thus, for practical calculations, (4.38) should be used.

We will examine one more component of antenna noises in (4.32) and (4.33), caused by resistance losses in an antenna:

$$T_{n.A} = T_0(L_m - 1)/L_m \tag{4.39}$$

where $T_0 = 290$ K and L_m are losses in the antenna reflector material. Modern metallic reflector antennas have very low losses, and therefore the values of $T_{n.A}$ are small. The values on different frequencies are given in Table 4.9. However, when metal-plated glass fiber reinforced plastic earth antennas are used in satellite broadcasting systems, the proportion of these losses can increase, and for this reason it is deserving of practical evaluation.

TABLE 4.9

F, GHz	0.3	1	3	10	30	60
$T_{n.A}$, K	0.018	0.04	0.06	0.09	0.18	0.3

In some cases the antennas of earth stations are protected from precipitation by a radio transparent dome. Signal losses and the corresponding increment of noises, caused by these losses, usually are minor and in practice need not be considered. However, in heavy rain, a water film forms on the surface of a fairing, which causes appreciable absorption of the signal and penetration of secondary noises. According to experiments, when $\epsilon = 1$ mm/hr, the increment of the noise temperature is 4–8 K, and when $\epsilon = 10$ mm/hr, it can reach 12–20 K; the lower limits correspond to small antenna elevation angles, and the upper limits correspond to $\beta = 90°$.

4.10 THE RECEPTION OF SIGNALS BY EARTH ANTENNAS WITH SMALL ELEVATION ANGLES

As follows from Sections 4.5 and 4.9, the reception of signals from a satellite by an earth station antenna at small angles of elevation is difficult because of negative atmospheric effects. Consequently the CCIR recommends that the minimum elevation angles be limited to 3°–5° on frequencies up to 6 GHz and to 10°–15° on frequencies above 10 GHz. However, in application to Soviet territory located on northern latitudes, these recommendations cannot hold up during operation through geostationary satellites. At small elevation angles, the equivalent signal path in the atmosphere increases sharply, and the negative influence of atmospheric effects is exacerbated accordingly; these effects were already examined in previous sections, and the results given there provide an opportunity for their qualitative estimation for small β.

Among new effects that need to be considered in the case examined here are the following: interference of the direct signal from a satellite with the signal reflected from the earth, gain losses of the receiving antenna, and the effect of noises of the earth on the antenna through the main lobe of the radiation pattern. Problems of interference of the direct and reflected signals for line of sight radio relay links [1.9] and for satellite links [4.13] are already described in the literature.

Let us look at Figure 4.22. The receiving antenna is installed at height h above the earth and is pointed at a satellite at angle β; the normal radiation pattern is

$$G'(\phi) = G(\phi)/G(0)$$

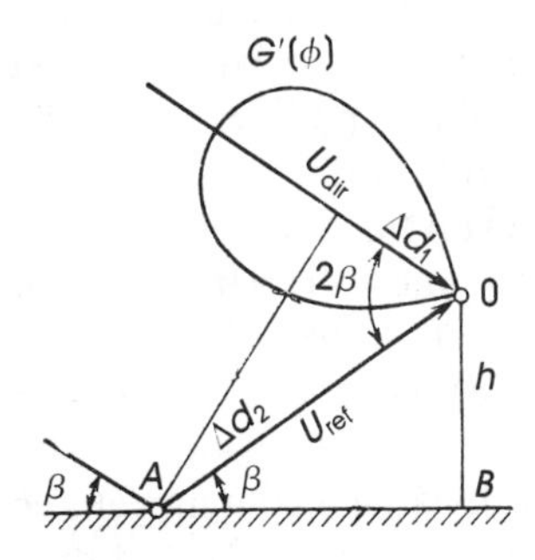

FIGURE 4.22. Calculation of attenuation factor during interference.

The direct wave, at the point of reception (on the antenna irradiator), produces the voltage:

$$U_{\text{dir}} = U_0 \exp\left[\mathrm{i}\left(\frac{2\pi}{\lambda}\right)d\right]$$

and the wave reflected from the earth:

$$U_{\text{ref}} = U_0\, \phi \sqrt{G'(\phi)/L_{\text{p}}} \exp\left[\mathrm{i}\left(\frac{2\pi}{\lambda}\right)(d + \Delta d) + \mathrm{i}\theta\right]$$

where ϕ and θ are the modulus and the phase of the reflectivity and L_{p} are polarization losses. The difference of the paths of these waves for small angles is $\Delta d \approx 2h\beta$. The resulting voltage is $U_{\Sigma} = U_{\text{dir}} + U_{\text{ref}}$, and the modulus of the attenuation factor is

$$|V| = \sqrt{1 + \phi^2 G'(2\beta)/L_n + 2\phi \sqrt{G'(2\beta)/L_p} \cos\left(\frac{4\pi h\beta}{\lambda} + \theta\right)} \quad (4.40)$$

For $4\pi h\beta/\lambda + \theta = 2m\pi$, where $m = 1, 2, 3, \ldots$, the resulting field will have peaks, and for $4\pi h\beta/\lambda + \theta = (2m + 1)\pi$ the field will have valleys. The difference of the heights of the antennas, corresponding to the phase distance between like extrema (for instance, peaks), is $\Delta h = \lambda/2\beta$, and the height corresponding to the first peak is

$$h_1 \approx \lambda/4\beta$$

By using the formulas for computing the modulus of the albedo for vertical and horizontal polarizations (Section 4.9) and the known directive gain $G'(2\beta)$ of a receiving antenna, it is possible to determine, on the basis of (4.40), the amplitudes of interference fluctuations of a signal for different elevation angles. Here, according to the materials in Section 4.9, $L_p \approx 1$ should be used for linear polarization, and $L_p \approx 10^2$–10^3 for circular polarization, because, on reflection from the earth at $\beta > \beta_B$, the direction of circular polarization reverses, and the reflected component is weakened effectively by the receiving antenna. Attention is invited to the fact that in the case of vertical polarization, the reflected component vanishes at some angle β_B (Brewster's angle), corresponding to the complete refraction of the incident beam; for circular polarization of signals from a satellite this means that the reflected signal will contain yet another horizontal component. At small elevation angles $\beta < \beta_B$, the direction of rotation of the polarization plane stays the same and $L_p \approx 1$.

The previous discussions are based on the assumption of purely mirror reflection, whereas in reality diffuse reflection also occurs at small elevation angles. The joint analysis of the mirror and diffuse components poses significant problems, and therefore experimental results usually are used, which show that the diffuse component has an appreciable effect, particularly on high frequencies.

The integral statistical distribution of the depth of fading due to the interference of the examined waves, on the assumption that the amplitudes of the interfering waves are constant in time, and that the phase difference between them is random and can acquire any values from 0 to π with equal probability, acquires the form [4.13]:

$$F(V) = 1 - \frac{1}{\pi} \arccos \frac{V^2 - 1 - \phi^2 G'(2\beta)/L_p}{2\phi\sqrt{G'(2\beta)/L_p}} \quad (4.41)$$

Results of calculation by formula (4.41) are given in Figure 4.23 for several amplitudes of a reflected wave at the point of reception. It follows from the analysis that, for linearly polarized radio waves, operation at small elevation angles will be accompanied by significant fluctuations of the signal level (from 6 to 10–15 dB) due to the influence of the beam reflected from the earth. The use of directional antennas with a wide pattern $\Delta\phi_{0.5} < \beta$ and circular polarization, such that the depth of fading in the worst case (for elevation angles close to the Brewster angles) will not exceed 6 dB, is an effective means of suppressing this effect. On high frequencies, it is also important to consider the effect of the averaging of the interference extrema by the antenna aperture. For instance, on frequencies above 10 GHz at elevation angles $\beta \approx 1°$, more than two extrema will fit in the aperture of an antenna with linear dimensions of 1–1.5 m, which, on the one hand, will result in the averaging of the result and, on the other hand, in dephasing of the field in the antenna aperture and a loss of gain of the antenna, associated therewith; on the basis of experimental data these losses can be estimated as approximately 0.5–1 dB.

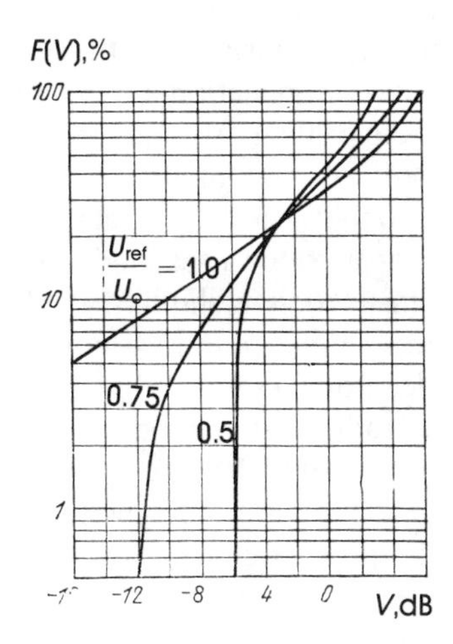

FIGURE 4.23. Integral distribution of attenuation factor during interference.

The change of the noise temperature of a receiving antenna, caused by the penetration of noises of the earth into the main lobe of the radiation pattern, can be determined by using formulas in Section 4.9. A quantitative estimation of this effect [4.13] shows that the increment of the temperature is approximately 60–80 K when $\beta \to 0$.

4.11 STATISTICAL SUMMING OF THE INFLUENCE OF ATMOSPHERIC FACTORS AND THE CALCULATION PROCEDURE

The materials of the previous sections provide an opportunity to calculate a satellite link. It will be assumed here that the basic constants that enter into the coupling equations are given*: the frequency bands, the altitude and type of orbit of a satellite, the gains of earth and space antennas, the efficiency of the waveguide paths, and also (in application to the calculations of this chapter) the kind, number, and parameters of the signals that are transmitted. The purpose of the calculation is to determine the variables that enter into the coupling equations (the signal energy losses due to the above-listed factors and noises of different origin), and to compute the signal-to-noise ratio at the end of a communications link (for given earth and space transmitter powers), or to determine the powers of these transmitters for the necessary signal-to-noise ratios.

Usually the initial calculation is done for the single-signal mode (for example, for the transmission of a TV signal); the parameters of the transponder of a communications satellite are determined. Then, if telephone messages must be transmitted, the carrying capacity of the transponder is calculated in application to the selected multiple access method and the parameters of the earth receiving-transmitting equipment are refined. The problem of determining (synthesizing) the optimum parameters of the transponder with the necessary carrying capacity for transmitting telephone messages for a selected multiple access method can also be solved directly; this problem should be solved on the basis of the materials presented in this chapter and in Chapter 7.

The problem of calculating a satellite link in the single-signal mode is most easily solved by the traditional method—by sections; the corresponding coupling equations (4.9) and (4.10) are given in Section 4.3. When the calculation is done by these equations, it is necessary first to determine the margins a and b and additional signal losses L_{1add}, L_{2add}, and equivalent noise temperature of the space and earth receiving systems $T_{\Sigma sat}$, $T_{\Sigma e}$. Coefficients a and b can be determined by (4.7) on the basis of a reasonable compromise. Usually $b = 1.1$–1.2, and accordingly $a = 11$–6. In some cases, in particular for systems with small earth stations, $a = b = 2$ may be used.

Additional losses are caused by the influence of factors, examined in Sections 4.5–4.9 and can be determined by the formula:

---

*These variables are selected on the basis of systems and technical-economic analyses or are determined by the technical parameters of the equipment.

$$L_{add} = L_a L_r L_{bp} L_p \tag{4.42}$$

where not only the quantitative values of the components of (4.42), but also their statistical characteristics, must be taken into consideration. The values of L_a for different frequency bands are given in Figure 4.5; as was pointed out in Section 4.5, these losses are constant and exist 100% of the time. The value of L_p is determined from the curves in Figure 4.15; these losses are also assumed constant. Losses in rain L_r are random. The value for a given climate region can be determined by (4.21), and also from Figures 4.6, 4.7, 4.9, and 4.10. The distribution of L_r is given in Section 4.5. Losses due to antenna beam-pointing L_{bp} also are random and are determined by (4.24), and the distribution function is given by (4.27).

Additional losses L_{add} can be estimated by determining the quantiles of the distribution of the values of L_r and L_{bp}, computed for a given percentage of the time (for example, 99 or 99.9%), and by adding them (in dB). The results of the calculation of the quantiles of L_r of the order of 99 and 99.9% (of the corresponding values T_r = 1 and 0.1%) were given earlier in Figure 4.11.

The quantiles of the distribution of L_{bp} can be computed in the same way by (4.24) and (4.26) for given $\Delta\phi_{0.5}$ and beam-pointing errors ϕ_Σ. Here, however, it is important to remember that the result of the determination of L_{add} will be overstated, because the summation of the maximum (quasipeak) values of L_r and L_{bp} presumes that both kinds of losses exist simultaneously, although actually they are independent.

Now it is necessary to determine $T_{\Sigma sat}$ and $T_{\Sigma e}$ by (4.29) in consideration of the variables entering in it, represented by (4.32)–(4.34) and (4.38), and also in Figures 4.19–4.21. The values of $T_{\Sigma sat}$ and $T_{\Sigma e}$ obtained also will be quasipeaks, because they are computed on the basis of the quantiles of the distribution of the rates of precipitation, used for the calculation of the functions given in Figure 4.20.

After substituting a, b, L_{add}, $T_{\Sigma sat}$ and $T_{\Sigma e}$ into coupling equations (4.8) and (4.9), we find the earth and space transmitter powers, necessary for achieving the required signal-to-noise ratio at the end of the link for a given percentage of the time (for example 99 or 99.9%).

For a more accurate calculation of a link and elimination of redundant energy potential due to the use in the preceding calculation of the quasipeak losses and noise, it is necessary to use generalized equations of a communications link, given in Section 4.4. However, it is first necessary to solve the problem of the statistical summation of the influence of all the components of loss and noise.

In accordance with the materials in Section 4.4, generalized equation (4.16) of a link may be written as

$$(P_n/P_s)\Sigma = A(BL_{r1}L_{bp1}T_{\Sigma 1} + CL_{r2}L_{bp2}T_{\Sigma 2}) \quad \textbf{(4.43)}$$

where A, B, and C are constants, determined by the parameters of a system; the subscript 1 pertains to the earth-satellite section, and the subscript 2 to the satellite-earth section. It is not hard to see that six variable parameters enter into (4.43), which, in turn, are functionally connected with the initial independent random variables ϵ, ϕ_1, and ϕ_2 by given distribution functions. The problem is to determine the integral distribution function of the resulting random variable $(P_n/P_s)_\Sigma$ and to compute the quantiles of the necessary order, for example 99 and 99.9%, i.e., to solve the equations:

$$F(x) = 0.99 \quad \text{and} \quad F(x) = 0.999 \quad \textbf{(4.44)}$$

for the purpose of finding the values of (P_n/P_s) that are not exceeded, for example, for 99 and 99.9% of the time (in accordance with CCIR standards).

The following is known relative to the initial random variables ϵ and ϕ:

- the distribution of ϵ (of the rate of precipitation) is logarithmically normal with the parameters m and σ_ϵ, determined by (4.23);
- the random variables ϕ_1 and ϕ_2 (the beam-pointing angle errors of the transmitting and receiving earth antennas) have a Rayleigh distribution, (4.26);
- ϵ, ϕ_1 and ϕ_2 are independent.

Secondary random variables are formed by the functional transform of the initial variables. According to (4.21) and (4.22),

$$L_{r1} = L_{01} \exp(\epsilon/\epsilon_0), \quad L_{r2} = L_{02} \exp(\epsilon/\epsilon_0) \quad \textbf{(4.45)}$$

in accordance with (4.24):

$$L_{r1} = 1 + (\phi_1/\Delta\phi_{01})^2, \quad L_{r2} = 1 + (\phi_2/\Delta\phi_{02})^2 \quad \textbf{(4.46)}$$

Then, from L_{r1} and L_{r2}, on the basis of (4.29), (4.30), and (4.34), the new random variables $T_{\Sigma 1}$ and $T_{\Sigma 2}$ are found:

$$T_{\Sigma 1} = D_1 + T_0(L_{r1}-1)/L_{r1}, \quad T_{\Sigma 2} = D_2 + T_0(L_{r2}-1)/L_{r2} \quad \textbf{(4.47)}$$

where D_1, D_2, T_0 are constants.

The problem as stated was solved in two ways: by computing the distribution of the resulting random variable $\nu = (P_n/P_s)_\Sigma$ in explicit form, followed by the approximate solution of (4.43), or by modeling the random variables with the aid of an electronic computer and estimating the quantiles on the basis of the selected value of the corresponding ordinal statistics. The latter method is the most effective, because it avoids the solutions of complicated integral equations and gives a quantitative result in explicit form. We will give the result of the energy calculation by electronic computer of a satellite TV broadcasting system in the 12.5/14-GHz band for the following initial data:

$0.1 \leqslant \epsilon \leqslant 40$ mm/hr, $m = 0.15$, $L_{01} = 2$, $\epsilon_0 = 17.5$ mm/hr, $\sigma_e = 0.55$

$L_{02} = 1.2$, $\phi_{01} = 0.06$ deg, $\sigma_{\phi 1\Sigma} = 0.017$ deg (antenna $D = 12$ m)

$\phi_{02} = 0.85$ deg, $\sigma_{\phi 2\Sigma} = 0.17$ deg (antenna $D - 1.2$ m)

$T_{n.sat} = 5000$ K, $T_{n.e} = 700$ K, $n_{sat} = 2$ dB, $n_e = 0.5$ dB

$D_1 = 8200$ K, $D_2 = 870$ K, $T_0 = 290$ K, $A = 1$, $B = 1$, $C = 10$

The electronic computer calculation by the program given above yields the following quantiles of losses:

$$F_{99\%}(BL_{r1}L_{bp1}T_{\Sigma 1}) = 2.3 \cdot 10^4, \quad F_{99\%}(CL_{r2}L_{bp2}T_{\Sigma 2}) = 1.6 \cdot 10^4 \tag{4.48}$$

Here the modeling error was 0.75% for the 99% quantile, and about 2.5% for the 99.9% quantile. Simple summation of quasipeak (determined for the same probabilities that they will not be exceeded) losses in accordance with the procedure described earlier for the same initial data yields the values:

$$F_{99\%}(BL_{r1}L_{bp1}T_{\Sigma 1}) = 3.6 \cdot 10^4, \quad F_{99\%}(CL_{r2}L_{bp2}T_{\Sigma 2}) = 2.47 \cdot 10^4 \tag{4.49}$$

By comparing (4.49) with (4.48) we establish that the energy gain is about 2 dB. It is obvious that for smaller probabilities (99.9%), and also for higher frequencies (20 and 30 GHz), where the proportion of losses in the atmosphere increases, the use of the above-described procedure of statistical analysis will yield even more effective results.

4.12 CALCULATION OF NOISE IN LOW-FREQUENCY CHANNELS

The final step of the power calculation of satellite links is the conversion of the ratio $(P_s/P_n)_\Sigma$ at the end of the link to a low-frequency channel, i.e., the determination of the signal-to-noise ratio at the demodulator output of the receiver system of the earth station. In the case of FM TV transmission, the signal-to-noise ratios at the input and output of the receiver are connected by the relation:

$$(P_s/P_n)_{out} = (P_s/P_n)_{in} B_{TV}(\text{FM})\, B_v \Delta k \tag{4.50}$$

where B_{TV} (FM) is the gain in the signal to fluctuation noise ratio, produced by an FM TV receiver; B_v is a videometric coefficient; Δ is the gain in terms of thermal noises due to the use of linear predistortions; and $k = 9$ dB is the factor for converting the amplitude of a sine wave signal to the effective voltage (this conversion factor must be used because of the procedure that is used for normalizing the signal-to-noise ratio on a TV channel; see Chapter 2).

In turn,

$$B_{TV}\ (\text{FM}) = \frac{3}{2}\frac{\Delta f_n \Delta f_d^2}{F_u^3} \tag{4.51}$$

where Δf_d is the peak frequency deviation, reserved for the TV signal itself without sync pulses and the accompanying signals (for example, sound accompaniment); and F_u is the upper frequency of the TV signal band (more accurately the upper frequency of the low-pass filter at the demodulator output). The numerical values of coefficients B_v and Δ are given in Chapter 2; for the standard linear predistortion curve, recommended by the CCIR, when an old videometric filter is used, $B_v\Delta = 18.1$ dB, and when a new filter is used, $B_v\Delta = 14.2$ dB.

If just a TV signal is transmitted in the transponder of a communications satellite (and of an earth station), the calculation of the signal-to-noise ratio on a TV channel is completed. But, if other signals are transmitted in the transponder for example, sound accompaniment, radio broadcasting, facsimile, *et cetera,* then it is necessary also to consider the interference on the TV channel, produced by these signals (see Chapter 19). When telephone messages are transmitted, the number of interference components on a telephone channel is significantly greater than in the case of the transmission of TV programs.

The general equation that connects the cumulative noise level on a telephone channel with the individual interference components is

$$P_{\text{n.c.}} = P_{\text{n.t}} + P_{\text{n.non}} + P_{\text{n}\phi} + P_{\text{n.AM-PM}} + P_{\text{i}} \quad (4.52)$$

where $P_{\text{n.t}}$ is the power of the thermal noises on a channel; $P_{\text{n.non}}$ is the power of noises caused by the nonlinearity of the amplitude responses; $P_{\text{n}\phi}$ is the power of noises produced by the nonlinearity of the phase responses; $P_{\text{n.AM-PM}}$ is the power of inaudible AM-PM noises; and P_{i} is the power of interference from neighboring transponders, satellites, and systems. These noise components will be analyzed quantitatively in application to different telephone message transmission methods and multiple access methods on the basis of materials in Chapter 7.

Chapter 5

Problems of the Electromagnetic Compatibility of Satellite Systems

5.1 A GENERAL DESCRIPTION OF COMPATIBILITY PROBLEMS OF SATELLITE SYSTEMS AND NORMALIZATION OF INTERFERENCE LEVELS

The frequency bands given in Section 3.1 are assigned for satellite services on a joint basis with terrestrial services. Fixed and mobile services operate in such frequency bands, as well as one or two other services. For example, radar or radio broadcasting are added to them in some bands. Because of this the problem of determining the electromagnetic compatibility conditions for satellite systems is broken into two separate problems: the compatibility of satellite systems with each other and compatibility of satellite systems with terrestrial systems (i.e., compatibility of different services).

5.1.1 Crosstalk Interference Between Satellite Systems

In the most general case, when the frequency bands of both systems coincide in each direction (Figure 5.1(a)), it is necessary to take into consideration interference that is created by the transmitter of the earth station of the interfering system with the space receiver system, which is subjected to the interference (line 1 in Figure 5.1(a)), and interference from the space transmitter with the receiver system of the earth station of the other system (line 2 in Figure 5.1(a)).

In some frequency bands, in which reverse (i.e., two-way) use is permitted, the situation shown in Figure 5.1(b) can arise, when the frequency band is used in one system for one direction (for instance earth-space), while in the other system it is used for the opposite direction

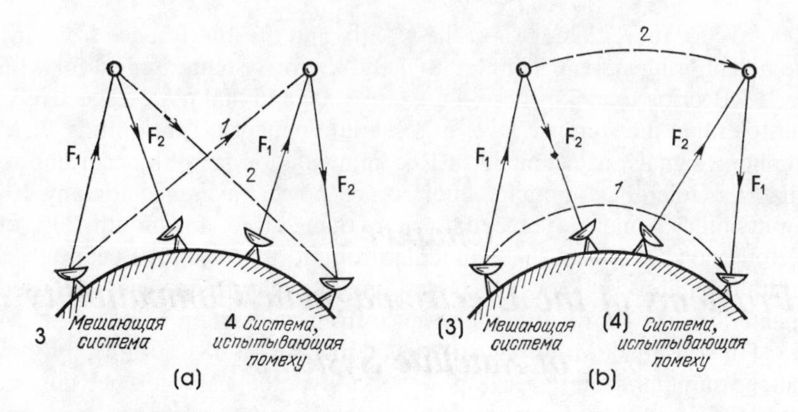

Key: **(3) Interfering system**
(4) System experiencing interference

FIGURE 5.1. Possible interference between satellite systems: (a) when transmission directions coincide; (b) when frequency bands are used in reverse.

(for example space-earth). In this case, it is necessary to examine the interference created by the transmitter of the earth station of the interfering system with the receiver of the earth station of the system that is subjected to the interference (1 in Figure 5.1(b)), and the interference from the space transmitter of the interfering system with the space receiver of the other system (2 in Figure 5.1(b)).

The coordination and registration procedure, described in Chapter 3, is used for guaranteeing the compatibility of satellite systems. During this process, refined calculations of the possible interference levels are executed (see Section 5.2), and the results are compared with the standards of the CCIR Recommendations or with standards established by agreement between the Administrations, if the appropriate CCIR Recommendations do not exist.

According to Recommendation 466-2 [8] (Mod-1)*, the mean minute psophometric noise power on any telephone channel at the zero relative level point, caused by interference from all other systems, may not exceed

*Recommendation 466 applies to systems operating in frequency bands below 10 GHz. There are no corresponding recommendations for higher frequency bands.

2,000 pW for more than 20% of the month, and the interference from any one neighboring system is limited to 600 pW. For systems applied for with the IFRB prior to 1978, these limits are 1000 and 400 pW, respectively. The tolerable interference level in a satellite network, transmitting PCM telephone signals, is examined in Recommendation 523 [8], according to which the tolerable summary interference power, averaged for any 10-minute interval, may not exceed for more than 20% of any month, 20% of the total noise power at the demodulator input, for which the guaranteed error probability is 10^{-6}, 15% in systems in which frequencies are used repeatedly, and 4% for the interference from one system.

For a useful FM TV signal, Recommendation 483-1 [8] establishes that the summary interference power in a hypothetical reference circuit, created by emissions of earth and space stations of other systems, may not exceed 10% of the tolerable noises on the video channel for more than 1% of any month. Here the interference from any one system should not exceed 4%. This recommendation must be taken into consideration during the design of systems, and must serve as a criterion for analyzing the levels of interference between systems. In practice, however, another approach often is used, based on the results of experimental studies and generalized curves of the tolerable safety demodulator input signal-to-noise ratios, depending on the separation of the carriers and the modulation parameters. These curves are given in Report 634 [10] in the concluding transactions of WARC-1977 [6] and in other sources, for example, the book [4].

In addition to the above-listed combinations of useful and interfering signals, one should remember the transmission of signals of one channel on a carrier, which experiences interference from a TV signal, because, in many cases, this combination is the determining factor for the computation of the tolerable angular distance between satellites. Some preliminary data on the protection of one channel on a carrier are given in the plan of Report AC/4 [8]. In this plan of the report, in particular, is given a criterion, which the Intelsat organization uses for coordinating one channel on a carrier of the SPADE type:

$$P_s/P_{int} \geq 27.5 + \log \delta \tag{5.1}$$

where P_s/P_{int} is the tolerable signal-to-interference ratio; $\delta = B/\Delta f$; B is the passband of one channel on a carrier; and Δf is the peak-to-peak frequency deviation of the TV carrier of the interfering signal, caused by artificial dispersion of the carrier.

The generalized expression for determining the required protection ratio:

$$P_s/P_{int}=a+b \log \delta \quad (5.2)$$

is given in suggestions of the Administration of Communications of France in WARC-1979 [5.1]. It was suggested that the coefficients a and b be determined in accordance with Table 5.1. There are some other suggestions, for example [5.2], pertaining to the protection criteria of the signals of one channel on a carrier, but the CCIR has not yet adopted the corresponding recommendations and additional studies are needed.

TABLE 5.1

Kind of modulation for transmission of one channel on a carrier	*Frequency of dispersion signal*	*a, dB*	*b, dB*
Digital	Frame frequency	20	3.5
Digital	Line frequency	20	9
Analog $\delta < 0.022$	Frame frequency	13.4	2.4
Analog $\delta > 0.022$	Frame frequency	22	7.6

5.1.2 Crosstalk Interference Between Satellite and Terrestrial Systems

During the determination of the conditions of compatibility of satellite and terrestrial services, the four following possible kinds of interference are examined:

1. from the transmitters of space stations, acting upon the receivers of terrestrial stations in frequency bands for space-earth links;
2. from the transmitters of terrestrial stations, acting upon receivers of earth stations in frequency bands for space-earth links;
3. from transmitters of terrestrial stations, acting upon receivers of space stations in bands for earth-space links;
4. from transmitters of earth stations, acting upon receivers of terrestrial stations in bands for earth-space links.

Shown in Figure 5.2, as an example, are these four kinds of interference for the jointly used 4- and 6-GHz bands of a fixed satellite service and radio relay terrestrial links. The CCIR adopted a particular model, closest to reality, for each of the combined services, and calculations of interference were done for different conditions. On the basis of these calculations specific limits for each service and ways of implementing the compatibility criteria were determined. The maximum tolerable interference on the channels of combined systems is used as a compatibility criterion. It is set equal to 10% of the total tolerable noises, produced by all sources.

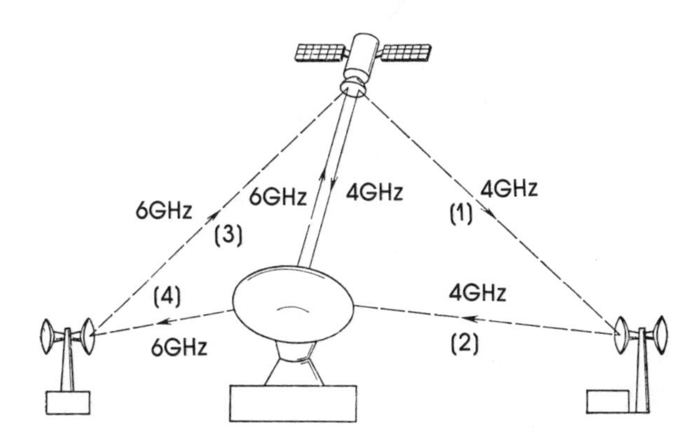

FIGURE 5.2. Possible interference between satellite systems and radio relay lines (6/4GHz).

5.1.3 Tolerable Interference Levels on Telephone Channels of Analog Radio Relay Links (RRL)

Recommendation 357-3 of the CCIR [9] defines the maximum tolerable noise levels on telephone channels of a 2500-km hypothetical reference line using FM and FDM, produced by interference signals from earth stations and space transmitters of satellite communications systems. In accordance with this recommendation, the mean minute psophometric noise power on any telephone channel at the zero relative level point for all sources of interference may not exceed:

1000 pW for more than 20% of any month;

50,000 pW for more than 0.01% of any month.

Mentioned in the CCIR Recommendation 357-3 is the possibility of determining the tolerable noise level for any percentage of time in the 0.01–20% range by interpolation, based on the assumption that the distribution of the noise power, caused by interference, is logarithmically normal in the mentioned range (Figure 5.3).

5.1.4 Tolerable Interference Levels on Telephone Channels of Analog Radio Relay Links

CCIR Recommendation 356-4 [9] specifies the maximum tolerable noise levels on telephone channels of a hypothetical reference satellite

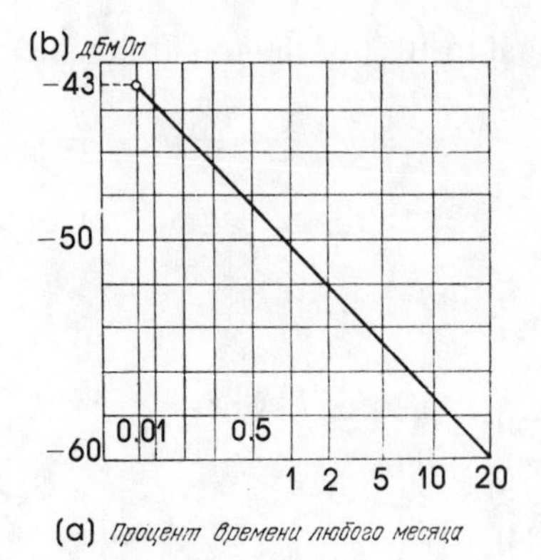

Key: **(a) Percentage of time of any month**
(b) dBp0ps

FIGURE 5.3. Maximum tolerable interference on telephone channel of radio relay line as function of percentage of time of any month.

circuit using FM and FDM, caused by interference from line of sight radio relay terrestrial links. In accordance with this recommendation, the mean minute psophometric noise power on any telephone channel at the zero relative level point for all sources of interference may not exceed:

1000 pW for more than 20% of any month;

50,000 pW for more than 0.03% any month.

Mentioned in Recommendation 356-4 is the possibility of determining the tolerable noise level for any percentage of the time in the 0.03–20% range by interpolation, based on the assumption that the distribution of the noise power, produced by interference, is logarithmically normal in the mentioned range (Figure 5.4).

5.1.5 Tolerable Levels of Interference in Satellite Communications Systems from Line of Sight Radio Relay Links during the Transmission of Eight-Bit PCM Telephone Signals

Recommendation 558 [9] specifies the maximum tolerable increments of the error probability in satellite communications systems, transmitting telephone signals using eight-bit PCM due to the action of interference from all terrestrial radio relay stations. The power of an interference signal at the receiver input, averaged over any 10-minute interval for 20% of any month may not exceed the level, at which the total noise power at the demodulator input is 0.1 of that which gives an error probability of 10^{-6}.

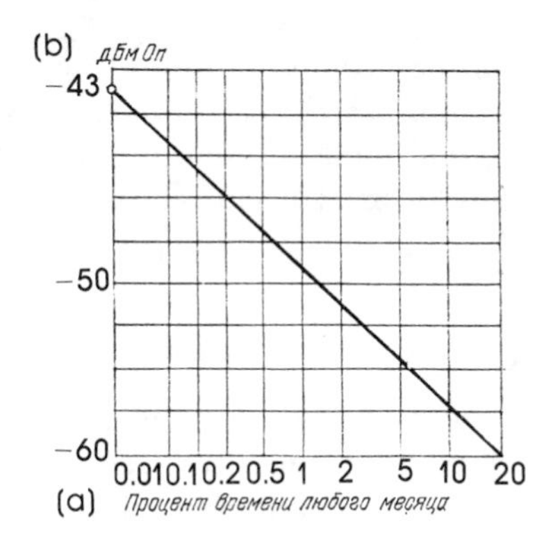

Key: **(a) Percentage of time of any month**
(b) dBp0ps

FIGURE 5.4. Maximum tolerable interference on telephone channel of satellite system as function of percentage of time of any month.

The receiver input interference power should not produce an error probability greater than 10^{-4} in and one-minute period for 0.03% of any month and greater than 10^{-3} in any one-second period for 0.001% of any year.

The CCIR Recommendations that are given in this section define the compatibility criteria of satellite and terrestrial systems. The comprehensive utilization of these criteria and some limits derived for them, imposed both on satellite, and on terrestrial services, are described in Sections 5.2 and 5.3.

5.2 CALCULATION OF CROSSTALK INTERFERENCE LEVELS FOR DETERMINING COMPATIBILITY OF SATELLITE SYSTEMS

The coordination procedure, described in Chapter 3, the technical part of which consists in the calculation of crosstalk interference between systems, comparison of the results obtained with the standards that are in effect (Section 5.1) and the development, when necessary, of measures to reduce interference, is used for compatibilizing satellite systems. The process by which interference is analyzed can be broken down into two steps: 1) calculation of the power ratio of the useful and interfering carriers at the input of the receiver system of an earth station $(P_2/P_{int})_{in}$, and 2) direct comparison of the value obtained with the standards (an FM TV signal, one channel on a carrier), or conversion to the noise power on a channel (FDM/FM) and then comparison with the standard. A detailed procedure for the calculation of $(P_s/P_{int})_{in}$ is given in Report 455-2 [8]. In the most general case, when the frequency bands in space-earth and earth-space links of both systems coincide (Figure 5.5),

$$(P_s/P_{int})_u = P_{e \cdot s} + G_1 - \Delta L_u - M_u - P_{e \cdot int} - G_1(\theta) + \Delta G_2 + Y_u \quad (5.3)$$

$$(P_s/P_{int})_d = E + G_4 - \Delta L_d - e - G_4(\theta) + Y_d \quad (5.4)$$

The following symbols are used in expressions (5.3) and (5.4): $(P_s/P_{int})_{u,d}$ —the signal to interference ratio in earth-space and space-earth links, respectively (dB); $P_{e.s}$, $P_{e.int}$—the powers of the useful and interfering carriers, supplied to the antenna of an earth station in (dBW); G_1, G_4—the gains of the transmitting and receiving antennas of the earth station of the useful system (dB); $G_1(\theta)$—the gain of the transmitting antenna of the interfering earth station in the direction of the satellite of the system that experiences interference, dB; $G_4(\theta)$—the gain of the receiving antenna of the earth station that experiences interference in the direction of the interfering satellite (dB); θ—the geocentric minimum angular separation between the useful and interfering satellites (deg); ΔG_2—the difference of the gains of the receiving antenna of the useful satellite in the direction of the interfering and useful earth stations (dB); $\Delta G_2 = G_{2s} - G_{2int}$; $\Delta L_{u,d}$—the difference of the losses to the propagation of the useful and interfering signals in the up-down link, respectively (dB); $\Delta L = L_s - L_{int}$; E, e—the equivalent isotropically radiated powers of the useful and interfering satellites in the direction of the earth station that experiences interference (dBW); $Y_{u,d}$—the minimum polarization isolation in the up and down links, respectively (dB); M_u—the net loss margin (dB); this margin needs to be taken into consideration only in the up link if automatic gain control is not used to maintain a constant level at the input of the satellite receiver. This margin has to be taken into consideration because the useful and interfering earth stations can be quite far apart in space and fluctuations of the useful and interfering signals will not be correlated in the up link.

This reserve is not taken into consideration in the calculation of $(P_s/P_{int})_d$ in the up link, because the paths of the useful and interfering signals through the atmosphere are very close and propagation conditions virtually coincide. By proceeding in the same way, it is also possible to derive expressions for the case of the reverse use of frequency bands, when the frequency band, used in one system in the down link, is used in the other system in the link and *vice versa* (see CCIR Report 455-2 [8]). Here, as was mentioned above (Section 5.1), two kinds of interference must be taken into consideration: from one satellite to another in the band used in the earth-space link in the protected system, and from one earth station to

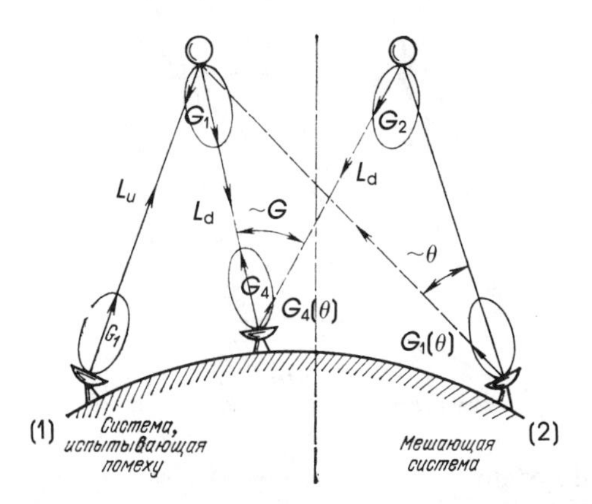

Key: **(1) System that experiences interference**
(2) Interfering system

FIGURE 5.5. Calculation of protection ratio at receiver input.

the other in the frequency band used in the space-earth link in the protected system. During the calculations, however, it is recommended that only interference that is created in the frequency band that is being used in the earth-space link in the protected system (Figure 5.6) be taken into consideration, because the interference that acts upon the receiver of an earth station from the transmitter of the other earth station can be reduced to zero by properly planning their relative locations. For this case:

$$(P_s/P_{int})_u = P_{e\cdot s} + G_1 - M_u + \Delta G'_2 - e' + Y + 20 \log \theta - 35.2 \qquad (5.5)$$

where $\Delta G'_2$ is the difference of the gains of receiving antenna of the satellite in the direction of the transmitting earth station and in the direction of the interfering satellite; $\Delta G'_2 = G_{2s} - G_{2int}$, dB; e' is the EIRP of the interfering satellite in the direction of the satellite that is experiencing the interference, dBW.

The equivalent ratio $(P_s/P_{int})_{eq}$ of a link can be found by determining the ratios $(P_s/P_{int})_{u,d}$, and then the contribution of the interference to the noises on a channel at the output of a system can be determined. The equivalent ratio is

$$(P_s/P_{int})_{eq} = 10 \log [1/(10^{-0.1(P_s/P_{int})_u} + 10^{-0.1(P_s/P_{int})_d})] \qquad (5.6)$$

The ratios P_s/P_{int} that are found can be converted to the noise level on a channel by using expressions given in the CCIR Report 388-4 [9].

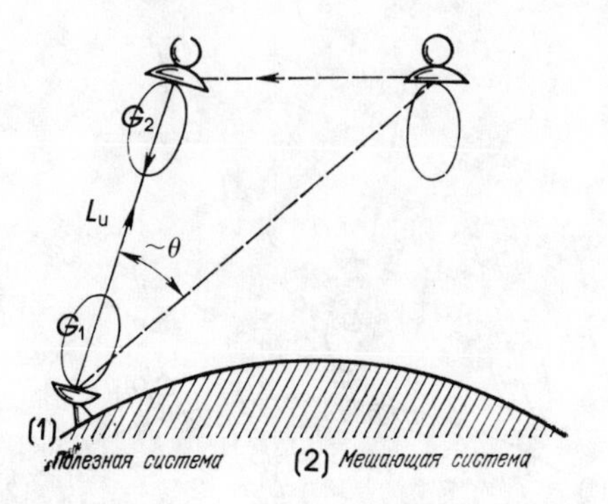

Key: **(1) Useful system**
(2) Interfering system

FIGURE 5.6. Calculation of protection ratio for reverse use of bands.

5.2.1 Analog Communications Systems

The ratio between the power of noise, created by interference on a telephone channel and P_s/P_{int} at the receiver input is characterized by the interference attenuation coefficient (in decibels)

$$B = 10 \log\left[\frac{(S/N_i)}{(P_s/P_{int})}\right] \tag{5.7}$$

where S is the power created on a channel by the test frequency signal, equal to 1 mW; N_i is the unweighed power of the interfering signal on a telephone channel; P_s is the power of the useful signal at the receiver input. P_{int} is the power of the interfering signal at the receiver input. The weighed power (in pW) of the interfacing signal on a telephone channel is determined by the expression:

$$10 \log N_p = 87.5 - B - 10 \log (P_s/P_{int}) \tag{5.8}$$

The interference attenuation coefficient is

$$B = 10 \log \frac{2(\delta f)^2 p\ (f/f_m)}{bf^2 D(f, f_v)} \tag{5.9}$$

where

$$\begin{aligned} D(f, f_0) = & \int_{-\infty}^{\infty} S(F)\, P_1\, (f + f_0 - F)\, dF + \int_{-\infty}^{\infty} S\,(F)\, P_1\, (f - f_0 - F)\, dF \\ & + S\,(f + f_0)\, P_{10} + S\,(f - f_0)\, P_{10} + S_0\, P_1\, (f + f_0) \\ & + S_0\, P_1\, (f - f_0) + S_0\, P_{10}\, \delta\, (f - f_0)/b \end{aligned} \tag{5.10}$$

$$P_1\,(f) = P\,(f)\,A^2\,(f) \tag{5.11}$$

$$P_{10} = P_0\,A^2\,(0) \tag{5.12}$$

δf is the mean-square deviation of the useful frequency by the test frequency (kHz); f is the central frequency of the examined channel in the group frequency spectrum (kHz); f_m is the upper frequency of the group frequency spectrum of the useful signal (kHz); $p(f/f_m)$ is the predistortion ratio for the central frequency of the examined channel in the group frequency spectrum of the useful signal; b is the telephone channel band (3.1 kHz); f_0 is the difference between the carriers of the useful and interfering signals (kHz); $S(f)$ and $P(f)$ are the normal spectral power densities of the useful and interfering signals, respectively (Hz^{-1}); S_0 and P_0 are the normal net powers of the carriers of the useful and interfering signals, respectively; and $A(f)$ is the amplitude-frequency response of the filter:

$$\delta\,(f - f_0) = 1 \quad \text{for } f = f_0, \;\; \delta(f - f_0) = 0 \quad \text{for } f \neq f_0$$

In order to calculate interference between two FDM/FM signals, it is necessary to compute the convolution of the spectra of the signal and interference by (5.10) and then to determine, by (5.8) and (5.9), the weighted power of the interference on the telephone channel. In certain specific cases the calculation procedure can be simplified. We will examine the following examples.

Interference between an FDM/FM Signal with Small Modulation Index and an FDM/FM Signal with a Large Modulation Index. This is interference from radio relay terrestrial links of a fixed satellite service. In this case the attenuation of the interference is determined by the simplified formula:

$$B = 10\log\frac{2\sqrt{2}\pi\,(\delta f^2)\,p(f/f_m)f_s}{\exp[-0.5\,(f_0 - f)^2/f_s^2] + \exp[-0.5\,(f_0 + f)^2/f_s^2]} \tag{5.13}$$

where f_s is the mean-square deviation of the useful multichannel signal (kHz).

Interference between Two FDM/FM Signals with Large Modulation Indexes ($m > 3$). The attenuation of interference is computed by (5.13), in which f_s is replaced with F_s:

$$F_s = \sqrt{f_{s1}^2 + f_{s2}^2} \tag{5.14}$$

where f_{s1} and f_{s2} are the mean-square deviation of the useful multichannel and interference signals, respectively (kHz).

Interference between Two FDM/FM Signals with Medium Modulation Indexes. To calculate the interference between the two specific signals with small modulation indexes for a certain carrier frequency separation, it is necessary to compute the convolution of the spectra of these signals by (5.10), and then to determine, by (5.8) and (5.9), the interference on the telephone channels. In certain particular cases the convolution of the spectra is very easy to compute using Figure 5.7. For example, when the upper frequencies of the group frequency spectra of the useful signal and interference are equal ($f_{m1} = f_{m2}$) the equivalent modulation index:

$$m = [m_1^2 + m_2^2]^{1/2} \quad \textbf{(5.15)}$$

is determined, and $f_mS(f_1)$ and $f_mS(f_2)$ (where $f_1 = (f_0 + b)/f_{m1}$ and $f_2 = (f_0 - b)/f_{m2}$) are found for this value of m from the curve in Figure 5.7. Then

$$D(f, f_0) = [f_mS(f_1) + f_mS(f_2)]/f_{m1} \quad \textbf{(5.16)}$$

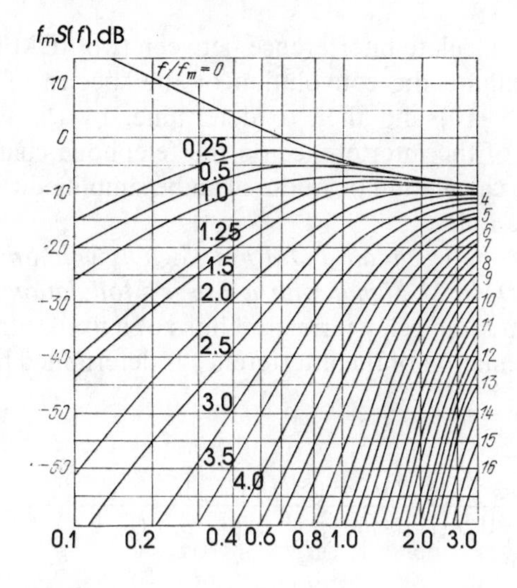

FIGURE 5.7. Dependence of $f_m(f)$ on equivalent modulation index.

The same procedure can be used for approximating $D(f, f_0)$ on the basis of the equivalent modulation index:

$$m = [m_1^2 + m_2^2(f_{m2}/f_{m1})^2]^{1/2} \quad \textbf{(5.17)}$$

In cases when $f_{m2} < f_{m1}$ and $m_2^2(f_{m2}/f_{m1})^2 << m_1^2$.

When the interference is an FM TV signal, modulated just by the dispersion signal, and the useful signal is an FDM/FM signal with a few telephone channels, the bandwidth of which is commensurable with that of the interfering signal, then, when the carrier frequencies are the same, the following simple expression from, CCIR Report 388-4 may be used [9]:

$$B = 10 \log \frac{(\delta f)^2 \Delta f p(f/f_m)}{f^2 b} \tag{5.18}$$

where Δf is the amplitude of the frequency deviation by the dispersion signal.

Interference between an FDM/FM Signal with a Large Modulation Index and an FDM/FM Signal with a Small Modulation Index. The following simple relation* may be used for calculating interference in the case that corresponds to interference between a fixed satellite service and radio relay terrestrial links:

$$\frac{\text{Power of interference on a telephone channel}}{\text{Power of thermal noise on a telephone channel}} = \frac{\text{Power of the interfering signal in two corresponding 4-kHz bands at the receiver input}}{\text{Power of thermal noises in the same two 4-kHz bands at the receiver input}}$$

5.2.2 Determination of the Error Probability per Symbol in Digital Communications Systems

The criterion of the quality of the output signal of a system for digital communications systems is the error probability per symbol in an information sequence. Therefore, for digital useful signals it is necessary to calculate error probability P_{er} at the output of a channel in the presence of interference. Here, in contrast to analog signals, it is not possible to separate the set noises of a system and interference, and P_{er}, caused by the summary action of the noise and interference, must be estimated.

---

*This expression is valid for calculating interference between an AM signal and an FDM/FM signal with a small modulation index.

In the general case, the probability of the erroneous reception of binary symbols in the presence of thermal noise for different kinds of discrete modulation for the perfect receiver is

$$P_{er} = 0.5 \text{ erfc } (\sqrt{E/N_0} \tag{5.19}$$

where

$$\text{erfc } (x) = \frac{2}{\sqrt{\pi}} \int_x^{\infty} e^{-t^2} \, dt$$

Here, $E/N_0 = P_s\tau/kT_n$ **(5.20)**

is the ratio of the energy of an elementary pulse to the power density of the receiver input noises, where τ is the elementary pulse duration, k is the Boltzmann constant, and T_n is the noise temperature of the receiver. The error probability per symbol in the presence of interference from other systems is, in general,

$$P_{er} = \frac{1}{2_\pi} \int_0^{\pi} \text{erfc} \left[\sqrt{\frac{E}{N_0}} \left(1 + \frac{\cos \phi}{\sqrt{\beta P_s/\alpha P_{int}}} \right) \right] d\phi \tag{5.21}$$

Here $\beta = 1$ if the phase of the signal is keyed by $\pm\pi/2$ (relative phase modulation) and $\beta = 1/2$ if the phase of the signal is keyed by $\pm\pi/4$ (double relative phase modulation). The coefficient α determines that fraction of the interference power that gets into the information band of the receiver of the useful system, and it can be found from the expression:

$$\alpha = \int_{-\infty}^{\infty} A(f) \, [P(f) + P_0] \, df \tag{5.22}$$

where $A(f)$ is the amplitude-frequency response of the receiving filter of the useful system; $P(f)$ is the normal spectral power density of the interference; and P_0 is the normal spectral component of the power of the residual carrier.

Expression (5.21) is valid for an interfering FM signal, but for the case of interference with amplitude modulation (with a suppressed carrier), it is necessary to use (5.19), in which N_0 is replaced by the sum of the power densities of thermal noise and interfering signal. Curves (calculated by the formulas given above), are given in Figure 5.8, where the tolerable ratios P_s/P_{int} are given as functions of P_s/P_n, for which different probabilities P_{er} are guaranteed. These curves are suitable for one or for several interference processes. A family of curves is given in each figure for different

$$PF = 10 \log(P_p/P_{mean}) \quad \textbf{(5.23)}$$

where P_p/P_{mean} is the ratio of the interference power in a pulse to the mean power of the interference spectrum. For interference with frequency modulation $PF = 0$.

5.3 METHODS OF ELECTROMAGNETIC COMPATIBILITY BETWEEN SATELLITE SYSTEMS AND TERRESTRIAL SERVICES

Various technical constraints on parameters, which guarantee that the compatibility criteria, formulated in Section 5.1, will be met, are used for achieving electromagnetic compatibility between satellite systems and terrestrial services. These constraints are written in the IRR and several CCIR recommendations and apply both to satellite and terrestrial services. Crosstalk interference between satellite and terrestrial services will be examined separately.

The possible level of the first-kind of interference (from the transmitters of space stations, into receivers of terrestrial stations) is limited to an acceptable level by limiting the maximum power flux density, created at the earth's surface by the space station's EIRP. These limits are found as a result of calculations using the appropriate models of systems. The model used for calculating the tolerable power flux density of a fixed satellite service is a 2500-km radio relay link, situated randomly on the smooth spherical surface of the earth and global coverage satellites, positioned uniformly in a geostationary orbit at intervals of 3° and 6°. It was assumed here that each satellite creaated at the earth's surface a power flux density that is identical for all angles of incidence. The maximum tolerable power flux density was selected such that the interference exceeded this level only in a small fraction of the highly sensitive radio relays. The limits on the power flux density, accepted as a result and written in the IRR, are given in Table 5.2 for a fixed satellite service. Similar limits also are imposed on emissions in other frequency bands, reserved for other satellite services. For example, according to Recommendation Space 2–10 of the IRR, the flux density at the earth's surface in the 620–790-MHz band, generated in other countries by emissions from a space station of a satellite broadcasting service, is limited temporarily to the following levels:

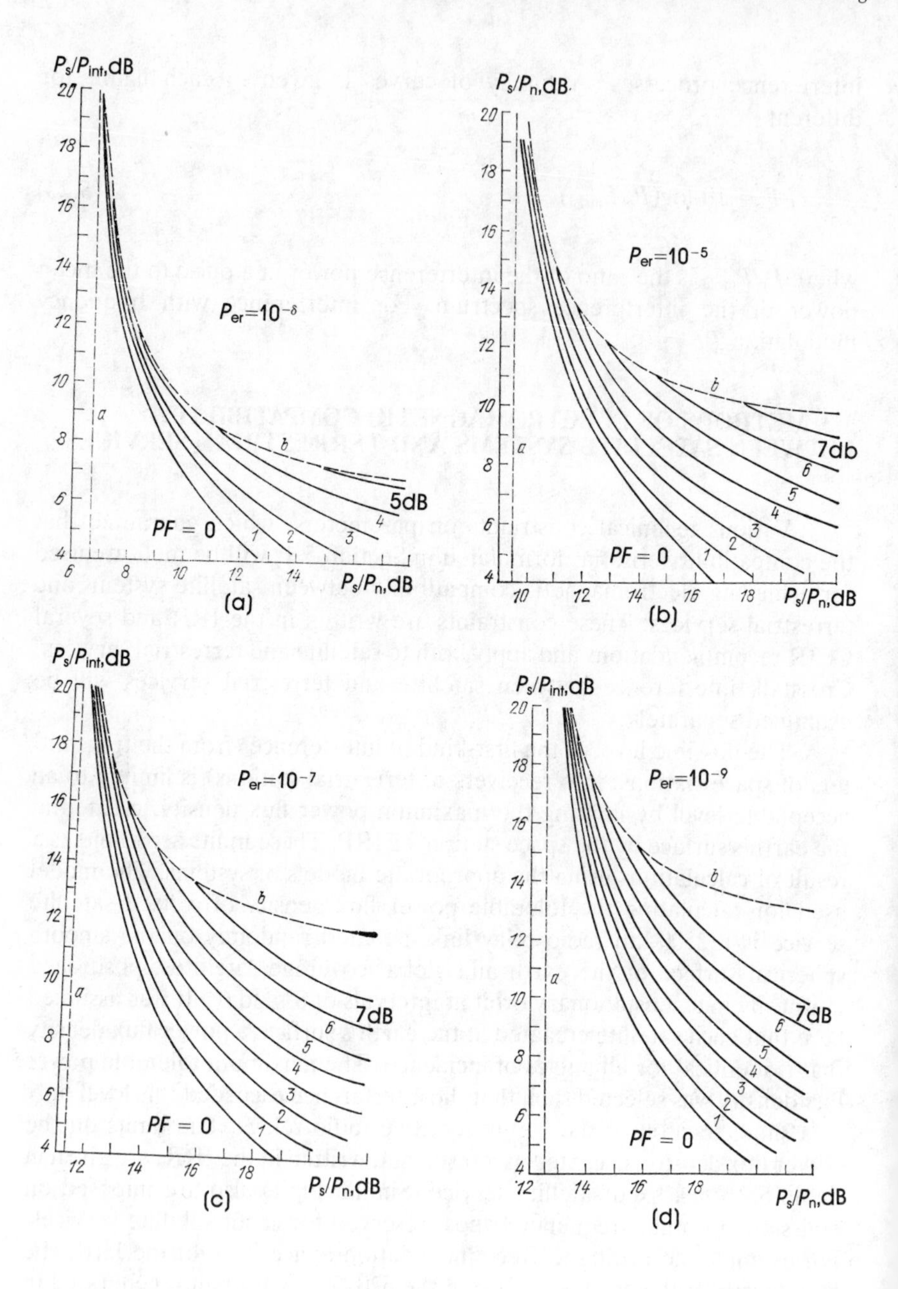

FIGURE 5.8. Tolerable interference level P_s/P_{int} as function of P_s/P_n for different P_{er}.

TABLE 5.2

Frequency band, GHz	*Power flux density limit, dBW/m²*			*Test bandwidth*
	$\theta \leqslant 5°$	$5° < \theta \leqslant 25°$	$25 < \theta \leqslant 90°$	
2.5–2.690	−152	$-152+0.75(\theta-5)$	−137	
3.4–7.750	−152	$-152+0.5(\theta-5)$	−142	
8.025–8.500 10.7–11.7	−150	$-150+0.5(\theta-5)$	−140	4kHz
12.2–12.75	−148	$-148+0.5(\theta-5)$	−138	
17.7–19.7	−115	$-115+0.5(\theta-5)$	−105	
31.0–40.5	−115	$-115+0.5(\theta-5)$	−105	1MHz

-129 dBW/m² for $\theta \leqslant 20°$

$-129 + 0.4(\theta-20$ dBW/m² for $20° < \theta \leqslant 60°$

-113 dBW/m² for $60° < \theta < 90°$

where θ is the angle of incidence of a wave above a horizontal plane (deg).

These limits were introduced for the purpose of protecting terrestrial TV broadcasting, for which given bands also are assigned. At the same time the IRR mentions an urgent need to analyze criteria that must be used in the case of the joint use of the 620–790-MHz band and to develop standards to replace the temporary standards, given above.

Studies [5.3] in the USSR have shown that the temporary standards that are in effect are unnecessarily rigid and require refinements. First and foremost, for determining the protection ratios for interference with reception of an AM single-sideband TV signal from an FM signal, it is necessary to consider its parameters. In particular, when the amplitude of the deviation of an interfering FM signal increases by 1 MHz, the protection ratio decreases by an average of 0.4 dB. Furthermore, the fact that the power of the carrier of FM interference experiences dispersion also weakens its interfering effect and should be taken into consideration.

In consideration of the material presented in the article [5.3], the following expression was derived for the tolerable power flux density in the examined band:

$$W = -77 - R_{oq} + \gamma \text{ dBW/m}^2$$

where R_{oq} is the protection ratio for the frequency deviation, taken as a reference, determined on the basis of curves given in the article [5.3]; and γ is a correction factor that takes into consideration the distribution of the power of FM interference in consideration of visual perception. Thus, the following standard on the tolerable power flux density may be used instead of the temporary standard:

$-77 - R_{oq} + \gamma$ dBW/m^2 for $\theta \leqslant 20°$

$-77 - R_{oq} + \gamma + 0.4\,(\theta - 20)$ dBW/m^2 for $20° < \theta \leqslant 60°$

$-61 - R_{oq} + \gamma$ dBW/m^2 for $60° < \theta \leqslant 90°$

We mention, as a particular example, the fact that, for an amplitude of the frequency deviation of 22 MHz in the absence of artificial dispersion, we have

-124 dBW/m^2 for $\theta \leqslant 20°$

$-124 + 0.4\,(\theta - 20)$ dBW/m^2 for $20° < \theta \leqslant 60°$

-109 dBW/m^2 for $60 < \theta \leqslant 90°$

In the 1530–1660.5-MHz band, reserved for mobile satellite services, the power flux density in the 4-kHz band should not exceed

-154 dBW/m^2 for $\theta \leqslant 5°$

$-154 + 0.5\,(\theta - 5)$ dBW/m^2 for $5° < \theta \leqslant 25°$

-144 dBW/m^2 for $25° < \theta \leqslant 90°$

Limits on the EIRPs of the transmitting stations of terrestrial services (55 dBW for all bands above 1 GHz, which are used jointly with space services) and on the power delivered to the antennas of transmitting terrestrial stations (13 dBW in joint frequency bands between 1 and 10 GHz and 10 dBW in joint frequency bands above 10 GHz) were introduced for the purpose of reducing second-kind interference (from the transmitters of terrestrial stations with the receivers of earth stations) and third-kind interference (from the transmitters of terrestrial stations with the receivers of space stations). There is also an additional limit, according to which the sites of the transmitters of terrestrial fixed and mobile services should be selected such that the peak radiation of an antenna be at least the following:

a. At 2° from the line to a geostationary orbit. This limit applies if the EIRP exceeds +35 dBW in bands between 1 and 10 GHz;
b. At 1.5° from the line to a geostationary orbit in frequency bands between 10 and 15 GHz. This limit applies if the EIRP exceeds +45 dBW.

There are no such limits at the present time in bands above 15 GHz. It is important to note that for cases when the limit given in a) is not acceptable from the practical standpoint, the maximum EIRP should not exceed the following:

+47 dBW in any direction within 0.5° of the line to a geostationary orbit, or

+47 to +55 dBW on a linear logarithmic scale within 0.5° to 1.5° (Figure 5.9).

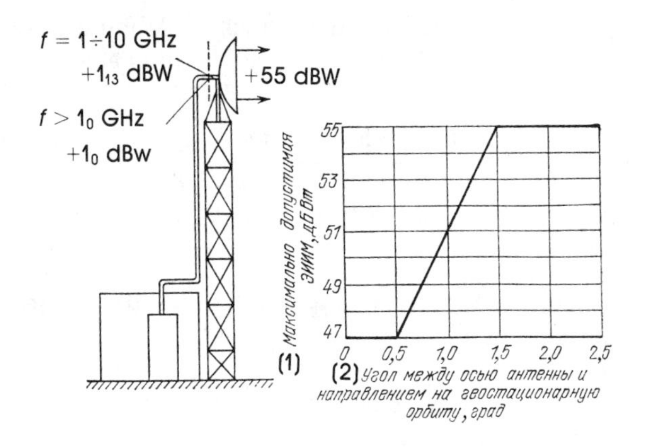

Key: **(1) Maximum tolerable EIRP, dBW**
(2) Angle between antenna axis and line to geostationary orbit, deg

FIGURE 5.9. Limits on the EIRP of a transmitting station of a radio relay link.

When these limits are used, it is necessary to consider a correction due to atmospheric refraction, which depends on a number of factors (see, for example, CCIR Report 393-3 [9]). Limits on the maximum EIRP of an earth station, radiated toward the local horizon (Table 5.3), and also a minimum elevation angle of the peak radiation of a transmitting antenna of an earth station, equal to 3° (Figure 5.10), were adopted for the purpose of protecting the receivers of earth stations from interference from earth stations (fourth-kind interference).

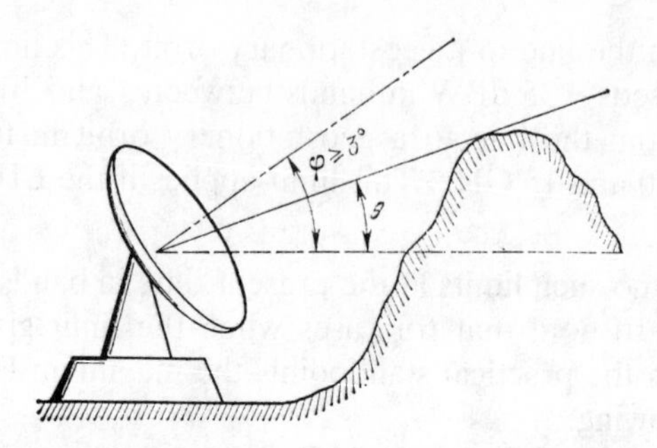

FIGURE 5.10. Limits on EIRP of a transmitting earth station.

TABLE 5.3

Frequency range, GHz	Limits on EIRP of earth station (dBW) in direction of local horizon, seen at elevation angle			
	$\theta \leq 0°$	$0 < \theta \leq 5°$	*Band*	*Comment*
1–15	40	$40+3\theta$	4 kHz	No limits for
>15	64	$64+3\theta$	1 MHz	elevation angles >5°

5.4 DETERMINATION OF THE COORDINATION AREA AROUND AN EARTH STATION

In addition to the above-mentioned limits, aimed at reducing the possibility of crosstalk interference between radio relay links and earth stations, a procedure is given in the CCIR Radio Regulations for coordinating frequency assignments to earth stations of a fixed satellite service in cases when the coordination area, drawn around an earth station, contains land of a different country. The coordination area is bounded by a contour, which connects the points at all angles of azimuth from an earth station, separated from it by distances, within which it can interfere with a radio relay link or experience intolerable interference from them [7, 9]. It should be pointed out that the existence of a radio relay station within the coordination area does not necessarily lead to intolerable interference, because the coordination area is determined on the basis of the worst prerequisites with respect to interference. In this case, it is necessary to do a detailed calculation of possible interference in consideration of the actual, not the conditionally worst parameters of radio relay links and earth stations, which will answer the question about the possibility of compatible operation.

Calculations of the coordination distance amounts to the following: the tolerable level of interference signal $P_r(p)$ at the output of the receiving antenna of a station that is experiencing interference is determined on the basis of matched hypothetical parameters of an earth station and a radio relay station, and then the distance beyond which this level is not exceeded for a given percentage of the time, $p\%$, is calculated. In the small fraction of time in which the two mechanisms of propagation are examined: troposphere propagation near an arc of the great circle, and propagation due to the scattering of a signal from hydrometeors.

The concept of minimum tolerable transmission losses $L(p)$ between the output of an interfering transmitter and the input of a receiver, subjected to interference, for $p\%$ of the time (i.e., the level that should be exceeded by the actual transmission losses for 100 - $p\%$ of the time), is used for determining the necessary coordination distance:

$$L(p) = P_{t'} - P_r(p) \tag{5.24}$$

where $P_{t'}$ is the power of the interfering transmitter at the input of the transmitting antenna, measured in a conditional frequency band (dBW); and $P_r(p)$ is the tolerable power level of interference at the input of a receiver, measured in a conditional frequency band, which should not be exceeded for more than p% of the time from each interference source (dBW). A detailed procedure for calculating the coordination area is given in Appendix 28 of the CCIR Radio Regulations [7].

Chapter 6
On the Efficient Utilization of a Geostationary Orbit

The material presented in Chapter 1 shows that in most cases a geostationary orbit is the best orbit in which to place a satellite; it actually is used most widely for communications and broadcasting satellites. Satellites of the Intelsat and Intersputnik international systems, of communications systems of the USSR, United States, Canada, Indonesia, Japan, and some other countries, and of satellite broadcasting systems of the USSR, Japan, Canada, France, and West Germany are in a geostationary orbit. A geostationary orbit, which has unique properties, is used so widely that in some cases it is difficult or impossible to deploy new satellites because of crosstalk interference. This applies primarily to certain sections of an orbit. Technical and organizational means are being used widely today for increasing the utilization efficiency of a geostationary orbit; the recommendations of CCIR are aimed at this objective.

We will examine, in succession, the technical factors that influence the utilization efficiency of a geostationary orbit. The relative sidelobe level of the radiation patterns of the antennas of earth stations, especially of the sidelobes that are close to the main beam of the radiation pattern (within approximately $\pm 10°$ of the axis of the main beam) are an important factor that determines the necessary angular distance between satellites. It is recommended at the present time to consider that the sidelobe gain of the antenna of an earth station (in decibels) be

$$G = 32 - 25 \log \theta \text{ dB} \tag{6.1}$$

at angles $\theta > 1°$ from the axis of the main lobe of the radiation pattern. Studies show that a 7-dB decrease in the gain of an antenna in the direction of the sidelobes, i.e., to $25 - 25 \log \theta$ dB, or an increase of the rate of decay of the lobes, such that $G = 32 - 30 \log \theta$ dB, will approximately double the total capacity of a geostationary orbit. It is considered feasible today to give all the antennas of an earth station a gain that not only satisfies (6.1), but also is better. The sidelobes of the radiation pattern of ordinary double reflector parabolic antennas (see Chapter 10) usually expand due to attenuation and diffraction, created by the counter-reflector and its supports that are in the path of the main flux of energy from the main reflector. The sidelobe level of the radiation pattern can be reduced by 5 dB and even by 10 dB by using axially asymmetric antennas with an extended irradiator (Chapter 11).

Crosstalk interference between satellite communications systems can be reduced significantly by virtue of the spatial selectivity of the space antennas of satellites, if the service areas of these two systems do not overlap. For this, the main lobe of the radiation pattern of a space antenna should cover as accurately as possible the service area and fall off rapidly outside of it. This problem is solved best of all by multibeam antennas, which make either several narrow beams, or on their basis a beam, the radiation pattern of which has a special shape, complex in cross section, which corresponds as accurately as possible with the territory being serviced. Multibeam antennas are being used in certain systems, for example in the Japanese BSE television broadcasting system.

The necessary distance between satellites increases by the sum of the errors with which both neighboring satellites are held on longitude. The CCIR Radio Regulations recommend that satellites be held in a geostationary orbit with an accuracy of not less than 0.5° and the CCIR recommendations contain the even more rigid limit of ±0.1° and express the wish that all efforts be made to hold satellites in orbit as accurately as possible. The error with which satellites are held in orbit in feasible satellite communications systems ranges from ±1.0 to ±0.1°.

Crosstalk interference can be reduced by radiating mutually orthogonally (circularly or linearly) polarized signals from satellites. Orthogonal polarizations can be used either for two signals of the same satellite, or for reducing interference between neighboring satellites. This method, in principle, makes it possible to double the carrying capacity of a geostationary orbit. However, because of the influence of factors that limit the possibility of separating orthogonally polarized signals, polarization difference alone is not enough for receiving signals in a common band with a low crosstalk interference level. Therefore, polarization division is always accompanied by space division due to the spatial selectivity of the antennas of satellites and earth stations, which was already discussed above. Polarization division of signals, radiated from satellites, in combination with space division, make it possible in certain satellite communications systems to reutilize a frequency band on a given satellite (or on a satellite in a given position). Further development of space division will make it possible to use the frequency spectrum over and over.

The efficiency with which a geostationary orbit is utilized can be increased by means of the reverse use of frequency bands, whereby frequency bands for earth-space and for space-earth links (see Chapter 3) are used in opposite directions. Crosstalk interference between systems with the usual and reverse utilization of frequencies occurs directly between satellites and between earth stations (Figure 6.1). Calculations have shown that for reducing the interference between satellites it is sufficient to separate them by fractions of a degree, while earth stations must be separated by a distance of the order of hundreds of kilometers, i.e., several earth stations cannot be built on the same site. Interference between satellites of reverse systems is most dangerous if the angular distance between them is somewhat less than 180° and they radiate signals directly toward the main lobe of the antenna of the other satellite (Figure 6.2). This limits the possibility of doubling the capacity of a geostationary orbit by means of the reverse utilization of frequencies.

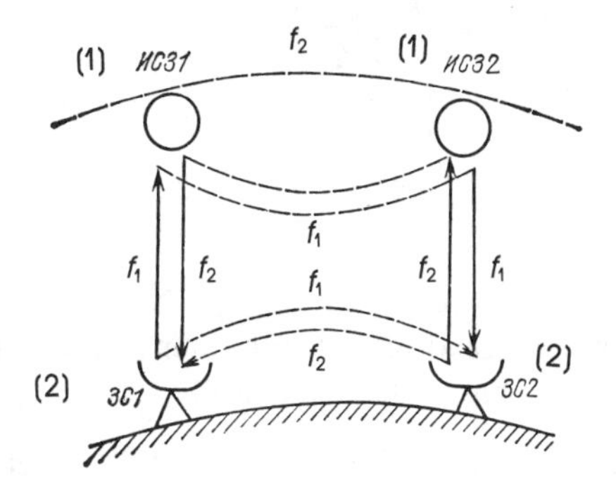

Key: **(1) Satellite**
(2) Earth station

FIGURE 6.1. Channels creating crosstalk interference between systems with conventional and reverse utilization of frequency bands.

Coordinating the frequency plans of neighboring systems can have a significant effect. For instance, if a communications system that is most vulnerable to interference (for example, a system with few channels, and accordingly with low-power transmitters) is located in some part of the frequency band, then a system with low-power transmitters also should be placed on an adjacent satellite in that same part of the band. A similar example of the coordination of the frequency plans of adjacent systems is the use on adjacent satellites of shifted or alternating frequency nets, whereby the carrier frequencies of the signals of one satellite are located exactly in-between the carrier frequencies of the signals of the neighboring satellite, i.e., where the spectral density of the signals of the neighboring satellite is minimal, and the filters of the receivers have a certain selectivity. For instance, when both mutually interfering systems are transmitting FM television programs, the necessary protection ratio for shifting the central frequency net by 20–25 MHz decreases by 13 dB; this was used for drafting the plan of a satellite broadcasting service on 12 GHz [4].

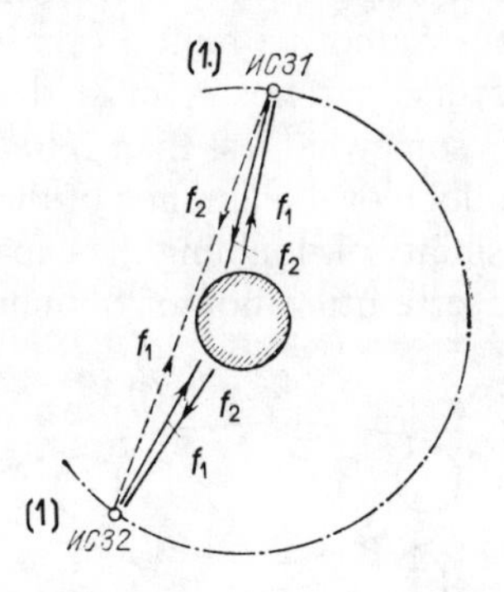

Key: **(1) Satellite**

FIGURE 6.2. Channels creating crosstalk interference with reverse utilization of frequency and opposed deployment of satellites.

The utilization efficiency of an orbit and of the spectrum depends to a great extent on the kind and parameters of modulation. Frequency modulation is used most often in satellite communications systems at the present time. As the frequency modulation index increases the frequency band occupied by each channel for a fixed number of transmitted channels also increases, and consequently the carrying capacity of a satellite decreases in some frequency band, which it occupies. But the noise immunity of signal reception in a given communications system increases; furthermore, the spectral density of the radiated signal decreases, with the result that interference with other satellite communications systems decreases. As a result, it becomes possible to place satellites closer together and, thus, to increase the carrying capacity of a geostationary orbit. An analysis makes it possible to determine the optimum frequency modulation index, which corresponds to the maximum carrying capacity of a geostationary orbit. This value will depend on the specific parameters of systems (the diameter of the antenna of an earth station, the frequency band, *et cetera*) [6.1]. When PSK is used and signals are transmitted in digital form, the optimum number of phase positions can be determined by the same procedure, because as the number of phases increases the carrying capacity of each individual communications link increases, but its noise immunity decreases [6.1]. However, the selection of the best modulation method always amounts to a compromise. Not only the carrying capacity of a communications system can be increased for a given noise immunity in relation to thermal noises, but noise immunity in relation to radio interference also can be increased, and consequently satellites can be placed closer together in orbit by using modulation methods that give the maximum noise immunity (for example, a combination of amplitude and PSK of the carrier for the purpose of increasing the distance between the vectors that correspond to the transmission of different code combinations). This obviously also applies to noise-immune reception methods, for example to tracking FM signal demodulators [16], the use of which makes it possible to increase noise immunity in relation both to thermal set noises in a given communications system, and to radio interference from neighboring systems.

Noise-immune coding methods have come to be used extensively in recent years for transmitting discrete signals, so that for a certain selectivity and by using the appropriate reception method, a substantial gain in noise immunity can be achieved, which makes it possible to increase the carrying capacity of a communications system, in spite of the need to transmit auxiliary signals. Here it is usually possible to increase noise immunity in relation to radio frequency interference. Thus, the total carrying capacity of a geostationary orbit increases by virtue of noise-immune coding. No matter what kind of modulation is used, it usually turns out that the system in which the spectral density is nearly uniform in the entire occupied frequency band creates the least interference with neighboring systems [2]. But in the case of narrow-band signals, the spectrum of which contains considerable discrete components, or of signals with a substantially nonuniform spectral density, it is necessary to resort to artificial dispersion of the transmitted signals [4].

Interference compensation is an interesting method [6.1, 6.2]. An interference compensator is a device that, by means of a special antenna with a radiation pattern that is different from that of the main antenna, receives an interfering radio signal (the useful signal and noises interfere with it to some extent) and then uses it for subtracting (compensating) interference from the sum of the useful signal, noise and interference, received by the main receiver path. Interference compensators are not used widely yet at earth stations, and the possibility of using compensation is not being considered in the calculations of crosstalk interference between satellite communications systems.

A different type of device, also usually a compensator, exists and is being used, in which interference is removed in the main receiver path by virtue of its *a priori* known differences from the useful signal, for example in terms of spectral density, or in terms of how the instantaneous values are distributed, and then is compensated to a certain accuracy in the receiver path, at the input or output of the demodulator.

The next method of increasing the carrying capacity of a geostationary orbit, which is very effective, but difficult to implement, is to make satellite communications systems uniform, if even in a certain arc of the geostationary orbit, in terms of the parameters that characterize crosstalk interference. In nonuniform systems, the angular separation obviously is determined by the need to protect the system that is more sensitive to interference, and in the opposite direction a noise immunity margin is formed. It is suggested in the CCIR study group for the purpose of promoting the attainment of systems uniformity, that certain limits be established on the following four parameters.

1. the spectral density of the EIRP of an earth station in the direction of a geostationary orbit (parameter A);
2. the limit of sensitivity to interference in an up link, defined as the spectral power flux density in a geostationary orbit, corresponding to the maximum tolerable interference on a channel (parameter B);
3. the maximum spectral power flux density at the earth's surface (parameter C);
4. the limit of sensitivity to interference in a down link, defined as the spectral power flux density (depending on the angle of incidence), corresponding to the maximum tolerable interference on a channel.

The first limit has already been adopted to some extent in the 6-GHz band in the form of the CCIR Recommendation 524, according to which the EIRP of an earth station in the direction of a geostationary orbit may not exceed $(35 - 25 \log \theta)$ dBW/4 kHz for the angles (relative to the axis of the main lobe of the radiation pattern of the antenna of an earth station) $48° \geq \theta \geq 2.5°$. The limits of the parameters require further study.

Planning of the use of an orbit and the spectrum is a more radical way to increase the utilization efficiency of a geostationary orbit. During planning, in addition to making a neighboring system uniform, it is possible to achieve the optimum matching of the frequency plans of adjacent systems and to select the optimum modulation parameters. During planning it is possible to deploy satellites advantageously (in such a way that satellites that service areas that are far apart from each other will be located at approximate points of the geostationary orbit, and signal separation will be accomplished here by virtue of the selectivity of the space antennas of the satellites; otherwise the satellites that service close and overlapping areas would be deployed as far apart as possible for the purpose of making the maximum utilization of the selectivity of the earth station antennas).

It is important to note that a plan can be drawn up only for certain parts of the frequency band, assigned to a given service, and that planning makes sense for an entire geostationary orbit, if only for long sections of the arc of a geostationary orbit, covering entire continents. It is for precisely this reason that the drafting and coordination of a plan for the utilization of an orbit are most difficult problems. An example of successful planning exists at the present time. A plan for the utilization of a geostationary orbit in the 11.7–12.5-GHz band (in Region 1) and the 11.7–12.2-GHz band (in Region 3) by satellite broadcasting systems was developed by the World Administrative Radio Conference (WARC) of the ITU, held in 1977 [4, 6]. The fact that, as time passes the technology of satellite communications systems is being developed and the demands of the countries that are covered by a plan change must be taken into consideration, of course. In this connection a plan must be corrected periodically; the plan, adopted by WARC-1977, provides an appropriate procedure.

The standard on the acceptable crosstalk interference between satellite communications systems has a significant effect on the utilization efficiency of an orbit and spectrum. As this standard becomes more rigid, the carrying capacity of each system will decrease, because it will be necessary to reduce the set (thermal) noises in a system, but in this case the satellites can be deployed closer together, so that the total carrying capacity of a geostationary orbit will increase. As a result of these conflicting factors, the maximum total capacity of a geostationary orbit will be achieved at a certain crosstalk interference level. According to a suggestion by the USSR, the recommended level of interference on a channel was increased from 1000 to 2000 pW for analog FM and FDM transmission systems (from all adjacent satellite communications systems).

The adoption of new, higher-frequency bands theoretically is a limitless means of increasing the utilization efficiency of a geostationary orbit. In anticipation of this, the CCIR Radio Regulations distributed among services the frequency band of up to 275 GHz; frequency bands up to 14–17 GHz had been adopted operationally by satellite communications and broadcasting systems; experimental systems have been built in the 20/30-GHz bands. It is important to mention that as frequency increases, the attenuation of the signals in the atmosphere increases sharply, particularly in clouds and precipitations (see Chapter 4), in which connection operation in the HF bands requires much greater energy potential on the earth and on satellites, and consequently the cost increases significantly. These losses are offset to some extent by the feasibility of building highly directional space antennas without increasing their size.

The indicator of the utilization efficiency of a geostationary orbit in some segment $\Delta\phi^\circ$ of its arc and in some frequency band Δf, MHz, is

$$E = C/(\Delta\phi^\circ \Delta f) \tag{6.2}$$

where C is the summary carrying capacity of all the satellite communications systems, operating in band Δf, the satellites of which are located in section $\Delta\phi^\circ$ of the geostationary orbit. More modern indicators are proposed and substantiated in the literature [6.3, 6.4]. It is worthwhile to estimate the limiting carrying capacity of a geostationary orbit, which can be achieved in the foreseeable future by using the above-examined techniques. This was done, for example, in the article [6.1] on the limiting assumption that the energy resources of communications links are great enough that the contribution of thermal noises was negligible in comparison with interference between the satellite communications systems.

Chapter 7
Multiple Access and Signal Separation Methods

7.1 THE PRINCIPLES AND FEATURES OF MULTIPLE ACCESS

Multiple access is a specific feature of satellite communications, which distinguishes it advantageously from other kinds of communications and makes it possible to significantly increase the utilization efficiency of the transponders of a communications satellite. Multiple access is defined as the capability of several earth stations to access one satellite—a repeater. Another characteristic of multiple access is the requirement of the simultaneous retransmission of several signals through a common satellite transponder, which places serious requirements on the methods of the transmission and separation of these signals. Because actual circuits have imperfect characteristics (the frequency band is limited, the amplitude and phase responses are nonlinear, *et cetera*), crosstalk interference between signals, which detracts from the quality of their separation and reception by earth stations, is unavoidable.

The problem of choosing the best multiple access technique in the general sense consists in finding a set of orthogonal or nearly orthogonal signals, for which the energy characteristics of a repeater (power and frequency band) will be used most efficiently, and the level of crosstalk interference between the signals will be the lowest and have the minimum effect on the separability and noise immunity of the reception of each signal. There are three basic ways to generate a set of orthogonal signals, based on the separation of signals by frequency, time, and waveform. Each of them has specific features and produces specific effects in multiple access. These effects ultimately detract from the carrying capacity of a repeater transponder in multiple access in comparison with the single-signal mode of operation.

The effectiveness of multiple access methods can be evaluated on the basis of the extent to which the carrying capacity (or capacity) of a repeater is utilized [17] as a function of the number n of signals:

$$\eta(n) = \sum^{n} C_i/C_0 \quad \text{or} \quad \eta(n) = \sum^{n} N'/N \tag{7.1}$$

where C_0 and N are, respectively, the carrying capacity of a repeater transponder in the single-signal mode and the number of telephone signals that can be transmitted through that transponder on one carrier when multiple access is not used; C_i and N' are the carrying capacity and the number of telephone signals that can be transmitted in part of a common repeater transponder, reserved for the ith earth station using multiple access. The characteristic $\eta(n)$ is a monotonically diminishing function of the number of transmitted signals, and the most effective method obviously is the one for which this function is expressed in the least way.

7.2 REPRESENTATION OF EQUIVALENT NONLINEAR CHARACTERISTICS OF A REPEATER

The first step in the quantitative analysis of specific effects for multiple access is to determine the equivalent nonlinear characteristics of a satellite repeater. The amplitude response can be found by practical measurements (Figure 7.1)*.

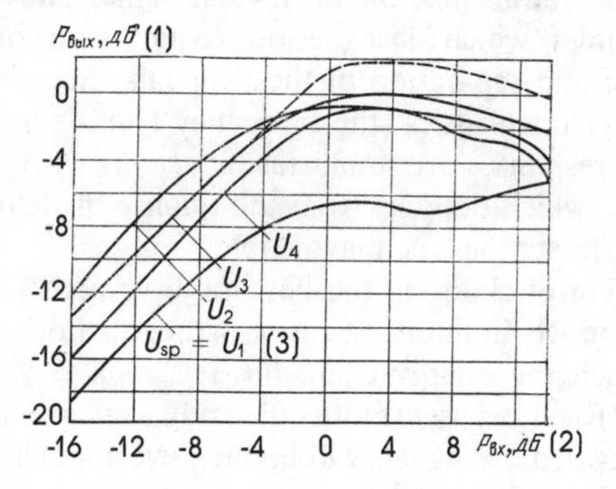

Key: (1) P_{out}, dB
(2) P_{in}, dB
(3) $U_{sat} = U_1$

FIGURE 7.1. Transfer functions of repeater with traveling wave tube operating in different modes.

*The family of curves in Figure 7.1 corresponds to different spiral voltages U_{sat} of a traveling wave tube.

For practical calculations it is first necessary to carry out the equivalent conversion from the amplitude response to the transfer function of instantaneous values, and then to process (approximate) the latter. The solution of this problem will depend on how the original amplitude response of a repeater is represented; we will assume that it is given as the first term of the odd harmonic polynomial

$$u_{out} = U_0 \sum_{n=1}^{m} A_n \cos n\omega t, \quad n = 1, 3, 5 \tag{7.2}$$

describing the output signal for a harmonic process:

$$u_{in} = U_0 \cos \omega t$$

Likewise, the output signal (its instantaneous value) can be determined through the transfer function $E(u)$ of instantaneous values: $u_{out} = E(U_0 \cos \omega t)$. By comparing these expressions we obtain

$$E(u) = U_0 \sum_{n=1}^{m} A_n \cos n\omega t \tag{7.3}$$

Stated as such, the problem of determining $E(u)$ can have a set of solutions, but on the condition that $E(u)$ may be represented as an odd power polynomial with the same number of terms, i.e.,

$$E(u) = a_1 u_{in} + a_3 u^3_{in} + \cdots + a_m u^m_{in} \tag{7.4}$$

The previous equation has one solution and acquires the form:

$$\sum_{n=1}^{m} a_n U_0^n \cos^n \omega t = U_0 \sum_{n=1}^{m} A_n \cos n\omega t$$

The relationship between the coefficients of the trigonometric polynomial that describes the transfer function of instantaneous values and the coefficients of the harmonic polynomial that describe the amplitude response can be found by means of harmonic synthesis. The formulas for computing them for seventh-power polynomials and for $U_0 = 1$ are

$$a_1 = 2^0 (A_1 - 3A_3 + 5A_5 - 7A_7), \qquad a_3 = 2^2(A_3 - 5A_5 + 14A_7)$$

$$a_5 = 2^4(A_5 - 7A_7), \qquad a_7 = 2^6(A_7)$$

Thus, if the measured response of a transponder is given as $P_{out} = f(P_{in})$, then it is first necessary to compute the amplitude response $A(U)$ and to represent it in analytical form, for example as a harmonic polynomial, and then to convert to characteristic $E(u)$ of instantaneous values. Shown in Figure 7.1 by the dashed line as an example is the computed response of instantaneous values; it is different from the initial amplitude response (for $U_{sat} = U_3$); their difference can reach several decibels.

Sometimes it is convenient to represent the transfer function of a nonlinear circuit in a different analytical form than the above trigonometric polynomial, specifically as the probability integral ("soft limiting"):

$$E(u) = \frac{A}{\sqrt{2\pi}} \int_0^{U_{in}/b_0} \exp(-z^2/2)\, dz \tag{7.5}$$

where b_0 is a normalizing parameter; or in exponential form:

$$E(u) = u_{in}^{\nu} \tag{7.6}$$

The last notation is most general; when $\nu = 1$ expression (7.6) describes a linear amplifier, and when $\nu = 0$ it describes a perfect limiter; when $0 < \nu < 1$ it represents intermediate forms of limiting.

The phase response of a repeater also is subjected to preliminary conversion; it should be represented as the AM-PM conversion factor (in deg/dB) as a function of the input power:

$$k_\theta = \Delta\theta/\Delta P_{in} = f(P_{in}) \tag{7.7}$$

The results of the corresponding conversion of the phase responses of a typical traveling wave tube to the form (7.7) are given in Figure 7.2, which shows that, generally speaking, k_θ is a complex function of P_{in}, but it may be assumed in the first approximation that this function is linear in the range of weak input signals, all the way to the saturation range. The equivalent responses of a repeater, thus found, are used as the basis for the quantitative estimation of effects for multiple access.

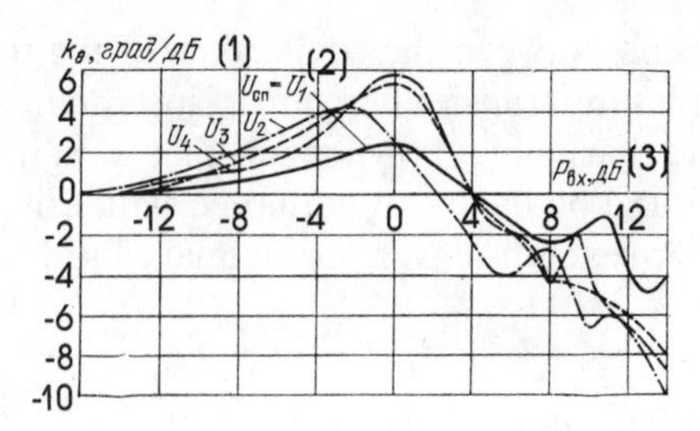

Key: (1) deg/dB
(2) $U_{sat} = U_1$
(3) P_{in}, dB

FIGURE 7.2. AM-PM conversion factor of repeater as function of operating mode of traveling wave tube.

7.3 FREQUENCY DIVISION MULTIPLE ACCESS. THE CHARACTERISTICS AND MODEL OF THE CUMULATIVE SIGNAL

The simplest and most common multiple access method is frequency division of channels, whereby each earth station transmits its own signals in a section of the frequency spectrum, reserved for it, with frequency band Δf (Figure 7.3). There are safety frequency intervals Δf_s between the signals, which on the receiving end make it possible to separate the signals with the necessary accuracy. Thus, n radio signals, each of which carries N' telephone (or other) messages, are transmitted in a common repeater transponder with frequency band W.

Great experience in the development and operation of frequency division systems, accumulated earlier during the design of other communications systems, and the comparative simplicity of the equipment are reasons why FDMA is used extensively today in nearly all existing satellite communications systems, including the Soviet national system and the Intelsat and Intersputnik systems.

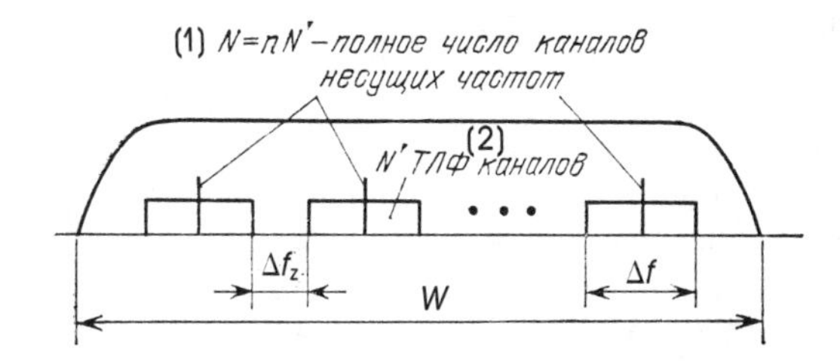

Key: **(1) $N = nN'$ the total number of carrier channels**
(2) N' telephone channels

FIGURE 7.3. Structure of frequency division signals.

In FDMA the signal that goes to the input of the repeater represents the sum of quasiharmonic partial signals of earth stations, each of which, in the general case, can be amplitude- or phase-modulated:

$$u(t) = \sum_{i=1}^{n} U_i(t)\cos\left[\omega_i t + \phi_i(t)\right] \qquad \textbf{(7.8)}$$

The correlation function of each partial signal in the case of modulation by random processes is

$$B_i(\tau) = \frac{U^2 i}{2}\, B_{ai}(\tau)\, B_{\phi i}(\tau)\cos\omega_i\tau \qquad \textbf{(7.9)}$$

where

$$B_{ai}(\tau) = \sigma^2 a \left[\frac{1}{\sigma_{2a}} + Ra(\tau)\right] \cos \omega_i \tau, \qquad B_{\phi i}(\tau) = \text{esp}\,\{-\sigma_\phi^2\,[1 - R_\phi(\tau)]\}$$

and $R_a(\tau)$, $R_\phi(\tau)$ are the correlation coefficients of the modulating processes.

Accordingly, the energy spectrum of the cumulative signal, i.e., of the signal that represents the sum of the signals at the repeater input (for identical partial signals) is:

- for phase modulation with a low index ($\sigma_\phi << 1$):

$$G_{\text{PM}}(\omega) = U^2{}_i \sum_{i=1}^{n} \pi\delta(\omega - \omega_i) + \frac{1}{2}\sigma^2{}_\phi\, G_{\text{m}}(\omega - \omega_i) \tag{7.10}$$

  where $G_{\text{m}}(\omega)$ is the energy spectrum of the modulating signal, and $\delta(\omega)$ is a delta-function;
- for frequency modulation with a high index ($\sigma_f >> 1$):

$$G_{\text{FM}}(\omega) = \frac{U^2{}_i}{2\sqrt{2\pi}\,\sigma_f} \sum_{i=1}^{n} \exp\left[-\frac{(\omega - \omega_i)^2}{2\sigma^2{}_f}\right] \tag{7.11}$$

An important parameter of the cumulative signal is the distribution function of the instantaneous values, which makes it possible to estimate the amount by which the peak signal level exceeds its mean square and the probability of excess. In the general case, the signals of earth stations arrive at the repeater input asynchronously and have a random phase, the value of which may be assumed to be uniformly distributed in the range 0–2π. The distribution function of the cumulative signal for n identical partial signals with constant envelopes (which corresponds to frequency or phase modulation) is

$$\omega(x) = \frac{1}{2nU_i}\left[1 + 2\sum_{r=1}^{\infty}\left(\cos\frac{\pi r x}{nU_i}\right) J_0^n\left(\frac{\pi r}{n}\right)\right] \quad \text{for} |x| < nU_i \tag{7.12}$$

where J_0^n is a zero-order Bessel function.

The peak factor of the cumulative signal, determined as the ratio of the peak power of that signal, not exceeded with probability $p(\lambda)$, to the mean cumulative power of n harmonic oscillations, is

$$\nu = \frac{2\lambda^2 U^2 i}{nU^2 i} = \frac{2\lambda^2}{n} \tag{7.13}$$

where λ is the peak cumulative signal level, not exceeded with probability $p(\lambda)$. Curves of the peak factor, determined by (7.13), are shown in Figure 7.4 as a function of the number of signals, from which it follows that, for n

≥ 10, the peak factor of the cumulative signal does not exceed 8 dB with probability $p = 0.99$, and for $p = 0.999$ it approaches 11 dB.

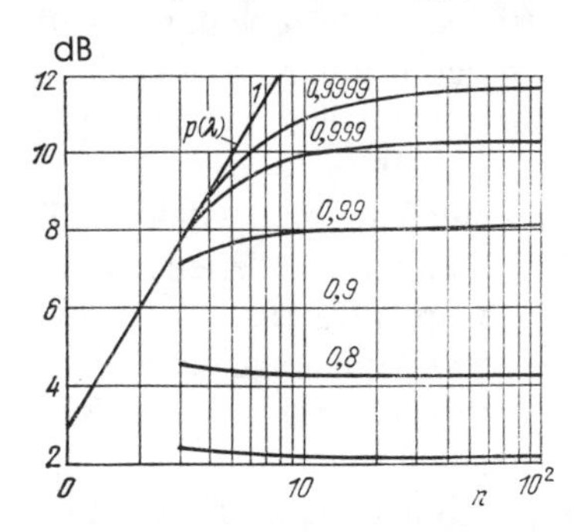

FIGURE 7.4. Peak factor of cumulative signal as function of number of retransmitted signals.

The pulse ratio is an important characteristic of a multistation signal; for frequency division this parameter characterizes the utilization of the frequency band of the repeater and can be determined by the formula:

$$q_f = \frac{1}{n} \sum_{i=1}^{n} \frac{\Delta f_i + \Delta f_e}{\Delta f_i} = \frac{W}{n \Delta f} \tag{7.14}$$

A model of the transponder signal is necessary for a quantitative analysis of the effects in frequency division; the parameters of the model should reflect well its amplitude features and spectral characteristics. However, these properties depend to a great extent on the number of signals of the earth stations, and for this reason it is not possible to choose a single model, which describes well enough the features of any cumulative signal. Therefore, two models are used in practice, one of which corresponds to a large, and the other to a small number of signals. When the number of signals is large ($n \geq 10$), the simplest and natural model is Gaussian noise with a spectral density that is uniform in frequency band W of a repeater:

$$\sigma^2_{\text{e.n}} = \begin{cases} \dfrac{nU_i^2}{2W} & \text{for } f_0 - \dfrac{W}{2} \leqslant f \leqslant f_0 + \dfrac{W}{2} \\ 0 & \text{for } f < f_0 - \dfrac{W}{2}, f > f_0 + \dfrac{W}{2} \end{cases} \tag{7.15}$$

It is important to remember here that the relative crosstalk interference level, determined with the aid of this model, is understated by 10 log

q_f dB, because the model does not take into account the frequency spaces between the signals of the earth stations. Consideration of this circumstance would require (the satellite repeater is assumed to have the same mean power) that the spectral power density be increased in accordance with (7.14), which, in turn, would lead to an increase of the spectral density of nonlinear products.

As a model of a cumulative signal with a small n, it is desirable to use an additive sum of some number of independent unmodulated sine waves, the frequencies of which are in a ratio such that the cross-modulation products of various orders will not coincide in terms of frequency, and can be separated by a filter; this requirement is satisfied, for example, by the frequencies f_0, $f_0 + \Delta$, $f_0 + 2\Delta$, $f_0 + 4\Delta$. The minimum number of sine waves, with which all the characteristic crosstalk interference is generated in the repeater, is three, and the total number in consideration of a calibrated oscillation should be four. Here the average power of the sum of the sine waves should be equal to the average power of the cumulative signal, and the amplitude of each of them is

$$U_e = \sqrt{nU_i^2/4} \tag{7.16}$$

We will present the ensuing examination in application to these two models.

7.4 NONLINEAR EFFECTS IN FREQUENCY DIVISION

7.4.1 General Description

In FDMA, the following effects occur: losses of the repeater output power in the multisignal mode; suppression of weak signals by strong ones; intermodulation (crosstalk) interference due to the nonlinear transfer function of a repeater; intermodulation (crosstalk) interference of AM-PM conversion, i.e., interference due to the conversion of amplitude modulation to phase modulation of signals. The effect of a reduction (loss) of the repeater output power is manifested most completely during operation in the saturation range. In application to the repeater characteristic, represented as hard limiting with a nonzero threshold [see (7.6), Figure 7.5 for $U_0 \neq 0$], it was shown in the item [7.1] that, when one harmonic signal is amplified, the output power, normalized relative to saturation power P_0, is

$$\frac{P_{1\,\mathrm{out}}}{P_0} = \frac{2}{\pi^2}\left[\left(1 - \frac{1}{2_\rho}\right)^{1/2} + (2_\rho)^{1/2}\sin^{-1}\left(\frac{1}{2_\rho}\right)^{1/2}\right]^2 \tag{7.17}$$

where $\rho = P_{1\,in}/P_0$, and when the input signal has a high power, corresponding to excursion into the saturation range ($P_{1\,in} >> P_0$),

$$P_{1\,out}/P_0 \approx 8/\pi^2 \approx 0.81 \tag{7.18}$$

When two harmonic signals are amplified at the same time, their cumulative output power is

$$P_{1\,out} + P_{2\,out}/P_0 = (8/\pi^2)^2 \approx 0.65 \quad \text{for } \rho \to \infty \tag{7.19}$$

When many signals are amplified (i.e., for the noise model):

$$\frac{\left(\sum^{n} P_{i\,out}\right)}{P_0} \approx \frac{2}{\pi} \approx 0.64 \quad \text{for } \rho \to \infty \tag{7.20}$$

The results of calculations by formulas (7.17)–(7.20) are given in Figure 7.5, which shows that output power losses in the multisignal mode are 1–1.5 dB relative to the single-signal mode. This can be explained physically by the expenditure of power for the generation of harmonic components and cross products.

As was pointed out above, the passage of signals of different levels through a nonlinear signal path is accompanied by the suppression of a weak signal by a strong one. The suppression coefficient is usually determined as the ratio of the mean powers of the signals at the input and output of a nonlinear path:

$$K_{sup} = (P_{1\,out}/P_{2\,out})/(P_{1\,in}/P_{2\,in}) \tag{7.21}$$

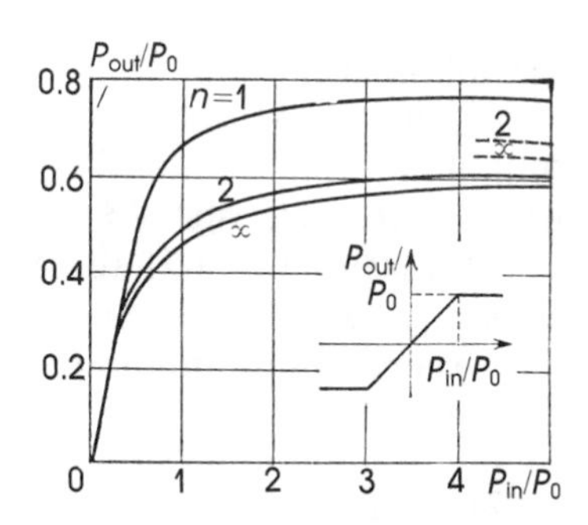

FIGURE 7.5. Determination of output power losses of repeater in multisignal mode.

This effect was analyzed in the classic work [7.2], where it is shown that for two harmonic signals (Figure 7.6):

$$K_{sup} \begin{cases} \to 4 & \text{for } P_{1\text{in}} >> P_{2\text{in}} \\ =1 & \text{for } P_{1\text{in}} = P_{2\text{in}} \\ \to 0.25 & \text{for } P_{1\text{in}} << P_{2\text{in}} \end{cases} \tag{7.22}$$

and for a harmonic signal and gaussian noise (Figure 7.6):

$$K_{sup} \begin{cases} \to 2 & \text{for } P_s >> P_n \\ =1 & \text{for } P_s = 0.35P_n \\ \to \pi/4 & \text{for } P_s << P_n \end{cases} \tag{7.23}$$

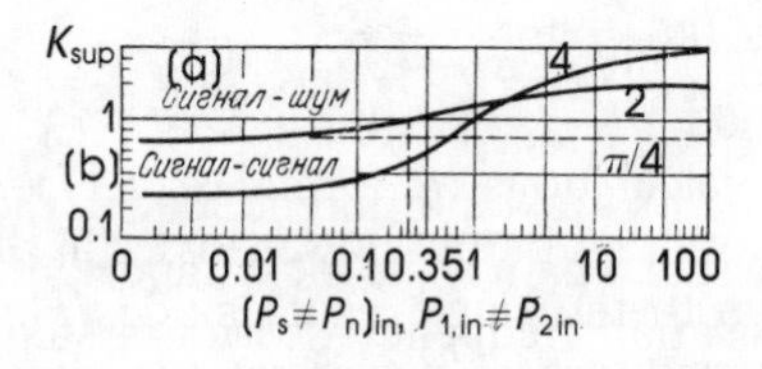

Key: **(a) Signal-noise**
(b) Signal-signal

FIGURE 7.6. Determination of signal suppression coefficient.

The above expressions show that, in the worst case (for harmonic signals), the suppression of a weak signal reaches 6 dB, and the suppression of a harmonic signal by noise does not exceed 1 dB. However, these figures are the limits, because they were found in the article [7.2] in application to the perfect limiter and to two extreme kinds of interference: single-signal and noise. For practical calculations, including for the above four-signal model, quantitative estimation of the suppression coefficient is necessary for any number of signals and for a nonlinear element with any characteristic, represented in the general form (7.6).

By means of expansion of characteristic (7.6) into a Fourier series, on the assumption that the input signal to interference ratio of a nonlinear signal path is $U_{s.in}/U_{sup.in} << 1$, was derived in the work [7.3] for the suppression coefficient in the general case (for any kind of harmonic signal and interference):

$$K_{sup} = 2\left(\frac{\nu + 1}{2}\right)^2 \frac{m^2\,[U^{\nu}_{int.in}(t)]D[U_{int.in}(t)]}{m^2[U^{\nu+1}_{int.in}(t)]} \tag{7.24}$$

where $D(x)$ is the dispersion of x; and $m(x)$ is the mean of random variable x, given by the distribution of $w(x)$. From (7.24) it is not hard to obtain partial results [7.2], represented previously by (7.22) and (7.23). For instance, for two harmonic signals, it is easy to directly find $K_{sup} = (\nu + 1)^2/4$, which generalizes the result of the article [7.2] for a signal path with any response and for the case of hard limiting ($\nu = 0$) yields $K_{sup} = 0.25$, just like (7.22). For the previously used four-signal model interference is the sum of three harmonic signals with amplitude U and with a random phase; according to (7.12), the distribution function of this interference is

$$w(x) = \frac{1}{6U}\left[1 + 2\sum_{r=1}^{\infty}\left(\cos\frac{\pi r x}{3U}\right) J_0^3\left(\frac{\pi r}{3}\right)\right]$$

and calculation by (7.24) for $\nu << 1$ yields $K_{sup} \approx$ -2.5 dB.

Thus, we have the following suppression coefficients for a future quantitative analysis of nonlinear products:

- for the four-signal model $K_{sup} \approx$ -2.5 dB;
- for the noise model K_{sup} = -1 dB.

The formation of crosstalk interference due to the nonlinearity of the transfer function of a signal path is the most important and crucial effect in an analysis of multistation frequency division systems. In practice, two basic methods of analysis are used:

- harmonic, based on the representation of the transfer function of a nonlinear element as odd power polynomial (7.4) of the input signal and calculation of combination products as the sum of the components of expansion of a power series;
- correlative, based on the representation of the transfer function as probability integral (7.5) and subsequent calculation of the correlation function and energy spectrum of the output signal.

The former is the simplest and physically obvious method ([7.4]–[7.6]), but it is suitable for only for a few input signals, because when $n >> 1$, the trigonometric calculations are considerably more complicated; this method will be used in this section for studies with the four-signal model. The latter method (correlative) (see the works [7.1], [7.7], [7.8]) is more general, because the correlation functions are computed for any number of signals, but the easiest solution is found for the noise model of an input signal, and that is what we will be using below.

7.4.2 The Harmonic Method

The harmonic method of calculating nonlinear interference for the four-signal model will be used. The transfer function of the instantaneous values of a nonlinear signal path by odd power polynomial (7.4) will be approximated. Here it will be necessary to take a large enough number of terms of the polynomial so that the approximation will be good in a range of input levels that exceed the conditional saturation point by 10–12 dB. The cumulative input signal is represented as (7.8).

By substituting (7.8) into (7.4) it is possible to find all the components of the signals and interference at the output of a nonlinear path. Here it is necessary to raise a harmonic polynomial to a power, i.e., to compute [7.4]

$$\begin{aligned} a_m u_{\text{in}}^m &= a_m \left\{ \sum_{i=1}^{n} U_i(t)\cos[\omega_i t + \phi_i(t)] \right\}^m \\ &= \sum \frac{m!}{p_1! p_2! \cdots p_n!} U_1^{p_1}(t)\, U_2^{p_2}(t) \cdots U_n^{p_n} \\ &\quad \times \cos^{p_1}[\omega_1 t + \phi_1(t)] \cos^{p_2}[\omega_2 t + \phi_2(t)] \cdots \cos^{p_n}[\omega_n t + \phi_n(t)] \end{aligned}$$

where $p_1, p_2, \ldots, p_n$ are arbitrary positive integers, connected by the relation $p_1 + p_2 + \cdots + p_n = m$. The direct determination of the amplitudes of components using the above formula involves unwieldy trigonometric conversions, because kth-order components of a given combination frequency are present not only in the expansion of the term with the power k, but also in the expansion of all terms with higher powers, and all these components must be added in order to determine the desired amplitude.

General expressions for the output components and cross products of a signal were derived in the article [7.4] for n identical signals with amplitudes U_i and with cumulative power $P_{\text{in}} = nU_i^2/2$. According to the article [7.5], the following enter the principal frequency band:

- useful signals with frequencies ω_i, the amplitude of each of which is

$$\begin{aligned} U_{\text{out}1}(n) = a_1 \sqrt{\frac{2P_{\text{in}}}{n}} \Bigg\{ & 1 + 3\frac{a_3}{a_1}\left(\frac{P_{\text{in}}}{n}\right)\left(n - \frac{1}{2}\right) \\ & + 15\frac{a_5}{a_1}\left(\frac{P_{\text{in}}}{n}\right)^2 \left[\frac{1}{6} + (n-1)(n-2) + \frac{3}{2}(n+1)\right] \\ & + 105\frac{a_7}{a_1}\left(\frac{P_{\text{in}}}{n}\right)^3 \Bigg[(n-1)(n-2)(n-3) + 3(n-1)(n-2) \\ & + \frac{34}{24}(n-1) + \frac{1}{24}\Bigg] + \cdots \Bigg\} \end{aligned}$$

- cross-modulation products of the type $2\omega_i - \omega_j$ with amplitudes:

$$\begin{aligned} U_{\text{out}2}(n) = \frac{3}{4} a_2 \left(\frac{2P_{\text{in}}}{n}\right)^{3/2} \Bigg\{ & 1 + \frac{2}{3}\frac{a_5}{a_3}\left(\frac{P_{\text{in}}}{n}\right)\Big[12.5 + 15(n-2)\Big] \\ & + 105\frac{a_7}{a_2}\left(\frac{P_{\text{in}}}{n}\right)^2\left[(n-2)(n-3) + \frac{13}{6}(n-2) + \frac{7}{12}\right] + \cdots \Bigg\} \end{aligned}$$

- cross-modulation products of the type $\omega_i + \omega_j - \omega_k$:

$$U_{out3}(n) = \frac{3}{2} a_3 \left(\frac{2P_{in}}{n}\right)^{3/2} \left\{ 1 + 10 \frac{a_5}{a_3} \left(\frac{P_{in}}{n}\right) \left[\frac{3}{2} + (n-3)\right] + 210 \frac{a_7}{a_3} \left(\frac{P_{in}}{n}\right)^2 \left[1 + \frac{7}{4}(n-3) + \frac{1}{2}(n-3)(n-4)\right] + \cdots \right\}$$

- cross-modulation products of the type $3\omega_i - 2\omega_j$:

$$U_{out5}(n) = \frac{5}{8} a_5 \left(\frac{2P_{in}}{n}\right)^{5/2} \left\{ 1 + \frac{49}{4} \frac{a_7}{a_5} \left(\frac{P_{in}}{n}\right) \left[1 + \frac{12}{7}(n-2)\right] + \cdots \right\}$$

The calculation of the combination products for the four-signal model with the nonlinear response of a traveling wave tube approximated by a seventh-power polynomial yields: main components with frequencies ω_i:

$$U_{out1}(n=4) = a_1 \sqrt{\frac{P_{in}}{2}} \left(1 + 26 \frac{a_3}{a_1} P_{in} + 10 \frac{a_5}{a_1} P_{in}^2 + 46.3 \frac{a_7}{a_1} P_{in}^3\right) \quad (7.25)$$

Third-order products with frequencies $2\omega_i - \omega_j$:

$$U_{out2}(n=4) = \frac{3}{4} a_3 \left(\frac{P_{in}}{2}\right)^{3/2} \left(1 + 7.1 \frac{a_5}{a_3} P_{in} + 45.4 \frac{a_7}{a_3} P_{in}^2\right) \quad (7.26)$$

Third-order products of the type $\omega_i + \omega_j - \omega_k$:

$$U_{out\,3}(n=4) = \frac{3}{2} a_3 \left(\frac{P_{in}}{2}\right)^{3/2} \left(1 + 6.25 \frac{a_5}{a_3} P_{in} + 36.1 \frac{a_7}{a_3} P_{in}^2\right) \quad (7.27)$$

Fifth-order products of the type $3\omega_i - 2\omega_j$:

$$U_{out\,5}(n=4) = \frac{5}{8} a_5 \left(\frac{P_{in}}{2}\right)^{5/2} \left(1 + 13.5 \frac{a_7}{a_5} P_{in}\right)$$

Expressions for fifth-order products of the type $3\omega_i - \omega_j - \omega_k$, and also of the type $\omega_i + 2\omega_j - 2\omega_k$ are not given in view of their complexity.

The results of calculations of the relative power of single nonlinear third-order products as a function of the input level, normalized relative to the saturation point, for a traveling wave tube with the typical response (see, for example, the curve $U_{sat} = U_3$ in Figure 7.1), converted to the response of instantaneous values in accordance with Section 7.2, are given in Figure 7.7 (the scales of the input and output powers are given on the horizontal axis in order to facilitate the use of these results).

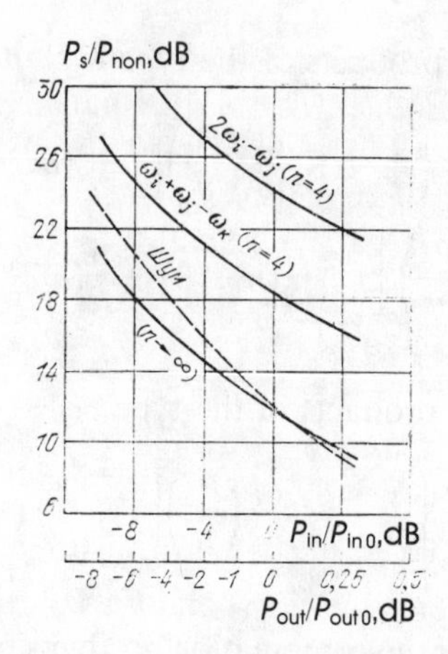

FIGURE 7.7. Relative power of nonlinear products as function of a load of a traveling wave tube.

Now it is necessary to determine the total number of products of each type and their distribution in the frequency band of a repeater. A general examination of this problem extends beyond the scope of this section. It is presented well in the articles [7.4] and [7.5]. The number of nonlinear product for some values of n is given in Table 7.1 as an example. As can be seen in the table, the number of fifth-order products is considerably higher than the number of third-order products, but as follows from (7.28), their amplitudes are exceedingly low; in practice it may be assumed, even for $n \geq 6$, that most of the interference is determined by products of the type $\omega_i + \omega_j - \omega_k$.

TABLE 7.1

n	Number of each type of product			
	$2\omega_i - \omega_j$	$\omega_i + \omega_j - \omega_k$	$3\omega_i - \omega_j - \omega_k$	$\omega_i + \omega_j - 2\omega_k$
4	12	12	12	24
10	90	360	360	720
20	380	3420	3420	6480

It should be pointed out that, when the frequencies of the input signals are uniformly distributed and separated by ΔW, the frequencies of all nonlinear products will be multiples of this separation, and consequently they will coincide with the frequencies of the signals. Here most of the interference gets into the central region, and the power of the interference on the edges of the frequency band of a transponder is less than in the center.

For practical estimation of the number of components of nonlinear products, we will use the results of the article [7.5], according to which the total number of single products $D(r, n)$ that fall on the frequency of a signal with the ordinal number r in the case of the transmission of n signals for third-order products of the type $2\omega_i - \omega_j$ is

$$D_2(r,n) = \frac{1}{2}\left\{n - 2 - \frac{1}{2}[1 - (-1)^n](-1)^r\right\} \tag{7.28}$$

and for third-order products of the type $\omega_i + \omega_j - \omega_k$:

$$D_3(r,n) = \frac{r}{2}(n - r + 1) + \frac{1}{4}[(n - 3)^2 - 5] - \frac{1}{8}[1 - (-1)^n(-1)^{n+r}] \tag{7.29}$$

For the four-signal model with $r = 2$, this yields $D_2(r,n) = 1$, $D_3(r,n) = 2$. The results of calculations by (7.28) and (7.29) are summarized in Tables 7.2 and 7.3 for different numbers n and r.

TABLE 7.2

n	Number of products of the type $2\omega_i - \omega_j$ on carriers with number r							
	1	*2*	*3*	*4*	*5*	*6*	*7*	*8*
1	0	—	—	—	—	—	—	—
2	0	0	—	—	—	—	—	—
3	1	0	1	—	—	—	—	—
4	1	1	1	1	—	—	—	—
5	2	1	2	1	2	—	—	—
6	2	2	2	2	2	2	—	—
7	3	2	3	2	3	2	3	—
8	3	3	3	3	3	3	3	3

TABLE 7.3

n	Number of products of the type $\omega_i + \omega_j - \omega_k$ on carriers with number r							
	1	*2*	*3*	*4*	*5*	*6*	*7*	*8*
1	0	—	—	—	—	—	—	—
2	0	0	—	—	—	—	—	—
3	0	1	0	—	—	—	—	—
4	1	2	2	1	—	—	—	—
5	2	4	4	4	2	—	—	—
6	4	6	7	7	6	4	—	—
7	6	9	10	11	10	9	6	—
8	9	12	14	15	15	14	12	9

It is necessary to consider one more factor: expansion of the frequency band of combination products in comparison with the frequency band of the useful signals, and the associated attenuation of interference by receiver filters. If the input signals are represented, for example, as FM signals, modeled by a process with a Gaussian distribution, the correlation function and power spectrum of which correspond to (7.9) and (7.11), then the power spectrum of their nonlinear products will be

$$G_2(\omega) = A \frac{U_i^4 U_j^2}{\sqrt{2\pi}\sigma_2} \exp\left[-\frac{(f - f_2)^2}{2\sigma_2^2}\right]$$

where

$$f_2^2 = 2f_i - f_j, \quad \sigma_2 = \sqrt{4\sigma_i^2 + \sigma_j^2} = \sigma\sqrt{5}$$

$$G_3(\omega) = B \frac{U_i^2 U_j^2 U_k^2}{\sqrt{2\pi}\sigma_3} \exp\left[-\frac{(f - f_3)^2}{2\sigma_3^2}\right] \tag{7.30}$$

where

$$f_3 = f_i + f_j - f_k, \quad \sigma_3 = \sqrt{\sigma_i^2 + \sigma_j^2 + \sigma_k^2} = \sigma\sqrt{3} \tag{7.31}$$

It follows from these expressions that the spectrum of nonlinear products of the type $2\omega_i - \omega_j$ is expanded by a factor of $\sqrt{5}$, and for products of the type $\omega_i + \omega_j - \omega_k$ by a factor of $\sqrt{3}$. The attenuation of the power of nonlinear products, inserted by the IF path of the receivers of earth stations, can be determined by integrating these spectra appropriately in frequency band Δf of the useful signals:

$$\alpha_2 = \int_{-\Delta f/2}^{\Delta f/2} G_2(\omega)\, d\omega \Big/ \int_{-\infty}^{\infty} G_2(\omega)\, d\omega = -1.8 \text{ dB}, \quad \alpha_3 = -1{,}2 \text{ dB} \tag{7.32}$$

Relations derived in the article [7.5], with correction factors that take into account the filtering of measured products (7.32) and the suppression of the signal by interference K_{sup} (for unequal signals) may be used for extrapolating the results, obtained for the four-signal model, to an arbitrary number of transmitted signals:

$$\frac{P_s}{P_{non}}(n) = \frac{U^2_{out\,1}(n = 4)K_{sup}}{\sum_t U^2_{out\,l}(n = 4)D_l(r,n)\alpha_l} \frac{n^2}{16} \tag{7.33}$$

where P_s is the useful signal power, and P_{non} is the power of nonlinear products in the band of the signal.

When there are many signals ($n \rightarrow \infty$), as was pointed out above, nonlinear products $\omega_i + \omega_j - \omega_k$ predominate; for them $D_3(r,n) \rightarrow 3n^2/8$ on the middle frequency of a transponder; $D_3(r,n) \rightarrow n^2/4$ on the edges of the band of a transponder. Accordingly, for $n \rightarrow \infty$, for the middle frequency of a transponder and for identical input signals, expression (7.33) acquires the form:

$$\frac{P_s}{P_{non}}(n \rightarrow \infty) \approx \frac{U^2_{out\,1}(n = 4)}{6U^2_{out\,3}(n = 4)\alpha_3} \tag{7.34}$$

Nonlinear interference is plotted in Figure 7.7 in accordance with this generalizing formula as a function of the input (output) power level of a repeater for $n \rightarrow \infty$.

7.4.3 The Correlative Method

As was shown in Section 7.3, when there are many signals a good model of the cumulative signal at the input and output of a repeater is gaussian noise with spectral density (7.15), uniform in the frequency band of the repeater; for calculating nonlinear interference here it is best to represent the response of a nonlinear element as probability integral (7.5). According to the article [7.7] the correlation function of the output process of a "soft" limiter with a response of the form (7.5), when a narrow-band normal process is applied to its input, is

$$B(\tau) = \frac{A^2}{2\pi} \arcsin\left[\frac{R(\tau)}{(1+\alpha^2)}\right]$$

where $\alpha = b_0/\sigma$ is a parameter that connects the responses of the nonlinear element and the input process.

Correlation function $B'(\tau)$, corresponding to the part of the process that gets into the frequency band of a repeater, can be extracted from $B(\tau)$ by means of expansion into a Taylor series and a series of transforms; then the spectra of the undistorted useful output signal $G_s(\omega)$ of the nonlinear element and the energy spectrum $G_{non}(\omega)$ of nonlinear interference can be found in accordance with Khinchin's theorem. Found in the book [2], after expansion of the integrals in the equation:

$$\frac{G_{non}(\omega)}{G_s(\omega)} = \sum_{n=2}^{\infty} \frac{[(2n-3)!!]^2 C_{2n-1}^{n-1}}{(2n-1)!2^{2n-2}} \frac{y_{2n-1}(\gamma)}{(1+\alpha^2)^{2n-2}} \tag{7.35}$$

where

$$y_{2n-1}(\gamma) = \sqrt{\frac{6}{\pi(2n-1)}} \exp\left[-\frac{3\gamma^2}{2(2n-1)}\right], \quad \gamma = 2(\omega - \omega_0)/W$$

The signal-to-interference ratio, computed on the basis of (7.35), is shown in Figure 7.7 as the dashed line as a function of the input signal power, normalized relative to the conditional saturation point $P_{out\,0} \approx 0.95P_{max}$. This function was computed for the middle frequency of a repeater transponder; the signal-to-interference ratio on the outer frequencies of a transponder is approximately 2 dB higher. A comparison of this function with the one computed for $n \to \infty$ by the harmonic method shows readily that they agree quite well.

Thus, for calculating nonlinear interference for a few signals ($n \leq 10$) it is better to use the harmonic method, but for many signals, the correlative method. This makes it possible to determine the output signal-to-interference ratio of the satellite repeater. This ratio can be converted to the power of nonlinear noises on a telephone channel, transmitting N'-channel group frequency division and frequency-modulated telephone signals on each carrier, by using formulas given in the work [7.5] and [2].

7.4.4 The AM-PM Conversion Effect

The AM-PM conversion effect is caused by the fact that the signal band of a repeater contains elements (the output traveling wave tube in particular), in which the insertion or phase shift depends on the signal level. Accordingly, these elements are amplitude-modulation to phase-modulation converters, and, when multiple access, is used they produce two types of crosstalk interference: audible and inaudible. Audible interference is caused by the fact that each signal (for example, frequency-modulated), passing through a signal path with a nonuniform frequency response (for example the transmitter of an earth station), acquires parasitic amplitude modulation in accordance with the principle of frequency modulation; conversion from FM to AM occurs. After passing through an element with AM-PM conversion, this parasitic amplitude modulation is converted to parasitic phase modulation of each of the amplified signals, and after demodulation it produces audible interference on a low-frequency channel. A quantitative assessment of this effect is given in the article [7.9]. Inaudible AM-PM conversion interference is explained by the fact that the envelope of the cumulative signal is not constant, but changes with the beat frequency between its components; accordingly, after passing through an element with AM-PM conversion, the phase of each signal will contain products of these beats. A quantitative assessment of this effect is given in the works [7.5] and [2].

The cumulative input signal of the form (7.8) is written as

$$u_{in}(t) = U_{in}(t) \cos [\omega_0 t + \phi_{in}(t)]$$

where the envelope is

$$U_{in}(t) = U_i \left\{ \left[\sum_{i=1}^{n} \cos (\omega_i - \omega_0)t \right]^2 + \left[\sum_{n=1}^{n} \sin (\omega_i - \omega_0)t \right]^2 \right\}^{1/2}$$

and the phase is

$$\phi_{in}(t) = \arctan \left[-\sum_{i=1}^{n} \sin (\omega_i - \omega_0)t \Big/ \sum_{n=1}^{n} \cos (\omega_i - \omega_0)t \right]$$

As was shown in the article [7.5], the phase shift, caused by amplitude modulation of a signal in a traveling wave tube, is proportional to the square of the envelope of the input signal. As a result of conversion of the expression for the square of the envelope and of the substitution of this expression into formula (7.3) in consideration of (7.7), an expression was derived for the phase shift of each signal after passing through an element with AM-PM conversion:

$$\phi_{out}(t) = \frac{0.1516 k_\theta}{2n} \sum_{j=1}^{n} \sum_{k=1}^{n} \cos (\omega_j - \omega_k)\, t$$

where k_θ is the AM-PM conversion factor (7.7).

For $k_\theta << 1$ and $\phi_{out}(t) << 1$, the general expression that describes the process at the output of an element with AM-PM conversion acquires the form:

$$u_{out}(t) = U_i \sum_{i=1}^{n} \cos \omega_i t - U_i \frac{0.1516 k_\theta}{2n} \sum_{\substack{i=1 \\ i \neq j}}^{n} \sum_{j=1}^{n} \sin (2\omega_i - \omega_j) - U_i \frac{0.1516 k_\theta}{n} \sum_{\substack{i=1 \\ i \neq j \neq k}}^{n} \sum_{j=1}^{n} \sum_{k=1}^{n} \sin (\omega_i + \omega_j - \omega_k) \quad \textbf{(7.36)}$$

The first sum in (7.36) determines the set of useful signals, the second AM-PM conversion products of the type $2\omega_i - \omega_j$, and the third products of the type $\omega_i + \omega_j - \omega_k$, and the amplitude of a product of the second type is 6 dB higher than of the first. It is interesting to note that AM-PM conversion products have the same structure as products of the nonlinearity of the transfer function of a signal path of the very same order, are in quadrature in relation to them, and consequently can be added with them in terms of power.

For the four-signal mode an expression can be derived from (7.36) for the signal-to-unit AM-PM conversion interference ratio:

- for products of the type $2\omega_i - \omega_j$

$$\frac{P_s}{P_{\text{AM-PM}}}(n = 4) = \left(\frac{2n}{0.1516k_\theta}\right)^2 \approx \frac{2785}{k_\theta^2} \tag{7.37}$$

- for products of the type $\omega_i + \omega_j - \omega_k$

$$\frac{P_s}{P_{\text{AM-PM}}}(n = 4) = \left(\frac{n}{0.1516k_\theta}\right)^2 \approx \frac{696}{k_\theta} \tag{7.38}$$

The general formula for the signal-to-unit AM-PM conversion interference ratio, which takes into account products of both types for an arbitrary number of signals, expressed through unit interference of the four-signal mode, by analogy with (7.33) may be written as

$$\frac{P_s}{P_{\text{AM-PM}}}(n) = \left[\frac{\alpha_2 k_\theta^2}{2785} D_2(r, n) + \frac{\alpha_3 k_\theta^2}{696} D_3(r, n)\right]^{-1} \frac{n^2}{16}$$

$$\approx \frac{174n^2}{k_\theta^2[\alpha_2 D_2(r, n) + 4\alpha_3 D_3(r, n)]} \tag{7.39}$$

For a larger number of signals ($n \to \infty$), as shown in Section 7.4, products of the type $\omega_i + \omega_j - \omega_k$ predominate, for which, according to (7.29), $D_3(r, n) \to 3n^2/8$. Accordingly, for $n \to \infty$,

$$\frac{P_s}{P_{\text{AM-PM}}}(n \to \infty) = \left(\frac{n}{0.1516k_\theta}\right)^2 \frac{8}{3n^2} \approx \frac{116}{k_\theta^2} \tag{7.40}$$

It follows from (7.40) that inaudible AM-PM conversion interference is significant and for $k_\theta \geq (2\text{–}3)°/\text{dB}$ can exceed the nonlinear interference level.

The formulas derived above express the output signal to AM-PM interference ratio of a repeater; it does not depend directly on the repeater output power, but the indirect function $k_\theta = f(P_{\text{out}})$ still exists, by virtue of which the level of interference of this origin also can be changed by choosing the right working point on the response of a repeater. The fact that, because the structures of nonlinear noises and AM-PM conversion noises are identical, they cannot be singled out during measurements on actual channels, is an important practical consideration; products, measured, for example, with the aid of the four-signal model, are produced by the combined action of both effects.

7.5 DETERMINATION OF THE OPTIMUM PARAMETERS OF A SYSTEM AND ESTIMATION OF THE EFFICIENCY OF FREQUENCY DIVISION MULTIPLE ACCESS

In order to assess the effectiveness of the examined multiple access method in accordance with criterion (7.1) which we used earlier, it is necessary to calculate the main parameters of a system, consisting of n earth stations with a total capacity of N telephone channels, and to analyze the function $N = f(n)$. It is assumed here that the capacity N' of all stations is identical; telephone channels within one station are transmitted by frequency modulation with the standard quality and are frequency-divided by means of standard channeling equipment, used on terrestrial links.

The first step of the solution of the problem as stated is to determine the total number of telephone channels N that can be set up through a repeater transponder, the propagation path of which has given parameters (Δf, P_{sat}, *et cetera*) or, conversely, to synthesize the parameters of the HF signal path that guarantee the transmission of a given number of telephone signals on one carrier.

The second step is to determine the number of telephone channels $N = N'n$ that can be set up through a link with the above parameters with two, three, and more carriers. The results of this analysis, carried out in the articles [7.10] and [7.11], are presented in Figure 7.8, which shows that for a repeater with constant parameters (P_{sat} and W) the capacity of a FDMA system decreases significantly as the number of signals increases. For instance, a repeater, designed for transmitting 600 telephone channels on one carrier (i.e., in the single-signal mode), is capable of transmitting only 360 channels with two carriers (180 channels per carrier), 270 channels with three carriers, *et cetera*. When $n > 50$ the capacity of a repeater is only 10–12% of the initial capacity. If the signal parameters are optimized for each number of carriers and the optimum working point is selected on the response of a repeater for the purpose of maximizing the carrying capacity on each of the carriers, then the capacity of a repeater can be increased; the corresponding function is shown by the dashed line in Figure 7.8.

If one telephone message is transmitted on each carrier, then the use of statistical suppression of the carriers in the pauses of modulation is effective. This method was investigated quantitatively in the book [7.14], where it is shown that, for a telephone channel with the actual activity coefficient $P_a = 0.3$, the average load on a repeater transponder with suppressed carriers decreases by 5 dB; this makes it possible to triple the number of transmitted carriers (see the dot-dashed lines in Figure 7.8). It is interesting to compare the results obtained with results achieved in real frequency division systems. Data on a US commercial system are given in the article [7.12], and data on the Intelsat system are given in the article [7.13]; these data are presented on Table 7.5. When comparing these data with Figure 7.8, it should be remembered that they correspond to previous estimates and confirm the conclusion that when FDMA is used, an increase of the number of signals leads to a decrease of the capacity of a system.

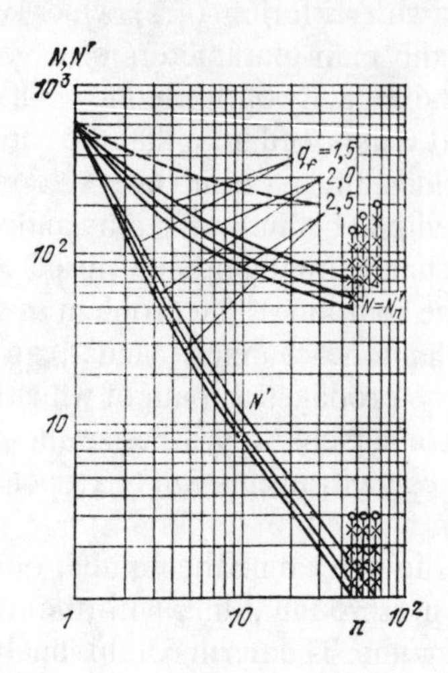

FIGURE 7.8. Carrying capacity of frequency division repeater as function of number of signals.

TABLE 7.5

Number of signals n	U.S. system		Intelsat system	
	N	$N/N(n=1)$	N	$N/N(n=1)$
1	700	1	900	1
2	540	0.77	—	—
4	360	0.51	456	0.5
8	240	0.34	400	0.44
16	160	0.23	300	0.3

7.6 TIME DIVISION MULTIPLE ACCESS

7.6.1 General Representation

As follows from the material presented above, FDMA, which has a number of indisputable advantages, nevertheless has a significant drawback, the need to operate the output power stage of the repeater in a quasilinear mode. Here the working point of this stage is usually 4–6 dB below the point that corresponds to the maximum power mode. Such an appreciable underutilization of the power potential of a link significantly reduces the carrying capacity of a communications system, and detracts from its economic characteristics accordingly. Time division multiple access (TDMA) is virtually devoid of this deficiency. The requirement in multiple access, that the signals of different stations be orthogonal, is satisfied by the fact that each station of a network, for radiating signals, is given a particular, periodically repeated time interval (the duration of which generally is determined by the traffic of a station). The transmission intervals of all stations are mutually synchronized, and therefore do not overlap. A time interval, in which all stations of a network radiate their signals one time each, is called a frame, and the duration of a packet, radiated by one station, is called a subframe. This kind of system makes it possible to use a repeater in close to the maximum power mode, because the signal of just one station passes through the repeater at any given time, and the problem of intermodulation (crosstalk) interference, being one of the main reasons why the carrying capacity of a multiple access system decreases, vanishes. Just as in FDMA, when the utilization efficiency of the passband of a transponder is determined by the need to insert certain frequency spaces between the individual modulated carriers, in TDMA, the utilization efficiency of the operating time η of a repeater is determined by the need to insert protection time spaces between the subframes, which keep them from overlapping during imperfect operation of the intersation synchronization system and by the need to insert a number of additional signals: $\eta = \Sigma_1^n \tau_{\text{inf}}/T$, where $\tau_{\text{inf}\,i}$ is a time interval, used for transmitting an information signal itself in the i th subframe; n is the number of subframes; and T is the frame duration. The structure of a frame of a TDMA system is described in greater detail in Chapter 18. As a rule, $\eta > 0.9$ in known existing or planned TDMA systems.

7.7 NOISE IMMUNITY ANALYSIS OF TIME DIVISION MULTIPLE ACCESS SYSTEMS

One of the most important characteristics during an analysis of the capabilities of a given multiple access system is the carrying capacity that is achieved. The main factors that influence this characteristic are the fluctuation noise level at the demodulator input of the receiver, the amount of distortion of the signal as it passes through a real signal path, and the imperfect operation of individual elements of the equipment. All these factors will be examined in greater detail. It should be pointed out first of all that PSK, combined with the transmission of signals in discrete form, is used in all known TDMA systems.

An analysis of the noise immunity of PSK (including relative PSK) communications systems has been carried out in numerous fundamental works [7.14, 7.15, 7.16, 7.17]. It is done, as a rule, on the assumption that interference is white noise with statistical properties, determined by a gaussian distribution, and there are no between-symbol distortions, determined by the limited passband of the signal path, and also by some additional signal distortions, which occur because of the imperfect characteristics of a signal path. The probability of the incorrect reception of PSK signals when coherent detection is used is

$$P \approx \frac{2}{m} F\,[\sqrt{2}\,h \sin(\pi/2m)] \tag{7.41}$$

where P is the probability of the incorrect reception of a bit in an m-times PSK system; h^2 is the ratio of the power of one pulse to the spectral power density of noise; and

$$F(x) = \frac{1}{\sqrt{2\pi}} \int_{-\infty}^{x} \exp^{(-t^2/2)}\, dt$$

is a Laplace function.

It follows from the above expression that the noise immunity of the examined systems deteriorates rapidly as the multiplicity of the signal increases. At the same time by increasing the multiplicity, it is possible to reduce the necessary passband of a communications channel. Thus, by varying the keying factor it is sometimes possible to match the energy potential of a link with its passband. For instance, a surplus of power in a band that is too narrow makes it possible to increase the carrying capacity by increasing the multiplicity and, conversely, a shortage of power in a surplus band can be offset by reducing the multiplicity.

Attention must be paid to the fact that in (7.41) the variable h^2 appears, which expresses the ratio of the power of one pulse to the spectral power density of noise. At the same time, during practical measurements on radio links, and also during their calculation, the signal-to-noise ratio, or P_s/P_n, is often used. These ratios are connected by the relation $P_s/P_n = h^2/FT$, where F is the passband of a signal path at the demodulator input, in which the ratio P_s/P_n is measured. When $F = 1/T$ (called a "matched band") these ratios coincide. Given in Table 7.6 as an example are the necessary signal-to-noise ratios for different keying ratios, information transmission rates and passbands of the linear part of a receiver.

TABLE 7.6

V, Mb/s	20			40			60		
m, multiplicity factor	1	2	3	1	2	3	1	2	3
Vch, Mb/s	20	10	6.7	40	20	13.3	60	30	20
T, ns	50	100	150	25	50	75	16.67	33.3	50
$F = 0.8/T$, MHz	16	8	5.4	32	16	10.6	48	24	16
P_s/P_n for p equal									
10^{-4}	9.8	12.8	17.8	9.8	12.8	17.8	9.8	12.8	17.8
10^{-5}	10.8	13.8	18.8	10.8	13.8	18.8	10.8	13.8	18.8
10^{-6}	11.8	14.8	19.5	11.8	14.8	19.5	11.8	14.8	19.5
$F = 1/T$, MHz	20	10	6.7	40	20	13.3	60	30	20
P_s/P_n for p equal									
10^{-4}	8.8	11.8	16.8	8.8	11.8	16.7	8.8	11.8	16.8
10^{-5}	9.8	12.8	17.8	9.8	12.8	17.8	9.8	12.8	17.8
10^{-6}	10.8	13.8	18.5	10.8	13.8	18.5	10.8	13.8	18.5
$F = 1.2/T$, MHz	24	12	8	48	24	16	72	36	24
P_s/P_n for p equal									
10^{-4}	8.0	11.0	16.0	8.0	11.0	16.0	8.0	11.0	16.0
10^{-5}	9.0	12.0	17.0	9.0	12.0	17.0	9.0	12.0	17.0
10^{-6}	10.0	13.0	17.7	10.0	13.0	17.7	10.0	13.0	17.7
$F = 1.35/T$, MHz	27	13.5	9.2	54	27	18	81	40.5	27
P_s/P_n for p equal									
10^{-4}	7.5	10.5	15.5	7.5	10.5	15.5	7.5	10.5	15.5
10^{-5}	8.5	11.5	16.5	8.5	11.5	16.5	8.5	11.5	16.5
10^{-6}	9.5	12.5	17.2	9.5	12.5	17.2	9.5	12.5	17.2

7.8 ANALYSIS OF THE INFLUENCE OF THE CHARACTERISTICS OF A SIGNAL PATH ON THE NOISE IMMUNITY OF TIME DIVISION MULTIPLE ACCESS SYSTEMS

It is known from an analysis of the influence of characteristics of the signal path of satellite links on the distortions of PSK signals that limitation of the passband of a signal path and additional deviations of its amplitude-frequency and phase-frequency responses from the ideal responses (uniform and linear, respectively) results in the appearance of between-symbol distortions. Between-symbol distortions can be estimated by the equivalent power losses in comparison with the noise immunity, determined by the theoretically calculated curves, given above [7.18].

The frequency response of an HF signal path can be represented as

$$\phi_{HF}(i\omega = (1 + b_1\Omega + b_2\Omega^2 + \cdots) \exp[i(a_1\Omega + a_2\Omega^2 + a_3\Omega^3 + \cdots)] \quad (7.42)$$

where $\Omega = \omega - \omega_0$; ω_0 is the central frequency of the signal spectrum. The coefficient b_1 expresses the bias of the amplitude-frequency response of a signal path, and b_2 expresses the parabolic component of the amplitude-frequency response; a_1 is a coefficient that determines the delay of the signal, and a_2 and a_3 are the linear and parabolic components, respectively of the group delay time.

The modulator input signal is

$$S_m(t) = \sum_{k=-\infty}^{k=\infty} C_k$$

where C_k through clock interval T randomly acquires the values ± 1, keeping them constant in time intervals $t_{k-1} < t \leqslant t_{k-1} + T$. The demodulator input signal in the case of single PSK is:

$$S_d(t) = \frac{\exp(i\omega_0 t)}{2\pi} \int_{-\infty}^{\infty} E(\Omega)\Phi_{HF}\,[i(\omega_0 + \Omega)] \exp(i\Omega t)\, d\Omega \quad (7.43)$$

where $E(\Omega) = \int_{-\infty}^{\infty} h(t)\exp(i\Omega t)\, dt$ is the spectral density of the envelope of PSK signal $h(t)$ after passing through the filters of the modulator and demodulator.

Trivially,

$$h(t) = \int_0^t S_m(\tau)g(t - \tau)\, d\tau \quad (7.44)$$

where $g(t - \tau)$ is the pulse response of modem filters, determined as

$$g(t) = \frac{1}{2\pi} \int_{-\infty}^{\infty} \Phi_m(i\omega)\Phi_d(i\omega) \exp(i\omega t)\, d\omega \quad (7.45)$$

$\Phi_m(i\omega)$ and $\Phi_d(i\omega)$ are the frequency responses of the modulator and demodulator.

It is assumed in the above analysis that the signal path of a communications link between the output filter of the modulator and input filters of the demodulator does not contain nonlinear elements, so that the generalized pulse response of both filters can be sought by multiplying their frequency responses. Response (7.42) of a signal path may be written as

$$\Phi_{HF}(i\omega) \approx \exp(ia_1\Omega)\,(1 + b_1\Omega + b_2\Omega^2 + \cdots)\,[1 + i(a_2\Omega^2 + a_3\Omega^3 + \cdots)] \tag{7.46}$$

If the responses of a signal path are not very different from the ideal and the distortions that are inserted by the path are minor, then only the first terms need to be retained in (7.46), whereupon

$$\Phi_{HF}(i\omega) \approx \exp(ia_1\Omega)\,[1 - ib_1(i\Omega) - b_2(i\Omega)^2 - 1a_2(i\Omega)^2 - a_3(i\Omega)^3 + \cdots] \tag{7.47}$$

Because the operation $(i\Omega)^k$ corresponds to the kth derivative of the envelope of PSK signal $h(t)$, then, in consideration of (7.44) and (7.45), it is possible to derive the following expression for the demodulator input signal:

$$S_d(t) = [h(t) - b_2h''(t) - a_3h'''(t)]\sin\omega_0 t + [-b_1h'(t) - a_2h''(t)]\cos\omega_0 t \tag{7.48}$$

We note that delay a_1 in (7.48) is assumed equal to zero, because it inserts only delay and does not cause distortions.

Expression (7.48) determines the distortions of the demodulator input signal as a function of the amplitude-frequency response and group delay time. It follows from this expression that the influence of the characteristics of a signal path is manifested in two ways: additional between-symbol distortions appear in a signal (b_2h'' and a_3h''), and an orthogonal component appears ($+b_1h'$ and a_2h'').

Calculations of the power losses, caused by the imperfect characteristics of a signal path for the case when the equivalent cumulative filter of the modem may be represented as a critically coupled second-order filter, follow. It is known from the work [7.19] that when the carrier signal jumps by π, the envelope of the output process is

$$h(t) = -\{1 - 2\exp(-t/\tau_k)[\sin(t/\tau_k) + \cos(t/\tau_k)]\}, \quad t \geqslant t_0$$

where τ_k is the time constant of the filter, and t_0 is the time of the phase jump at the filter input. If the filter passband is $\Pi = 1/T$, then $\tau_k = T\sqrt{2}/\pi$.

Under the assumptions made above, the transient process at the filter output decays rapidly when $t > 2T$, so that between-symbol distortions may be taken into consideration only in interval $2T$. Given in Table 7.7 are the relative values of the first three derivatives at sample times $t_k = t_0 + kT$, where $k = 1, 2, 3$ for a meander.

The linear component of the group delay time is attributed to the parabolic component of the phase response $a_2\Omega^2$. By definition of the group delay time, $\tau = 2a_2\Omega$. If the nonuniformity of the group delay time is normalized on the edges of the passband, and the deviation of the linear component of the group delay time on frequency $\omega_0 \pm 2\pi F$ is τ_{lin}, then $a_2 = \tau_{\text{lin}}/4\pi F$. The influence of the linear component of the group delay time is manifested as the appearance of an additional orthogonal component, which is most significant in double PSK. If one of the quadrature channels is assumed not to be keyed, i.e., $S(t) = h(t)\sin\omega_0 t + \cos\omega_0 t$, the error probability in this channel is

$$P_{\text{er}} = \left\langle \frac{1}{2}\left[1 - \Phi\left(\frac{1 - a_2 h''(t_k)}{\sqrt{2}\sigma}\right)\right]\right\rangle \tag{7.49}$$

where $< >$ is the sign of statistical averaging of the distribution of the random variable $h''(f_k)$ at the sample times, and σ^2 is the dispersion of noise.

TABLE 7.7

$h^{(k)}(kT)$	Sample time			$h^{(k)}(kT)$	Sample time		
	T	$2T$	$3T$		T	$2T$	$3T$
$h(t)$	−0.96	0.93	−0.93	$h''(t)\tau_k^2$	0.63	−0.66	0.66
$h'(t)\tau_k$	−0.36	0.41	−0.41	$h'''(t)\tau_k^3$	−0.51	0.47	−0.46

Shown in Figure 7.9(a) are additional losses for double PSK as functions of the linear component of the group delay time for the case when the nonuniformity of the group delay time is normalized in the band $F = +0.5/T$ and $P_{\text{er}} = 10^{-6}$. The parabolic component $3a_3\Omega^2$ of the group delay time leads to the appearance of between-symbol distortions. When the nonuniformity of the parabolic component τ_p is given in the band $F = 0.5/T$, it is possible to find $a_3 = \tau_p(2T)^2/3(2\pi)^2$. Shown in Figure 7.9(b) is the influence of the parabolic component of the nonuniformity of the group delay time on additional energy losses under the assumptions made above. The influence of the bias of the amplitude-frequency response can be calculated in the same way, and the result of the calculation is given in Figure 7.9(c), where bias ϵ of the amplitude-frequency response in the band $F \pm 0.5/T$ is plotted on the abscissa axis.

The information given above on the calculation of power losses, attributed to the characteristics of a signal path, are based on the prerequisite of perfect synchronization of the demodulator, both on the carrier, and on the clock frequency. At the same time, studies, conducted by V. M. Dorofeev, M. L. Payanskaya, and M. D. Grebel'skij, disclosed that the influence of the imperfection of the characteristics of a signal path produces an additional phase error in the reference signal regenerator, which creates additional power losses, commensurable with those inserted by the information channel itself.

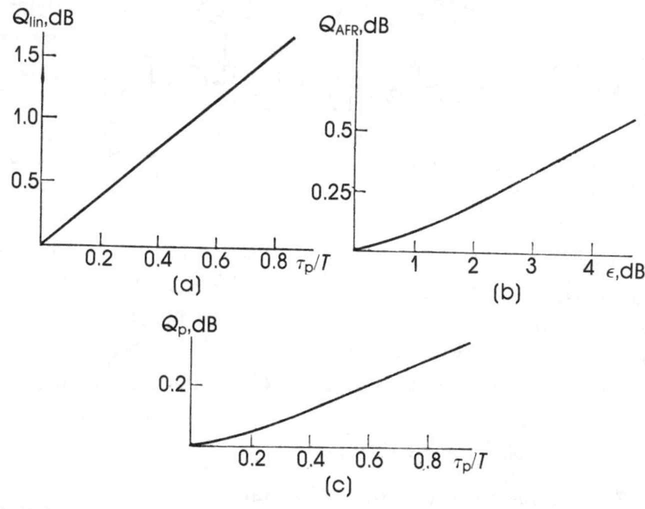

FIGURE 7.9. **Losses of noise immunity as functions of: (a) linear component of group delay time; (b) parabolic component of group delay time; (c) nonuniformity (bias) of amplitude-frequency response.**

7.9 CALCULATION OF THE NOISE IMMUNITY OF TIME DIVISION MULTIPLE ACCESS SYSTEMS BY ELECTRONIC COMPUTER

The inability to express the actual characteristics of a signal path through simple enough formulas, and the extreme mathematical difficulty of taking into consideration nonlinear effects, caused both by the nonlinearity of the amplitude response of some of the elements of a signal path, especially the power stage of a satellite repeater, and by the fact that the phase shift of the output signal of a repeater and power transmitter of an earth station depends on the output signal (AM-PM conversion), prevent the exact analytical calculation of losses, caused by the imperfection of the parameters of a signal path. The exact consideration of all these factors is possible only with the use of an electronic computer. To analyze distortions, inserted by an actual signal path, it is necessary to choose a model of a channel, equivalent to a satellite link. A diagram of a real satellite channel is shown in Figure 7.10. It includes transmitting and receiving filters of the earth station and repeater F1, F2, F3, F4, nonlinear amplifiers of the transmitters of the repeater and earth station, modulator and demodulator. Additive Gaussian white noise acts upon the signal path at two points: at the inputs of the receivers of the repeater and earth station.

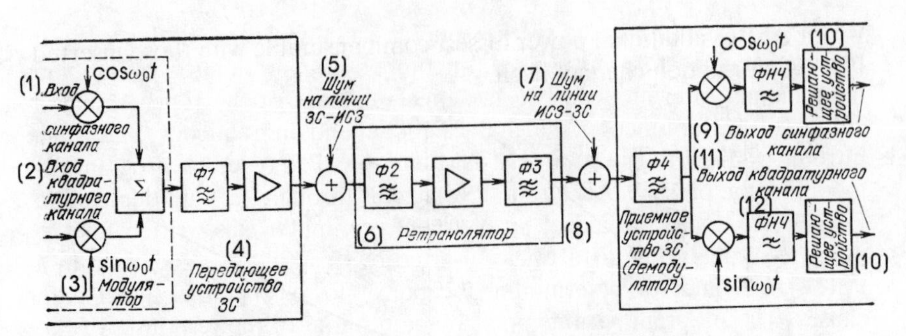

Key:
(1) In-phase channel input
(2) Quadrature channel input
(3) Modulator
(4) Earth station transmitter
(5) Noise in earth-satellite link
(6) Transponder
(7) Noise in satellite-earth link
(8) Earth station receiver (demodulator)
(9) In-phase channel output
(10) Computer
(11) Quadrature channel output
(12) Low-pass filter

FIGURE 7.10. Functional diagram of signal path of TDMA system.

Considering that the energy potential of an earth station-satellite link usually is considerably greater than that of a satellite-earth station link, and the transmitter of an earth station usually can be used in a nearly linear mode, the nonlinearity of the transmitter of the earth station and the repeater input noise may be ignored. In this case, the section of the link between the modulator output and the input of the nonlinear element may be assumed linear and may be represented by some equivalent transmitting filter. The section of the link between the output of the nonlinear element of the repeater and the demodulator input can be represented in exactly the same way by an equivalent receiving filter. It is fairly convenient to model, with the aid of an electronic computer, precisely this kind of model of a satellite link, sacrificing only a little calculation accuracy in comparison with the modeling of a complete real signal path. The transmitting and receiving filters cause linear intersymbol interference, and the combined nonlinear element of the repeater [inserting distortions of the AM-AM type (a nonlinear amplitude response), and of the AM-PM type (a conversion of amplitude modulation to phase modulation] causes power losses to out-of-band products and nonlinear intersymbol interference. The low-frequency equivalent of a satellite link of TDMA systems is depicted in Figure 7.11. The combined transfer functions $M_{\text{tr}}(\omega)$ and $M_{\text{rec}}(\omega)$ correspond to the transmitting and receiving filters in in-phase and quadrature channels, and the filters with the combined transfer functions $\pm N_{\text{tr}}(\omega)$ and $\pm N_{\text{rec}}(\omega)$ characterize crosstalk interference from one channel to another. The transversal corrector, the use of which often is justified in an actual signal path, is represented in the general case by a combination of four transversal correctors and is used for compensating both intersymbol, and interchannel interference. The nonlinear element of the repeater has a complex transfer function, which does not depend on frequency, but which introduces distortions of the AM-AM and AM-PM types. A program for modeling the passage of multiple PSK signals through the path depicted in Figure 7.11* with the aid of an electronic computer was written for the purpose of analyzing distortions, inserted by this kind of signal path in a PSK signal, and for determining the power losses that this signal path causes. The program permits point-by-point input of the responses of the transmitting and receiving filters, measured on an intermediate frequency, and automatic conversion of them to a low-frequency equivalent, and all the computations are carried out on a discrete time channel. The responses of the nonlinear element also are put in point by point.

In order to develop the discrete time channel, it is necessary to replace all of its existing analog filters with their digital equivalents, and to find the optimum sample time the time digitization frequency should be several times higher than the clock frequency. If $F_c = 1/T$ is the clock frequency of the transmission of information in a signal path, then the digitization frequency is $F_d = m_d F_c$, where m_d is the number of samples per symbol. m_d should be large enough, first, for guaranteeing the necessary calculation precision and, second, for virtually precluding the possibility of overlapping adjacent frequency responses, the danger of which arises when an analog filter is converted to a digital filter with the frequency response repeated periodically on the frequency axis [7.20]. A random number generator, which generates a pseudorandom sequence of sufficient length, is used as a simulator of the information source. The filter output signal is computed by means of convolution of the input signal by a filter with a digital pulse response with number $N_{ef} = N_f/m_d$ of sampling points. The pulse response is determined by a discrete Fourier transform with the aid of a standard fast Fourier transform subroutine [7.20]. The number of samples of the response N_f of the filter is selected such as to provide sufficient attenuation of the response. The gains at the taps of the transversal corrector are selected such as to assure zero between-symbol and between-channel interference within N_f symbols.

Input and output data of the program, written in Fortan IV language, are given in Table 7.8. The basic subroutines, used in the program, and a brief description of them, are given in Table 7.9. A flow chart of the program is shown in Figure 7.12. The program permits the exclusion of any filter, transversal corrector and nonlinear element from the signal path

---

*The material was provided by S. L. Portnoj and D. R. Ankudinov.

during the modeling process. By modifying the program somewhat, it is possible to take into consideration the influence of synchronization circuits, and also the effect of using noise-immune coding. There are also methods of modeling distortions of a PSK signal in an actual signal path without replacing the analog responses of the filters with their digital equivalents on the basis of the successive use of a pair of Fourier transforms [7.20].

Both the results of modeling by the program described above, and results presented in the article [7.21], and also the results of experimental studies and an analysis, contained in the article [7.22], lead to the conclusion that the power losses attributed to the linear elements of a signal path are 1.5–2 dB. The writing of the mentioned modeling program made it possible to analyze the influence of a nonlinear element on additional power losses, and to determine the optimum working point of the power stage of a satellite repeater. Modeling and subsequent experimental investigations disclosed that, when the input PSK signal of a nonlinear element decreases, the cumulative losses first decrease, and then they increase again. The cumulative power losses (in consideration of linear elements) are given in Figure 7.13 as a function of the input signal level of a nonlinear element. When the input signal level of a nonlinear element is $U_s = 1$, the mean signal power determines the working point, located at the point of inflection of the amplitude response of a nonlinear element. Here the peak signal power at individual moments of time is in the saturation range. The relative input level of $U_s = 0.7$ corresponds to the minimum power losses ($\approx$ 2.5 dB). Numerous calculations and tests on actual links show that, when the parameters of the linear part are optimized and the optimum input signal level of a nonlinear element is chosen, the cumulative power losses cannot exceed 3 dB.

7.10 SYNCHRONIZATION IN A TIME DIVISION MULTIPLE ACCESS SYSTEM

The synchronization equipment of earth stations is one of the main elements of TDMA systems. It eliminates crosstalk interference between the signals of earth stations at the repeater input. This equipment also solves the problem of the initial synchronization of earth stations, i.e., it connects them to an already functioning synchronous system. To synchronize the operation of earth stations in one transponder of a satellite repeater, some of the carrying capacity of the transponder is reserved for the transmission of frame synchronization signals, which the earth stations of a TDMA network use for establishing the necessary time relations. The optimum frame synchronization methods provide the necessary synchronization reliability and precision at the minimum cost of carrying capacity.

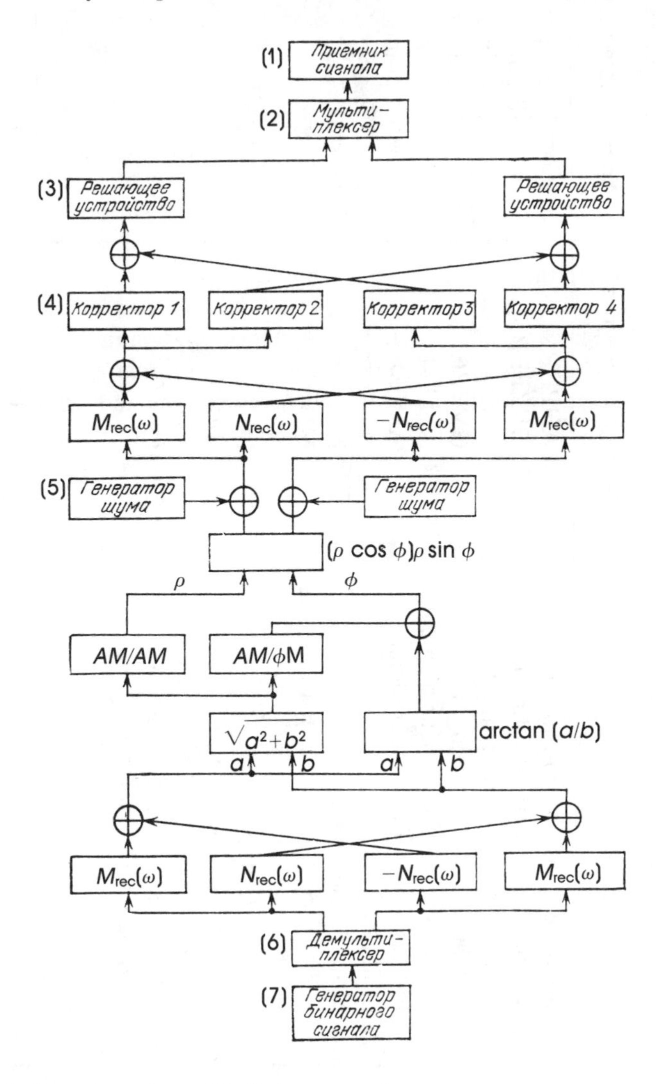

Key: (1) Signal receiver
(2) Multiplexer
(3) Computer
(4) Corrector
(5) Noise generator
(6) Demultiplexer
(7) Binary signal generator

FIGURE 7.11. Low-frequency equivalent of satellite link of TDMA system.

TABLE 7.8

	Parameter	Symbol in program	Approximate range
Input	Amplitude response of transmitting filter	AHPER	≤512 points
	Group delay time of transmitting filter	GVZPER	≤513 points
	Amplitude response of receiving filter	AHPR	≤512 points
	Group delay time of receiving filter	GVZPR	≤513 points
	AM-AM response of nonlinear element	AMAM	≤100 points
	AM-PM response of nonlinear element	AMFM	≤100 points
	Size of filter arrays	N	≤512
	Size of nonlinear converter arrays	N1	≤100
	Number of gradations per symbol in time	MD	≤8
	Length of signal	NY	$\leq 10^3$
	Phase-modulation factor	M	4–8
	Transponder input power loss	BACK	≤1
	Normal clock frequency (the ratio of the clock frequency to the passband of the receiving and transmitting filters at the 3-dB level)	FT/FPER FT/FPR	0.5–3
	Signal-to-noise ratio	SND	10–30 dB
Output	Power losses	POTER	>0

TABLE 7.9

Subroutine	*Brief description*
PTCHF	Subroutine converts group delay time arrays to phase characteristics of filters and generates combined array of filter characteristics
FDFT and DPF	Fast Fourier transform and ordinary Fourier transform of combined arrays
PEREUP	Subroutine rearranges combined pulse response of filter so that the element with the most active part has the index $(N/2 - 1)$
FORMIO	Subroutine makes multiple access pulse responses for different clock shifts of each filter and normalizes pulse response in accordance with unit energy
TRANSK	Pulse response of transversal corrector is found on basis of given combined pulse responses of transmitting and receiving filters
DATCH	Generates numbers 0 and 1 with equal probability
PSP	Generates pseudorandom sequence with length $\log_2 M$, where M is the number of phases, with the aid of DATCH subroutine
MODUL	Subroutine depends on phase modulation factor and generates NY complex numbers, the modulus of which is equal to $\sqrt{2}$ BACK and which acquire M values
SVERT	Input sequence with length NY with pulse response of digital filter is convoluted
NEL	Output signal of nonlinear element is determined with the aid of its combined response
OBRAT	Inverse probability integral subroutine
PRYAM	Probability integral subroutine
ANOSH	Power losses for given signal-to-noise ratio are found for the output signal of the receiving filter

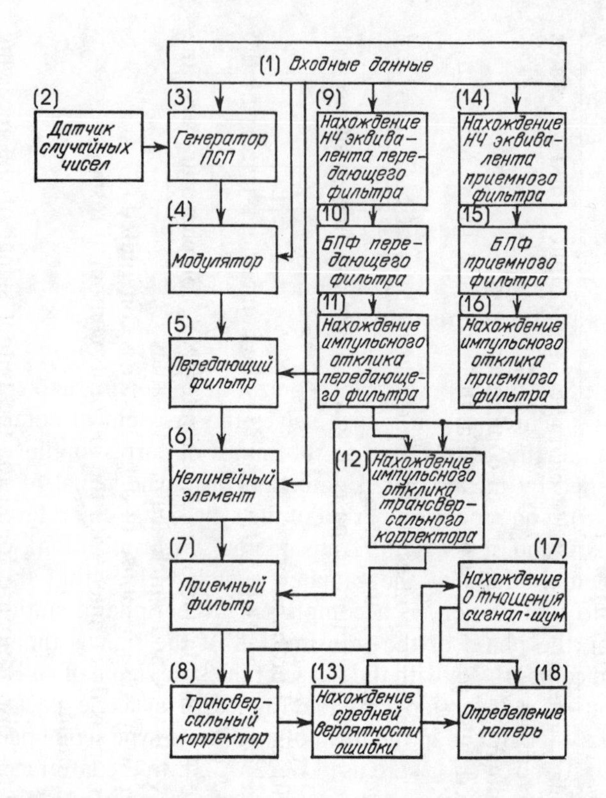

Key:
(1) Input Data
(2) Random number generator
(3) Pseudorandom sequence generator
(4) Modulator
(5) Transmitting filter
(6) Nonlinear elements
(7) Receiving filter
(8) Transversal corrector
(9) Low-frequency equivalent of transmitting filter found
(10) Fast Fourier transform of transmitting filter
(11) Pulse response of transmitting filter found
(12) Pulse response of transversal corrector found
(13) Mean error probability determined
(14) Low-frequency equivalent of receiving filter found
(15) Fast Fourier transform of receiving filter
(16) Pulse response of receiving filter found
(17) Signal-to-noise ratio determined
(18) Losses determined

FIGURE 7.12. Flow chart of program that models signal path of TDMA system (the numbers correspond to number of subroutine).

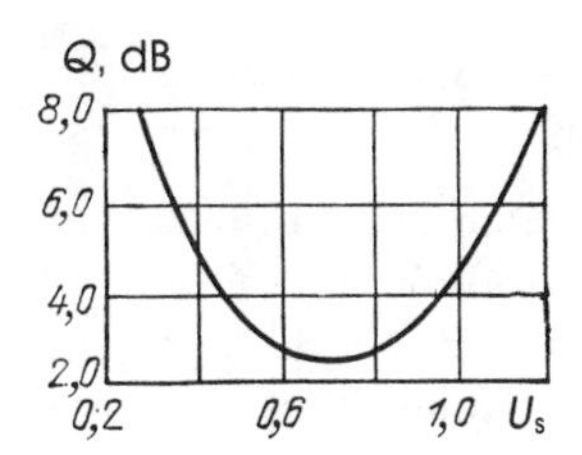

FIGURE 7.13. Power losses as function of choice of working point of traveling wave tube of repeater.

There are two types of synchronization: receiving and transmitting. Receiving synchronization, which solves the problem of determining the time intervals that correspond to subframes of corresponding stations, is accomplished by means of the detection of a sync signal of a reference station (RS). The repetition period of the sync signal of a reference station assigns the frame of a system. Transmission synchronization, which solves the problem of holding the radiated signal itself within the subframe assigned to that station, is accomplished by peripheral stations (PS) by controlling the phase of the transmission of the signals themselves in a frame and comparing it with the phase of the sync signal of an RS. Both the generation of a sync signal in the frames of a single packet with the information signal, and in the form of a separate sync signal packet (called synchronization selection) are used [7.22, 7.23]. In the latter case, the sync signals of all earth stations are transmitted in a frame in fixed time positions apart from the information packets. The structure and duration of frame sync signals are constant, whereas the position and duration of information packets can change in accordance with the traffic of an earth station.

When selected frame synchronization is used, the frame efficiency of a system obviously decreases a little [7.23] in view of the need to transmit sync signals and the preamble of the demodulator, both in a sync packet and in an information packet. It is possible here, however, to separate completely the functions of the system and components of the equipment, which offers significant advantages during operation.

Entry into synchronism is accomplished by means of the initial connection of stations to the network after an interruption of communications. The following requirements are imposed on the entry procedure:

- the entry signal must have the minimum effect on the information signals of operating earth stations;
- the information signals must have the minimum effect on the quality of the reception of the entry signal;
- entry into synchronism must be quick.

The following methods for entering into synchronism are used in different TDMA systems:

1. a low-level entry signal is transmitted during the time of establishment of transmission frame synchronization;
2. the time positions of peripheral earth stations are predicted on the basis of results of calculation of the trajectory of the geostationary satellite;
3. pulse entry as a result of the one-time transmission of a short sync signal (with a duration of one or a few frames) of the nominal level, followed by phase correction of the transmitting frame signal generator.

The first method, which creates the minimum interference with other stations, has a significant entry time of several seconds. The second method requires a special network of measurement stations and has high satellite slant range measurement precision. Nevertheless, the precision with which the distance to a satellite is measured, sufficient for the entry mode in a TDMA network, can be achieved only when satellites are held in orbit with high precision. The third method, which creates brief interference with all the stations that are operating in a network, is the easiest to implement from the standpoint of hardware, and is exceedingly convenient when the entry procedure is rarely needed. The most detailed information on synchronization systems can be found in the already-mentioned work [7.23], which also contains a voluminous bibliography.

Chapter 8
Design Features of Receiving Equipment of Earth Stations

8.1 INPUT RECEIVERS

The limit on satellite communication EIRP requires that the sensitivity of earth stations be maximized. One way to do this is to use low-noise input equipment. This equipment, being a characteristic feature of earth stations, is described in this chapter. The basic design principles of individual elements of receiving-transmitting equipment, just like an IF signal path, modulators and demodulators, do not have specific features, and a description of them can be found in numerous works that have been published (see the articles [8.3, 8.5, 19]). The parameter that characterizes the sensitivity of input receivers is the effective noise temperature T_{rec}, which is connected with the noise coefficient n of a receiver by the relation $T_{rec} = (n - 1)T$, where T is the effective noise temperature of the signal source under normal conditions, $T \approx 300$ K.

Different types of low-noise amplifiers are used in the equipment of satellite communications earth stations, of which the main ones are the following: parametric amplifiers, cooled and uncooled, transistor amplifiers, tunnel diode amplifiers, cascade amplifiers (amplifiers with several series-connected stages). The noise characteristics of certain types of low-noise amplifiers are given in Figure 8.1 as functions of frequency. The choice of the necessary type of low-noise amplifier (along with noise characteristics) is also determined by other electrical parameters: the passband, necessary tuning range, performance stability, saturation level, energy input, and also cost, size, and weight.

Basic Requirements on Quiet Amplifiers

1. The amplification band should be maximum (for amplifying signals of several transponder at the same time). The amplification band in the

decimeter range, for example, in many cases can be expanded to an octave, and in a solid state version of a low-noise amplifier, the band reaches 15–20% relative to the middle frequency.

2. The noise temperature of an amplifier should be minimal.
3. The gain of a low-noise amplifier should be sufficient for the effective suppression of the noises of amplifier-converter stages that follow it. As is known [8.1], the noise coefficient of a receiver is determined with acceptable accuracy by the noises of the first two stages of the input receiver:

$$n_{\text{rec}} = n_{\text{q.a}} + (n_{2\text{s}} - 1)/K_{\text{g}}$$

where $n_{\text{q.a}}$ is the noise coefficient of a low-noise amplifier; K_{g} is the gain of a quiet amplifier; and $n_{2\text{s}}$ is the noise coefficient of the second stage. When K_{g} is large enough, the second term of the equation may be ignored. Then n_{rec} will be determined only by the noise of a low-noise amplifier.

4. Low-noise amplifiers should have a high enough saturation level, otherwise crosstalk distortions can occur, and the receiver even can become saturated. Usually the saturation level in modern low-noise amplifiers falls in the range of −40–70 dBW.
5. Low-noise amplifiers should have linear amplitude and uniform phase responses.

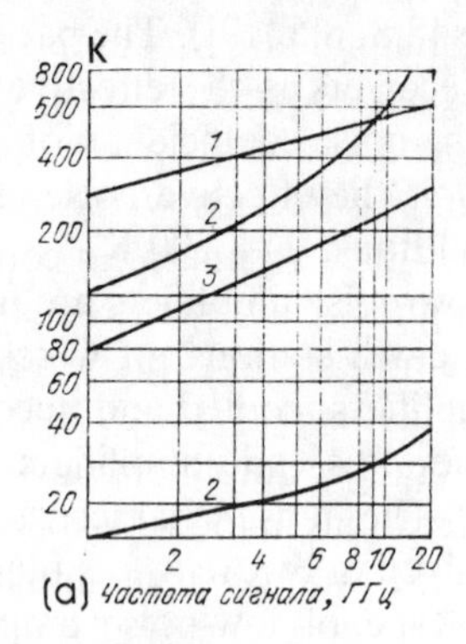

Key: **(a) Frequency of signal, GHz**

FIGURE 8.1. Noise characteristics of certain types of low-noise amplifiers as functions of frequency: 1-transistor amplifiers; 2-tunnel diode amplifiers; 3-parametric amplifiers.

The basic parameters of different types of low-noise amplifiers, used in the equipment of satellite communications earth stations, are listed in Table 8.1. We will examine these amplifiers.

Parametric amplifiers belong to a class of instruments, in which the negative impedance effect is used for amplification. There are different kinds of parametric amplifiers at the present time, with different designs and parametric (reactive) elements, which may be capacitors and induction coils. Parametric (varactor) diode amplifiers, which have the properties of nonlinear capacitance and which change their reactance at the cost of external power sources, are most often used in the equipment of satellite communications earth stations. Because purely reactive elements do not produce noises, the parametric amplifiers that are based on them generate low set noises, to 100–150 K.

In terms of design parametric amplifiers are classes as single-loop, double-loop and traveling wave amplifiers.

Double-loop amplifiers, in turn, may and may not have a frequency converter.

Single-loop parametric amplifiers generate rather low noises in the synchronous mode, but in the case of synchronous operation noises get into the system, both on the signal frequency, and on the different frequency, which increases the noise temperature considerably. Because it is practically impossible to guarantee synchronous operation, it is better to use double-loop parametric amplifiers.

A diagram that explains the operating principle of a double-loop parametric amplifier is given in Figure 8.2. In this amplifier, two resonators are tuned, one to the signal frequency (working), and the other to the difference (idle) frequency, and a varicap serves as a coupler. Additional gain can be achieved, and consequently the noise temperature can be reduced by connecting an "idle" resonator in combination with the signal circuit. Double-loop amplifiers can operate in the frequency conversion mode. In this case, the output signal is taken from a difference loop and there is no need for a circulator.

Until recently, a parametric amplifier-converter was used extensively in the equipment of earth stations in the 800–1000-MHz band at the Orbita station for operation with the Molniya-1 satellite [8.2]. A typical schematic diagram of this kind of parametric amplifier is shown in Figure 8.3. Here the signal with frequency f_s, received by the antenna, passes through an input bandpass filter and a ferrite valve to the signal frequency loop, coupled with a difference frequency circuit through the nonlinear capacitor of a parametric diode. At the same time, pumping oscillations with frequency f_p from the pumping generator are supplied to the parametric

TABLE 8.1

Type of amplifier	Basic parameters of low-noise amplifiers for range							
	Decimeter (0.3–3) GHz				Centimeter (3–30) GHz			
	Gain, dB	*Passband, %, of carrier*	*Noise coeffi-cient, dB*	*Noise temper-ature, K*	*Gain, dB*	*Passband, %, of carrier*	*Noise coeffi-cient, dB*	*Noise temper-ature, K*
Parametric amplifiers with semiconductor diodes at temperature, K:								
290	25	3–5	1.8–3	150–290	20	1–2	2.7	250
77	25	3–5	1–1.8	75–150	20	1–2	1.07	80
Tunnel diode amplifier	15	2–3	4–6	435–865	15	1–1.5	7–9	1160–2000
HF transistor amplifier	5	3–5	5	625	4	3	7	1160
Balanced SHF mixer	—	—	5–8	623–1540	—	—	510	625–2600

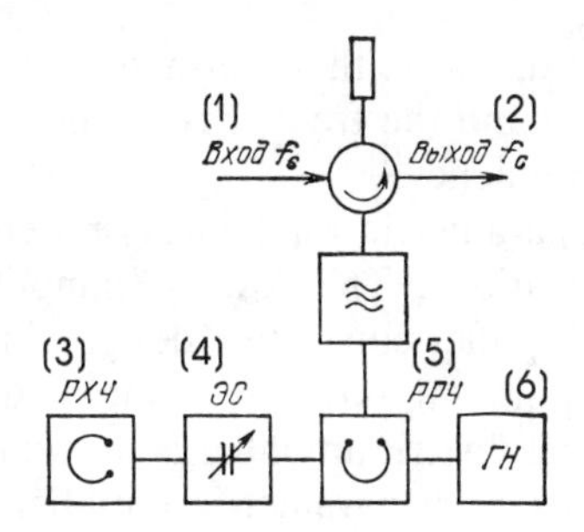

Key: (1) Input (4) Coupler
(2) Output (5) Working frequency resonator
(3) Idle frequency resonator (6) Pumping generator

FIGURE 8.2. Schematic diagram of double-loop parametric amplifier.

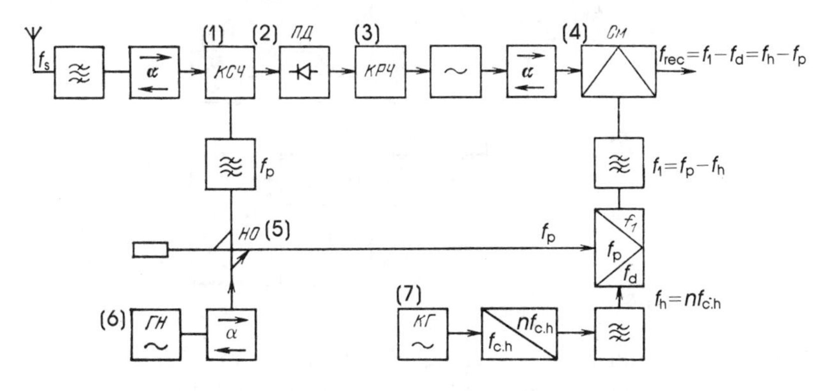

Key: (1) Signal frequency loop (5) Directional coupler
(2) Parametric diode (6) Pumping generator
(3) Difference frequency circuit (7) Crystal heterodyne
(4) Mixer

FIGURE 8.3. Schematic diagram of parametric amplifier-converter.

diode through a ferrite valve, directional coupler, and bandpass filter. The amplified oscillations of difference frequency $f_d = f_p - f_s$ from the difference frequency circuit pass through the filter and ferrite valve to the mixer, in which frequency f_{con} is reconverted to intermediate frequency f_{IF}. The same mixer receives oscillations with frequency f_1 from a shift mixer, where pumping frequency f_p is converted to frequency f_1, shifted relative to f_p by the amount equal to frequency f_h of the crystal heterodyne, i.e., $f_1 = f_p$

$- f_h$. When converted this way, the intermediate frequency depends only on the signal frequency and the frequency of the crystal heterodyne, and not on the pumping frequency.

Primarily, reflex parametric amplifiers are used for designing modern equipment of earth stations. This kind of amplifier has a number of advantages over an amplifier-converter: design simplicity, a lower noise temperature and less pumping power. A simplified schematic diagram of an input with a reflex parametric amplifier is shown in Figure 8.4. Here the output power is taken from the parametric amplifier itself not on difference frequency f_d, but on signal frequency f_s, and the amplified signal goes into the mixer of the receiver (signal frequency loop and difference frequency circuit are the signal and difference frequency loops).

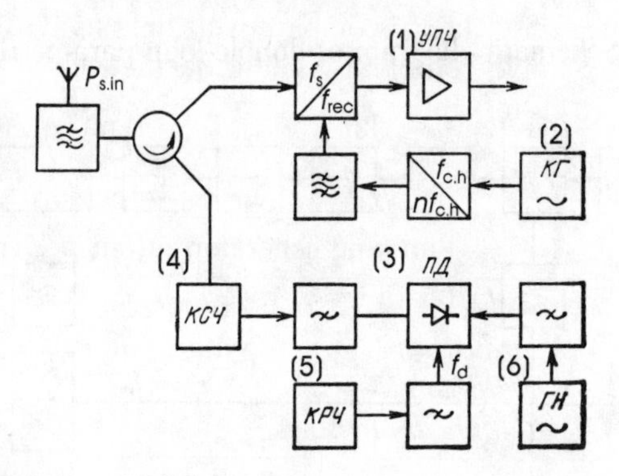

Key: (1) IF amplifier (4) Signal frequency loop
(2) Crystal heterodyne (5) Difference frequency loop
(3) Parametric diode (6) Pumping generator

FIGURE 8.4. Simplified schematic diagram of input with reflex parametric amplifier.

Pumping generators are one of the main elements of a parametric amplifier. Their frequency f_p is 10 to 15 times higher than signal frequency f_s. For a decimeter parametric amplifier, for example, the pumping frequency is 10–12 GHz. Both vacuum super high frequency (SHF) instruments (klystrons and magnetrons), and solid state instruments (Gunn diodes, avalanche transit diodes or transistor generators with multiple frequency multiplication) are used as pumping generators. Reflex klystrons sometimes are used as these generators. However, they have a number of drawbacks: frequency and power are strong functions of the power voltages, there is a long nomenclature of these voltages, thermal instability, difficult to operate, high cost, *et cetera*. Special sophisticated automatic frequency control systems, and sometimes thermostats have to be used.

Magnetron pumps are unreliable in operation and usually have a short service life. Their parameters change significantly as time goes by and as the ambient temperature changes.

Crystal transistor generators that utilize frequency multiplication, avalanche transit diode generators, and generators based on the Gunn effect are considered promising. Inexpensive, reliable and easy to operate pumps can be made with solid state SHF generators. Also most suitable are Gunn diode generators (for example AA 703B) because of their frequency stability and performance reliability, and transistor generators that utilize multiple frequency multiplication. The latter type has high stability, stable parameters, long service life, is comparatively light, and uses comparatively little power. Sample comparative data on the output power of these generators are given in Table 8.2.

TABLE 8.2

Frequency, GHz	**Pumping generator output power, W**		
	with crystal transistor utilizing multiplication	*with avalanche transit diode*	*with Gunn diodes*
10	2	1	0.1
20	0.3	0.2	0.01

An example of a typical schematic diagram of a solid state pumping generator that utilizes multiple frequency multiplication for a decimeter parametric amplifier is shown in Figure 8.5. The multiplication factor is determined by the carrier frequency and by the necessary pumping frequency.

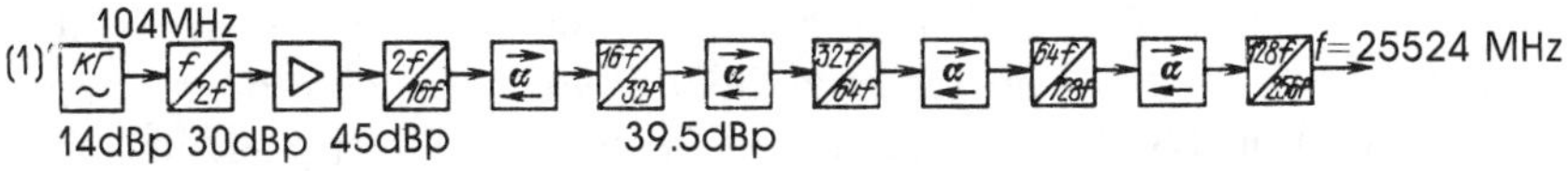

Key: **(1) Crystal heterodyne**

FIGURE 8.5. Typical schematic diagram of solid state pump utilizing frequency multiplication.

A parametric diode needs a power of from 10 to 100 mW from a pump. Varactor diodes made of gallium arsenide are used most widely as parametric diodes, although up until recently they were also made on a silicon base. Basically diffused (on the basis of mesostructure), welded and point varactor diodes were manufactured in the initial stages. Schottky barrier diodes with better parameters began to be manufactured recently.

Parametric amplifiers have the best noise properties of all SHF amplifiers, working without forced cooling. However, in some cases, for example at Orbita stations, it turned out to be advantageous to lower the noise temperature of parametric amplifiers additionally by cooling the diode and other elements to the liquid nitrogen point. For example [8.3], three-stage cooled parametric amplifiers will be used at some of the Orbita-2 earth stations. These parametric amplifiers have the following parameters: gain about 30 dB, range 4 GHz, passband 600 MHz, pumping frequency 20.5–22.5 GHz; pump power 20–60 mV. These amplifiers, when cooled to 18 K, have a noise temperature of about 29 K. A parametric amplifier is cooled by placing a cavity resonator with a parametric diode in a cryostat, filled with liquid nitrogen (T = 77 K). Curves that characterize parametric amplifiers with different properties are given in Figure 8.6.

Low-noise SHF transistor amplifiers have begun to be used extensively recently in the equipment of satellite communications and TV earth stations. For example, transistor amplifiers with a noise temperature of 500–800° in the decimeter range are used in the receivers of the Ehkran TV broadcasting system. The development of new types of transistors, comparable in terms of efficiency with tunnel diode amplifiers and traveling wave tubes, made it possible to begin the adoption of even higher bands. Now up to 4-GHz and even 8-GHz transistor amplifiers are being used at earth stations as input low-noise amplifiers. Nevertheless they are used as the first stages of low-noise amplifiers only in the 1–3-GHz band, and in higher bands, for example 4 GHz, they are connected after the parametric amplifier.

As a rule, the minimum noise coefficient of a multistage transistor amplifier is achieved when each of its stages is built as a common emitter circuit. In order to reduce the set noises in a multistage transistor amplifier, it is necessary to satisfactorily match the total input and output impedances of adjacent stages. The following are important properties of transistor amplifiers: capability of operating without a circulator, ease of matching with conventional SHF oscillator circuits, integrated and strip circuits, and high reliability.

A sample diagram of a transistor amplifier with coupled stages on semiconcentrated elements [8.4] is shown in Figure 8.7. Here the signal to be amplified is selected on the collector of tansistor T1 with the aid of a

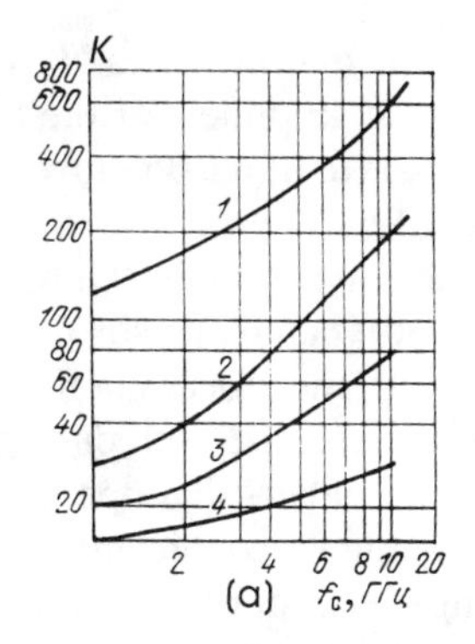

Key: (a) f_s, GHz

FIGURE 8.6. Curves of noise properties of different parametric amplifiers: 1-ordinary; 2-better uncooled; 3-better cooled to 77 K; 4-better cooled to 20 K.

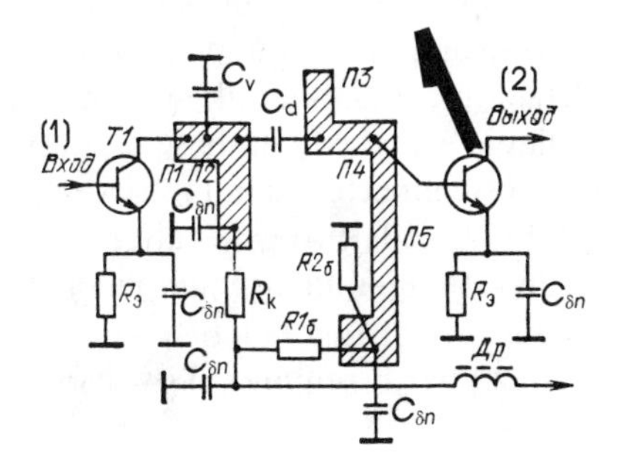

Key: (1) Input
(2) Output

FIGURE 8.7. Diagram of transistor amplifier.

circuit, made by shorted quarter-wave strip lines St1 and St2 and varicap C_v. Then the signal passes through dividing capacitor C_d to the base circuit of transistor T2. This circuit consists of high-impedance quarter-wave strip line St5 and special strip line St4 with a shorted loop St3, which make an equivalent circuit, tuned to the upper frequencies of the working band.

Low-noise three-stage transistor amplifiers, operating on frequencies of 700 and 1000 MHz, of integrated design on KT329A and KT341A transistors, have a gain of 18–25 dB (depending on frequency) with a noise coefficient n of about 3.5 dB and an up to 50 MHz band. The maximum attainable gain of this kind of amplifier for the SHF band is $K_g = (f_{max}/f)^2$, i.e., the gain decreases at a rate of 6 dB/octave. Here f_{max} is the frequency on which the gain is equal to 1.

Tunnel diode amplifiers are comparatively wide band amplifiers; their passband at a gain of 15–17 dB can reach 40%, and they make it possible to improve by at least 3–4 dB the noise coefficient of modern low-noise amplifiers without adding weight, size and, energy consumption. However, the design of tunnel diode amplifiers that operate in the centimeter band is limited by the state of the art noise coefficient and by the great scattering of parameters. Their strong dependence on ambient conditions and the difficulty of guaranteeing stable operation are significant deficiencies of tunnel diode amplifiers.

It is advantageous to use tunnel diode amplifiers as the second stage after a parametric amplifier, and also in cases when weight and size are the determining factors for moderate noise coefficients (6–10 dB), narrow dynamic range, *et cetera*. This applies to all transportable and mobile earth stations, and also to the space equipment of satellites.

Combined low-noise amplifiers found application in the earth station equipment of the Orbita and Intersputnik systems. For instance, a four-stage low-noise amplifier is being used at these earth stations at the present time, the first two series-connected parametric amplifiers of which are cryogenically cooled by liquid nitrogen, after which additionally are connected two more uncooled parametric amplifier stages.

Basic Parameters of a Combined Low-Noise Amplifier	
Passband, MHz	250
Gain, dB	40
Noise temperature, K	80–90

Two other types of combined low-noise amplifiers have begun to be used recently at earth stations [8.5]: the MShU-30 [low-noise amplifier] with cooled parametric amplifiers, and the MShU-60 with uncooled parametric amplifiers. The MShU-30 consists of three stages: two liquid nitrogen-cooled parametric amplifier stages with a subsequent (third) transistor stage. The amplifier is designed as follows: a coaxial waveguide using coaxial waveguide circulators and gallium arsenide diodes, based on the Schottky barrier effect. The pumps operate on Gunn diodes and on frequencies of about 33 GHz with an automatic power control system. This amplifier is a little inferior in terms of parameters to low-noise amplifiers that utilize cooling to the liquid helium point with T_n = 16–20 K, but then they have a simple cooling system, are inexpensive and, importantly, they are simple and reliable in operation. The cooled stages of a parametric amplifier are installed in a special case, which is immersed in a liquid nitrogen cryostat. The pumps are placed in a thermostat. The power, control, and monitor units are installed in a separate rack.

Basic Parameters of MShU-30 Low-Noise Amplifier	
Passband at the 1-dB level, MHz, not less than	500
Gain of two stages, dB	30
Gain with transistor amplifier, dB	30–50
T_n on output flange, K	30

The MShU-60 low-noise amplifier is an uncooled amplifier, consisting of a thermostatted and airtight parametric amplifier unit and a power unit. The parametric amplifier unit has one or two parametric amplifier stages and a transistor amplifier. As in the previous case, gallium arsenide Schottky barrier diodes are used in the parametric amplifier, and the pump is solid state and operates on a frequency of 50 GHz. Thermostatting is accomplished by using the Peltier effect, whereby heat is released or absorbed at the point of contact of two substances (a metal and a semiconductor) as electric current passes through the contact. When the circuit is closed, one of the contacts is heated and the other is cooled. When the current changes directions, the effect is reversed. The amount of heat that is released (or absorbed) is proportional to the electric charge on the contact. An important advantage of the MShU-60 low-noise amplifier is the fact that it can be installed not only in the equipment room, but also in the antenna shack, which can offer an additional gain in the G/T of a station, whereas the MShU-30 low-noise amplifier can be installed only in the equipment room.

Basic Parameters of MShU-60 Low-Noise Amplifier	
Passband, not less than, MHz	500
Gain, dB	35–40
T_n on output flange, K	60

Given in Table 8.3. as an example are the G/T ratios of earth stations using the last two versions of low-noise amplifiers for different installations. The G/T of an earth station is determined on 3.8 GHz with a 12 m antenna, effective area 0.7.

TABLE 8.3

Type of amplifier	*Noise temperature, K*	G/T[1]			
		Quiet amplifier in equipment room		*Quiet amplifier in antenna shack*	
		$L_{ABT}=0.8+0.9$ dB		$L_{ABT} \leq 0.3$ dB	
		$\beta=5°$	$\beta \geq 15°$	$\beta=5°$	$\beta \geq 15°$
MShU-30 low-noise amplifier	30	30.5–30.2	31.7–31.2	32.2	33.8
MShU-60 low-noise amplifier	60	30.1–29.7	30.7–30.4	31.3	32.5

[1] β is the angle of elevation of the antenna, deg; L_{ABT} are losses of the signal path

8.2. RECEIVER MIXERS

Receiver mixers of earth stations are intended for converting a received SHF signal, amplified in the input low-noise amplifier, to an IF signal (usually 70 MHz).

The basic requirements of receiver mixers are as follows:

1. Low noise coefficient and conversion losses.
2. Conversion ratio with the minimum nonuniformity.
3. Group delay time response in a wideband with the minimum nonuniformity.

The first two requirements are basically determined by the parameters of the selected mixer element, which, in addition to conventional mixer diodes, may be transistors, tunnel and reverse diodes, or Schottky barrier diodes. The third requirement is met by a receiver with high-quality characteristics in order to guarantee undistorted reception of information.

In the receiver mixers of earth stations, a weak signal is multiplied by stronger heterodyne oscillations, and therefore they operate, as a rule, in the linear conversion mode and the mixer parameters are virtually invariant to possible changes of the input signal level. Two types of semiconductor diodes are most widely used in earth station equipment as mixer diodes: point contact silicon diodes and gallium arsenide Schottky diodes. Schottky diodes are increasingly being used because of their high noise parameters and the stability of their characteristics at various parameters and under various mechanical conditions. The parameters of low-noise mixer diodes, used in earth station equipment, are given in Table 8.4 [19].

TABLE 8.4

Parameters	*For diode*		
	D405B, D408BP	*D408, D408P*	*AA111A*
Conversion losses, dB	6	—	6
Standing wave ratio	1.4	1.3	1.5
Heterodyne power, MW	1	0.5	3
Rectified current, mA	1	0.8	2.5
Output impedance, Ω	300–450	290–390	300–500
Noise temperature, K	7000	5500	2000–3000

A mixer diode should be matched in the frequency band of signals received from transmission lines with the optimum heterodyne power, and diodes must also be loaded on the frequencies of extraneous conversion components. Diodes are usually matched to the input in the entire working frequency band without tuning the mixer. The standing wave ratio of a mixer does not exceed 1.5 in consideration of the scattering of the parameters of diodes. That is good enough for achieving mixer parameters that are necessary for earth station equipment. The heterodyne power is optimized for minimizing the noise temperature of a mixer. However, the power is not critical and can vary in certain limits. The heterodyne power is usually set at about 3 mW for single-cycle Schottky diode mixers. In some cases, bias is supplied to the diode from a dc source, as was done in the first versions of the Orbita equipment. The bias voltage in this case is regulated by changing the diode and is optimized for minimizing the noises and the nonuniformity of the amplitude-frequency response of the receiver.

A mixer, along with the input stages and circuits that combine the received signal, and a heterodyne comprise the frequency converter of a receiver. It can be designed both as a single-cycle, and as a two-cycle (balanced) converter.

A simplified schematic diagram of a single-cycle frequency converter is shown in Figure 8.8. The signal, passing through a bandpass filter and ferrite valve, is combined with the aid of the fork of the filters with the heterodyne oscillations, and both signals pass through the harmonic filter to the mixer. The fork consists of two low-pass filters, a bandpass and a bandstop, tuned to the heterodyne frequency. To the heterodyne circuit is connected a low-pass filter, which passes the heterodyne energy to the T-junction of the filter fork and prevents the received signal from entering the heterodyne circuit to avoid losses. A bandstop filter is connected to the signal circuit, which passes the received signal and reflects the heterodyne energy to the mixer. It prevents the heterodyne signal from passing to the receiver input. The distance between the filters in the fork is usually selected for each transponder to ensure a low return loss, both in the signal circuit and in the heterodyne circuit.

A ferrite circulator is often used instead of a filter for combining a received signal with heterodyne oscillations in a single-cycle converter. A schematic diagram of this kind of converter is shown in Figure 8.9. In this case, the received signal passes from the bandpass filter through the circulator (arms 1 and 2) and the mixer. The heterodyne signal is connected through a low-pass filter and ferrite valve to arm 3 of the circulator. In accordance with the direction in which energy circulates (signified by the arrow), it enters arm 1. The heterodyne energy, reflected

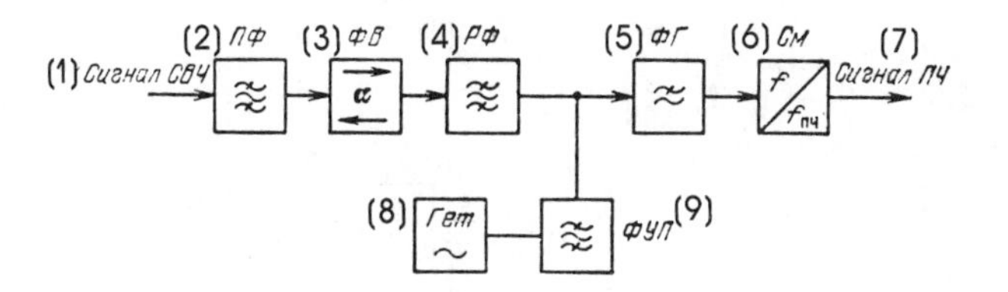

Key: (1) SHF signal
(2) Bandpass filter
(3) Ferrite valve
(4) Bandstop filter
(5) Harmonic filter
(6) Mixer
(7) IF signal
(8) Heterodyne
(9) Low-pass filter

FIGURE 8.8. Schematic diagram of single-cycle converter.

from the bandpass filter, tuned to the signal frequency, returns to the circulator and is supplied through arms 1 and 2 to a harmonic filter and mixer. In this design, the circulator functions as a valve and a signal circuit at the same time, because the wave reflected from the mixer is sent by the circulator to the heterodyne circuit, where it is absorbed in the ferrite valve and does not reach the bandpass filter. Precisely the same thing is done with the reflection that occurs in a mixer diode. It is also absorbed in a ferrite valve, by which the loads of the bandpass filter and the diode are matched at the frequency of the reflection. A schematic diagram of a balanced frequency converter is shown in Figure 8.10. Converters of this kind are used in cases when the heterodyne power must be lowered. If the mixer is well matched and the right diodes are selected the isolation of the signal and heterodyne circuits in it will reach 20 dB, and the power from the heterodyne accordingly will decrease by the same amount. In this converter, a bridge (slotted bridge) is used for combining the received signal and heterodyne signal. This bridge divides in halves the energy of the SHF signal, supplied to arms 1 and 2. As a result, beats of the signal and heterodyne frequencies pass to both diodes of a balanced mixer, connected through harmonic filters to output arms 3 and 4 of the bridge.

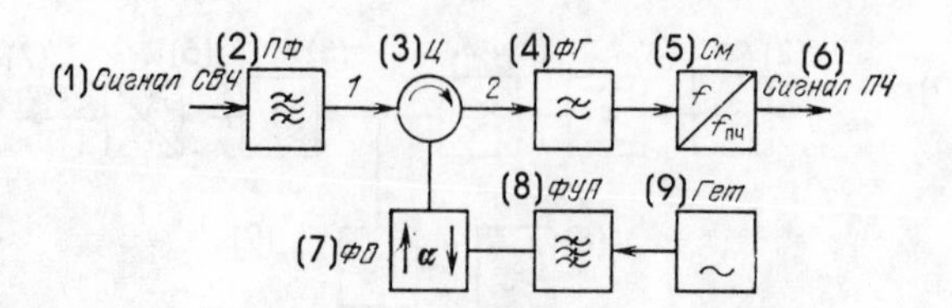

Key: (1) SHF signal
(2) Bandpass filter
(3) Circulator
(4) Harmonic filter
(5) Mixer
(6) IF signal
(7) Ferrite valve
(8) Low-pass filter
(9) Heterodyne

FIGURE 8.9. Schematic diagram of a converter with a circulator.

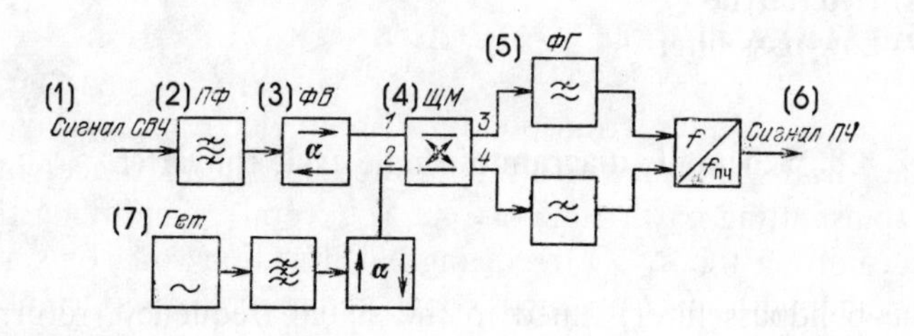

Key: (1) SHF signal
(2) Bandpass filter
(3) Ferrite valve
(4) Slotted bridge
(5) Harmonic filter
(6) IF signal
(7) Heterodyne

FIGURE 8.10. Balanced converter.

Chapter 9

Transponders for Satellite Communications and TV Broadcasting Links

9.1 DEFINITION, PURPOSE, AND MAIN FEATURES

A space repeater (transponder) is an electronic receiving-transmitting apparatus, installed on a satellite and intended for receiving signals from a transmitting earth station (one or several), amplifying them and then transmitting them to a receiving earth station (one or several). A space repeater with space antennas connected to it is an important part of a satellite communications and broadcasting system. Space repeaters, depending on the purpose of the system of which they are a part, differ with respect to how they convert the frequency of a received signal to the output frequency of the signal to be transmitted, with respect to the working frequency band, the number of transponders, the transponder-generation method, the kind of transmitted information (the operating mode of the transponders), the kind of processing of signals in space, and with respect to how they are structurally installed on a satellite, *et cetera*.

Of course, the above differences are conditional, because most communications and broadcasting satellites are multifunctional vehicles, each containing several repeaters, connected to several antennas (spot-beam or global) for solving different problems. For example, there are the following possible space repeater designs on the basis of how the received signal is converted:

Space Repeaters of the Heterodyne Type. These are most often encountered in satellite communications and broadcasting practice. The passband of this kind of space repeater usually does not exceed 40 MHz, and the main amplification takes place in the IF circuit, which in some cases is in the 70–120-MHz range. In this kind of space repeater, there are usually two frequency converters: a step-down and a step-up.

Space Repeaters with Single Frequency Conversion (the terms linear space repeater and direct amplification space repeater are encountered in the literature). In this kind of space repeater, the frequency of a received signal is converted only one time, and, as a result, the spectrum of the signal is transferred to the frequency range of the signals transmitted to earth. The advantages of this design are its simplicity and wide band. The passband can reach up to 80 MHz. However, technical design difficulties, related to the need to achieve considerable gain on one of the frequencies, are a drawback of this design. For instance, when the output power on the signal frequency is f_{out} – 20 W (+13 dBW) and the input signal is received on frequency f_{rec} = –100 dBW (i.e., for values that are typical of most communications links) the necessary gain is K = 120 dB, i.e., it is necessary to achieve high gain in a space repeater, which is difficult from the standpoint of guaranteeing stable performance.

Space Repeaters Utilizing Demodulation (or Processing) *of a Signal in Space.* These were used, as a rule, for transmitting special kinds of information. As satellite communications systems underwent development (the conversion to digital techniques, signal processing in space, a change of the kind of modulation, message switching on beams, *et cetera*), these systems also began to be used for two-way communications through stationary satellites utilizing detection (demodulation). For example, in the case of operation by digital techniques in space, received signals sometimes are detected and regenerated.

9.1.1 Structural Design Features of Space Repeaters

The selection and the development of the structural design of a space repeater are connected with the continuous and extended operation of it under the specific conditions of outer space as a part of a satellite. Among these conditions are a deep vacuum (10^{-4} – 10^{-14} mmHg/sr), weightlessness, the possibility of entering meteor showers, intense solar radiation, cosmic rays, and x-rays.

Space repeaters must be designed so that they can operate independently and reliably under the complicated conditions of all these factors for their entire service life and meet the following requirements:

- the least possible weight for given reliability and energy requirements;
- the best shape for installation on a satellite in order to minimize the stresses on the booster;
- the optimum utilization of the interior of a space repeater for the

purpose of meeting heat control requirements, for convenient access to assemblies and units, and for the prelaunch replacement of units;

- the influence of dynamic stresses and of the uneven distribution of weight in flight on other satellite systems, chiefly on the steering and stabilization systems, must be minimized;
- the capability of withstanding various vibrations, accelerations and impacts that occur during launch and trajectory correction, the main sources of which are the operating motor of the spacecraft itself and the booster, with vibration frequencies ranging from a few to thousands of hertz;
- the capability of withstanding a sudden change of temperature (+60 to −150° C). As can be seen, rather rigid, and often conflicting requirements are imposed on space repeaters. For instance, space repeaters must have the minimum weight and size, have high reliability and be economical. At the same time they must have the maximum possible power, and their parameters must be highly stable over a long period of operation.

For the above reasons, the design of the space repeater equipment has a number of significant differences from similar equipment installed on land. They include, primarily, the use of special wiring and spraying techniques, i.e., the use of special manufacturing technology.

9.1.2 The Main Energy Features of Space Repeaters

The power that must be obtained at the transmitter output of a space repeater (or of a space repeater transponder) depends on the following: the functional purpose of the system in which a given space repeater functions, the kind of information being transmitted, the prescribed carrying capacity, energy resources that are available in space, the prescribed service life, *et cetera*. To reduce the power that is consumed from space energy sources, efforts are made to improve the so-called *industrial efficiency* η_{ind} of a space repeater, which is the ratio of the useful oscillation power P_{use}, delivered to the antenna, to the total power P_{sou} that is consumed by a space repeater from power sources:

$$\eta_{ind} = P_{use}/P_{sou}$$

The effort to improve (increase) the industrial efficiency requires the use of economical electronic instruments in the output stage (transmitter). This stage uses the most energy and is the largest unit of a space repeater. These electronic instruments are traveling wave tubes, klystrons, transistors, *et cetera*. The parameters of these, and the characteristics of the amplifiers are examined in greater detail in Section 9.5.

9.1.3 Features of the Operation of Space Repeater Assemblies in Space

As is known [18], some of the elements of a space repeater may be installed, not only in the sealed compartment of a satellite, but also on the outside in open space. As a rule, these assemblies are either passive elements (for instance, SHF filters), or powerful output units, which cannot always be installed inside the sealed capsule. The specific feature of space is the existence of a high vacuum or the virtual absence of atmospheric pressure. The average pressure is plotted in Figure 9.1 as a function of altitude. As can be seen in the figure, at an altitude of 1000 km, the pressure is 10^{-12} times less than at sea level. The transition from the earth's atmosphere to interplanetary space takes place at an altitude of about 10,000 km, where the pressure reaches 10^{-19} mmHg/sr, and from which point the pressure does not change for practical purposes up to the altitudes of satellites in geostationary orbits (about 35,800 km).

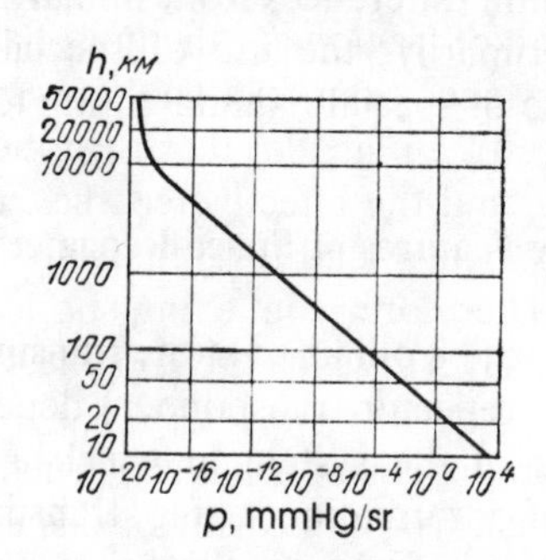

FIGURE 9.1. Average pressure as function of altitude.

The influence of a high vacuum on the elements of a space repeater in space is manifested in the following effects:

- the lack of natural (free) convection;
- coronal discharge;
- sublimation (evaporation) of material;
- a change of the strength of materials;
- a loss of lubricants.

The absence of free convection detracts from the conditions of heat exchange of the working surfaces of transponders in space, and normal heat conditions of the equipment can be maintained only by controlled radiation of heat in space. According to Pashen's law, the breakdown potential of a coronal discharge depends on the pressure of the gas, the gap between electrodes, and the geometry of electrodes. This law is valid only for certain products of the electron gap and pressure. For instance, when the gap between strip electrodes is 1 cm, the minimum breakdown potential is 173 V/mm at a pressure of 2 mmHg/sr, which corresponds (see Figure 9.1) to an altitude of about 30 km, which is not characteristic of satellites. However, there is also a danger of this effect in the high-vacuum environment of a satellite due to the sublimation of materials. For this reason, conditions of a "microatmosphere" are created around a satellite itself, and special measures must be taken to prevent breakdown.

The evaporation of materials in a vacuum is related to the molecular weight of a material, the ambient temperature, and the vapor pressure of materials. In a deep vacuum, especially when a material is hot, intense evaporation can occur. In a high vacuum the mechanical properties of many materials change. In certain metals, for example steel and molybdenum, they improve, but in other materials (aluminum and magnesium) they deteriorate. Losses of lubricant, the volatilization of lubricants and an increase of the coefficient of friction are the most harmful results of a high vacuum. Under ordinary conditions friction, between surfaces coated with metal decreases because of an air film that is formed. In a high vacuum, this film does not exist, and the effective coefficient of friction increases because the surfaces of metals tend to diffuse into each other during the cold welding process. Therefore, contacting surfaces should be made of malleable metals (gold, silver), which are capable of acting as nonvolatile lubricating materials.

The influence of irradiation, which is construed as the influence of the flux of corpuscular particles (protons, neutrons, *et cetera*) and electromagnetic waves (gamma rays, x-rays, *et cetera*) on components of space repeaters that are located in open space, is manifested as a change of structurally sensitive characteristics of materials: their resistivity increases, plasticity and viscosity decrease, and the strength and creep resistance increase. Ionization effects are the main kind of radiation damage to organic materials.

9.1.4 Heat Control Features of Space Repeaters [9.1]

The heat control system of a space repeater, or the heat regulation system is, as a rule, a part of this kind of satellite system. There are heat control systems of space repeaters that automatically maintain a prescribed temperature mode inside. The need to solve this problem is explained by the nonuniform heating, both of a satellite itself in flight on the sun side and in the earth's shadow, and of individual subsystems. In contrast to conditions on the earth, in space only radiative heat exchange takes place, during which a satellite is affected by radiation from the sun and from the earth. A satellite, heated to a certain temperature, also radiates heat into space, the amount depending both on external heat fluxes absorbed by a satellite, and on internal heating due to the operation of space equipment, chiefly the space repeater. Therefore, even if a satellite could be assumed to be isolated from external radiation, a stable temperature still could not be maintained in a vehicle without the use of special measures. The temperature inside a vehicle would begin to rise continuously, soon to exceed the tolerable temperature. Because a satellite cannot be completely isolated from external radiation, and internal heating is unavoidable, the purpose of a heat control system is to maintain a balance between the absorption and radiation of heat. The overall thermal balance usually is maintained by creating a radiating surface (a radiator) on the skin—the nosecone of a satellite, capable of radiating a lot of heat without absorbing much from the outside. Heat from the space repeater is also pumped to the skin. The temperature mode of a satellite can be controlled by regulating the amount of heat supplied to the radiator. Because there is no convection inside of a satellite, heat is transferred among its elements by means of the thermal conductivity of a structure by transmitting the heat from the elements inside of the capsule by means of the circulation of gas in the compartments, and also by making special sealed liquid coolant heat exchangers, capable of cooling both the elements inside of a satellite, and elements installed in open space.

Modern heat control systems include the following parts: sensors that measure the temperature inside of a satellite, and sensors installed on the assemblies of space repeaters that produce much heat; electronic units, which generate control signals for final controls; the final controls themselves, which influence the heating process; the surplus interior heat radiator; and vacuum-shield insulation. The final controls in the simplest heat control systems are fans, which circulate gas inside of the compartments of a satellite, and a mechanism that changes the active surface of a radiator (louvers). More sophisticated heat control systems have a fluid loop that carries heat from sources to the radiator. The circulation of the liquid coolant can be changed in this loop and the amount of heat that is released from a satellite thus can be influenced.

9.2 TYPICAL STRUCTURAL DESIGNS OF SPACE REPEATERS

A simplified schematic diagram of a heterodyne space repeater is shown in Figure 9.2(a). The signal on frequency f_{rec}, picked up by the antenna, goes to the input of the space repeater and is converted to IF signal f_{IF}. The main amplification of a space repeater in the given frequency band takes place on frequency f_{IF} in an IF amplifier. In the next converter, the amplified IF signal is stepped up to the signal on transmission frequency f_{tr}, which, after additional amplification in the output power stage of a power amplifier (usually assembled on a traveling wave tube or a klystron), is transmitted to earth. The space repeaters of the Soviet Ehkran TV broadcasting satellite and the transponders of the Molniya and Raduga communications satellites are built on this classical design.

A simplified schematic diagram of a space repeater that utilizes single-frequency conversion is shown in Figure 9.2(b). The signal with central frequency f_{rec}, which passes to the input of the space repeater from the antenna, appears at the output of the space repeater in the transmitted frequency band with central frequency f_{tr} as a result of single-step down-conversion. The output frequency band is removed far enough from the input band to avoid possible self-excitation of the repeater as prescribed by the Radio Regulations [1]. After conversion, the signal on frequency f_{tr} is amplified in one or several output amplifiers for increasing the power to the prescribed level, and is sent through the transmitting antenna to earth.

A simplified schematic diagram of a space repeater that demodulates (or processes) signals in space is shown in Figure 9.2(c). Here the signal, received on frequency f_{rec}, is converted to IF signal f_{IF}, is amplified in the IF signal path, and is demodulated. The demodulated low-frequency signal goes to a modulator through a converter that alters the structure of the signal, and from there to the output power amplifier (OPA) and to the antenna. This kind of system is used most often in cases when the method or rate of modulation in the satellite-earth link must be changed, for signal regeneration and for beam switching.

In principle, the demodulation of signals of the up link in space makes it possible to improve the characteristics of the link. For instance, in the case of the transmission of digital signals, if the signal-to-noise ratio in the up-links and down-links is identical, then the regeneration of pulses can give a gain of about 2.5 dB [9.2], in comparison with a linear space repeater for an earth station with the same output error probability. In many cases, however, the signal-to-noise ratio in up links usually is high, and therefore the advantages that regeneration offers are insignificant. Cases when there is interference in the up-link, or when signals must be separated in the up-link and a new group signal generated in the down-link, are exceptions.

In a repeater that processes signals in space, the types of signals that can be used are limited by the specific kind of modulation that takes place in the repeater. Therefore, the potential advantages of a regenerating repeater must be compared with the limits that are imposed on the signal modulation parameters, and consequently, with the ability to change modulation after the satellite has been launched, which renders a system much less versatile and deprives it of its universality.

A simplified schematic diagram of a space repeater with a PSK signal regenerator is shown in Figure 9.2(d) for the case of the perfect regenerator [9.2]. Sequential packets of signals from different earth stations, picked up by the space repeater antenna, are incoherent to the regenerator, although their carriers and clock frequencies differ little from each other. Thus, for a modulating signal regenerator, the carrier and clock frequencies of all packets must be regenerated. Therefore, a space repeater with a regenerator for processing a PSK signal, in the general case, must perform the following functions: coherent detection with the aid of the regenerated carrier; clock frequency regeneration; digital resolution; relative phase to absolute phase conversion (relative PSK/PSK decoding); digital data processing; absolute phase to relative phase reconversion (coding); carrier frequency generation; and modulation. If a space repeater performs all these functions, then it is hard to find a compromise between these requirements on the one hand, and the need to guarantee light weight, low power consumption, high reliability, *et cetera*, at the same time. High-speed regenerated carrier and clock frequency regeneration units are especially hard to design and manufacture.

The exclusion of some of the elements of a regenerator simplifies a space repeater. For instance, the design of highly stable OPA units is an exceedingly difficult task, in which connection OPAs are excluded from a space repeater and are replaced with an autocorrelative relative PSK detector, which simplifies a space repeater (Figure 9.3) and increases its reliability, although it does detract a little from its energy capabilities by approximately 0.5, 2.5 3 dB (for two-, four- and eight-phase signals, respectively) in comparison with coherent detection. However, when the reliability of received information is determined mainly by the satellite-earth link, the use of autocorrelative detection in space has virtually no effect on the overall reliability.

As is known [9.2], the use of spot-beam space antennas in a satellite-earth link makes it possible to use the power of a space repeater most economically. This is most convenient in the 11–14- and 20–30-GHz bands. In this case, n separate transponders are generated in a space repeater, which either operate on different frequencies, or a frequency is reutilized by means of space and polarization separation.

For operation in a communications system with TDMA, an $n \times n$ transponder switch is installed in the space repeater, which connects stations on the "each to each" principle. The space switch performs cross

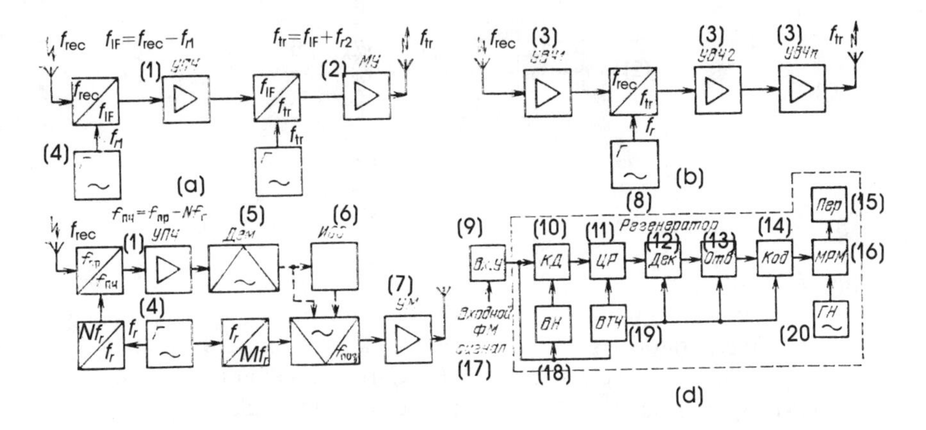

Key: (1) IF amplifier
(2) Power amplifier
(3) HF amplifier
(4) Heterodyne
(5) Demodulator
(6) Signal converter
(7) Power amplifier
(8) Regenerator
(9) Input amplifier
(10) Coherent detection
(11) Digital resolution
(12) Decoder
(13) OPA
(14) Coder
(15) Transmitter
(16) Modulation rate monitor
(17) Output PSK signal
(18) Regenerated carrier
(19) Clock frequency regeneration
(20) Carrier generator

FIGURE 9.2. Simplified schematic diagrams of space repeaters: (a) heterodyne type; (b) with single frequency conversion; (c) signal demodulated in space; (d) perfect regenerator.

switching of transponders in synchronization with the time packets that reach its input. Space repeater transponder switching on beams is the easiest (this is the main kind of space signal processing at the present time), and can be accomplished by choosing the right heterodyne frequencies of the transponder space repeater converters.

9.3 SCHEMATIC DIAGRAM OF THE MOLNIYA SATELLITE SPACE REPEATER

The first design principle (see Figure 9.1(a)) was used as the basis in the first Soviet Molniya-1 satellites. The space repeater of this satellite operates in the 800–1000-MHz band and has two modes: retransmission of the signals of one television program at an output power of 40 W; and

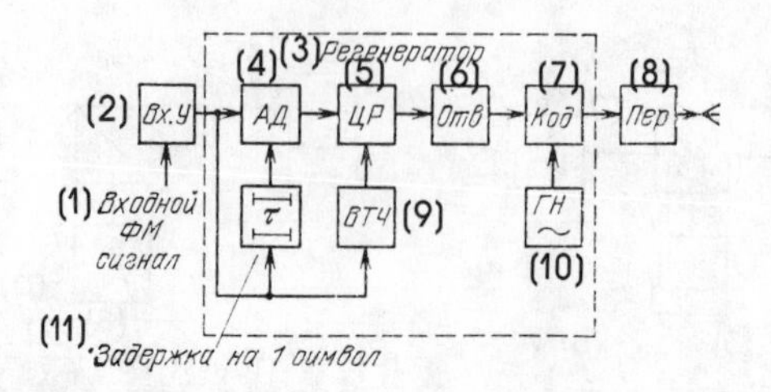

Key: **(1) PM input singal**
(2) Amplifier input
(3) Regenerator
(4) Autocorrelative detector
(5) Digital resolution
(6) OPA
(7) Coder
(8) Transmitter
(9) Clock frequency regeneration
(10) Carrier generator
(11) One symbol delay

FIGURE 9.3. Schematic diagram of space repeater with regenerator and autocorrelative detector.

two-way multi-channel communications with each transponder having a power of 14 W. A schematic diagram of the space repeater of the Molniya-1 satellite is shown in Figure 9.4 [9.3]. For improved reliability the space repeater of the Molniya-1 satellite has three identical receiving-transmitting sets, one of which is a working set, and the other two are in the so-called "cold" standby mode. Each receiver-transmitter set contains two receivers, tuned to frequencies f_{rec1} and f_{rec2} of one part of the 800–1000-MHz band, and one common transmitter, transmitting on frequencies f_{tr1} and f_{tr2} of the other part of that band.

Signals from earth stations on frequencies f_{rec1} and f_{rec2}, picked up by the satellite antenna, which the receiver and transmitter share in common, pass through bandpass filter 1 of a combiner which isolates the weak input signal from the strong output signal, and through a set switch to the receiving path of the set, selected for operation. The signal on frequency f_{rec1} is selected by bandpass filter 1 of the dividing network and the signal on frequency f_{rec2} is selected by bandpass filter 2, and these signals are supplied to the appropriate inputs of the receivers. That is how a space repeater transponder is divided into two paths. The receiver of each path contains a step-down converter (mixer with a common heterodyne), IF amplifier (IF amplifier 1 and IF amplifier 2), limiter, and a step-up converter (the second mixer with a common heterodyne).

In the receivers, the signals received on frequencies f_{rec1} and f_{rec2} are converted to IF signals f_{IF1} and f_{IF2} and are amplified in IF amplifiers 1 and

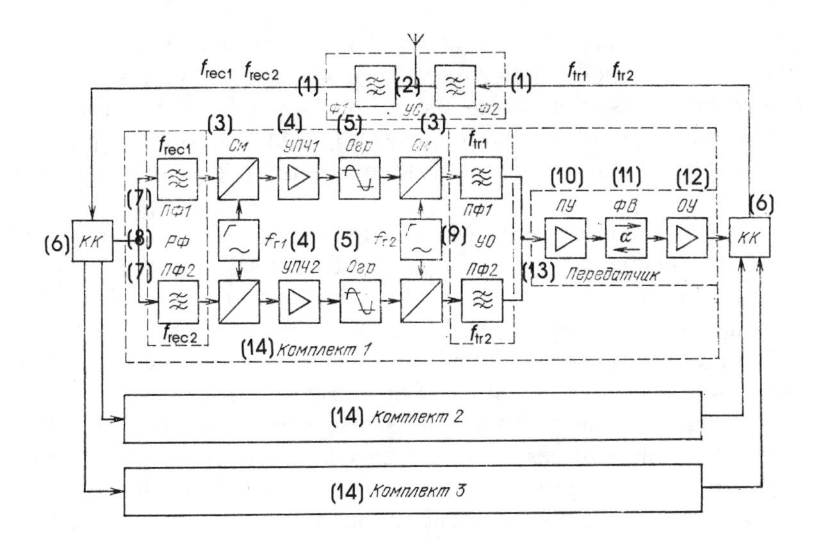

Key:

(1) Filter	(8) Dividing network
(2) Combiner	(9) Combiner
(3) Mixer	(10) Preamplifier
(4) IF amplifier	(11) Ferrite valve
(5) Limiter	(12) Final amplifier
(6) Set switch	(13) Transmitter
(7) Bandpass filter	(14) Set

FIGURE 9.4. Schematic diagram of transponder of Molniya-1 satellite.

2, which have the passband $\Delta f = 12$ MHz, to a level sufficient for the effective operation of the amplitude limiters. The limiters suppress parasitic amplitude modulation and stabilize the signal level, because the input signals can experience changes, both due to changes of propagation conditions, and due to changes of satellite flight, orientation and other conditions. From the limiters the signals pass to output mixers, where the signal on frequency f_{rec1} is stepped up by means of a heterodyne signal on frequency f_{h2} to f_{tr1}, and signal f_{rec2} is converted up to f_{tr2}. The difference between frequencies f_{h1} and f_{h2} of the heterodynes provides the necessary frequency shift.

The signals of the two frequencies f_{tr1} and f_{tr2} are combined through bandpass filters 1 and 2 of the combiner and then travel to the common transmitter, which consists of a two-stage traveling wave tube amplifier. In the first stage, which is a preamplifier, the traveling wave tube operates in the linear mode, and in the second, which is the final amplifier, it operates in the saturation mode. A ferrite valve, connected between the stages, provides the necessary matching and perfomance stability. From the final amplifier the signals of the two frequencies f_{tr1} and f_{tr2} pass through output bandpass filter 2 and the antenna and are transmitted to the correspondents.

The space repeater of the Molniya-1 satellite is used for transmitting both TV and telephone signals. During TV transmission, work is conducted in the simplex mode; in this case one of the receivers (depending on the direction of transmission) is turned off by command from the earth; the working receiver is turned off and the other is turned on to reverse the transponder (to change the direction of transmission).

During telephone transmission, two signals on the two frequencies f_{rec1} and f_{rec2} from two earth stations pass to the input of the receiver. Because the signal spectrum is considerably narrower during the transmission of these signals using frequency modulation with 60 telephone channels, than during the transmission of TV signals, common elements of the SHF receiving path (in front of the mixer) and one common transmitter are used for amplifying the signals of both links. For this, the transmission frequencies of the correspondents are selected such that the signal spectrum will not overlap, and at the same time will fit entirely in the passband of both signal paths.

Basic Technical Characteristics of the Space Repeater of the Molniya-1 Satellite

Number of sets	3
working	1
standby	2
Transmitting frequency band, MHz	About 1000
Receiving frequency band, MHz	800
Operating mode	TV[1]
	telephone[2]
Kind of modulation	FM
Passband, MHz	12[3]
Input signal level, dBW	90
Output signal level[4], dBW, in mode:	
TV	16
telephone	12

[1]Simplex, reversible.
[2]Duplex.
[3]At the 3 dB level.
[4]At the antenna input.

A similar heterodyne principle of transponder design is taken as the basis for the space repeater designs of satellites of the following generations: Molniya-2, Molniya-3, Raduga, and Gorizont. The only difference is, first, that these space repeaters became multitransponder repeaters, and second, they began to operate in the centimeter band. For instance, the Molniya-2 and Molniya-3 space repeaters have three transponders, and the space repeater installed on the Raduga satellite has six.

By using the centimeter band (6 GHz for receiving and 4 GHz for transmitting) it was possible to give repeater paths wider bands with better characteristics than the Molniya-1 satellite. For instance, an FM signal with considerably greater deviation can pass through a wideband signal path, and this significantly improves the performance characteristics of an entire communications link. In principle, transponders are designed for transmitting different kinds of information.

9.4 TYPICAL STRUCTURAL DESIGNS OF MULTITRANSPONDER SPACE REPEATERS

The schematic diagram of a typical multitransponder space repeater with frequency division and with two beams in the down link [9.2] is shown in Figure 9.5. Here two orthogonally polarized signals in frequency band Δf_u pass to the inputs of redundant receivers through a polarizer and signal divider. The signals of six different frequency transponders in this band of the up link are divided by frequency-selective filters of the frequency division multiplexers connected to the receiver outputs, and are transposed to the appropriate output frequency band, and are then supplied to different traveling wave tube amplifiers. The amplified signals from the outputs of the traveling wave tubes, in turn, are combined with the aid of frequency multiplexers and are fed through polarizers 3 and 1 to spot-beam antennas A3 and A4 of the down link in the band Δf_d. The frequency multiplexers that are shown in the diagram consist of parallel-connected bandpass filters and frequency converters. They separate the signals in the input frequency band to the necessary frequency transponders and transpose their frequencies to the appropriate output frequency band.

A schematic diagram of a space repeater with frequency division and with two beams in the down link is shown in Figure 9.6. Space repeaters of this kind, as a rule, are multitransponder repeaters. In them, transponders are generated by means of frequency-selective filters, which direct signals of different frequencies to separate amplifiers and antennas.

In the space repeater shown in Figure 9.6 [9.2], the signals received from each antenna, with orthogonally polarized transponders 1 and 2, are filtered and amplified separately. Used here are a global antenna with

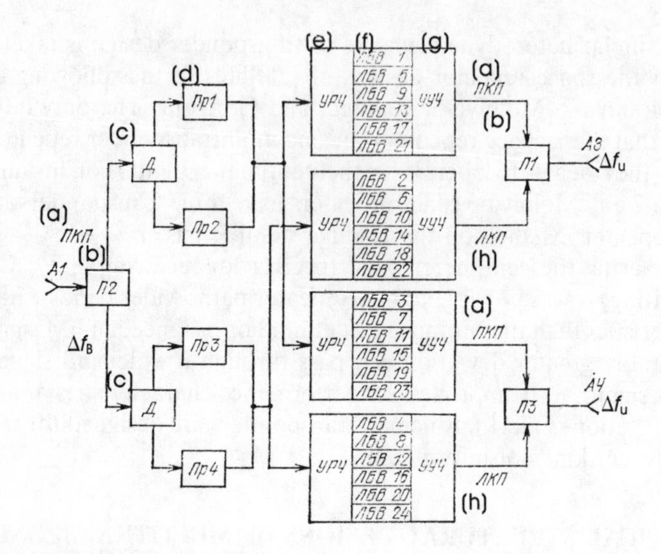

Key: **(a) Right-hand circular polarization**
(b) Polarizer
(c) Divider
(d) Receiver
(e) FDMA
(f) Traveling wave tube
(g) Frequency multiplexer
(h) Left-hand circular polarization

FIGURE 9.5. Schematic diagram of typical multitransponder space repeater with FDM.

right-hand circular polarization and a global antenna with left-hand circular polarization. The signals pass through a low-noise preamplifier and transponder bandpass filters 1 and 2, which isolate the sensitive receiving path from strong signals in the down link and divide the transponders. Additional amplification with or without AGC, or limiting is used to amplify the signals in each transponder to the necessary level. Gain control in a space repeater often is accomplished by commands from earth stations. Preference sometimes is given to this method of control over the use of AGC in a satellite, because the latter introduces some additional uncertainty in the control of the transmitted power of earth stations. Thus,

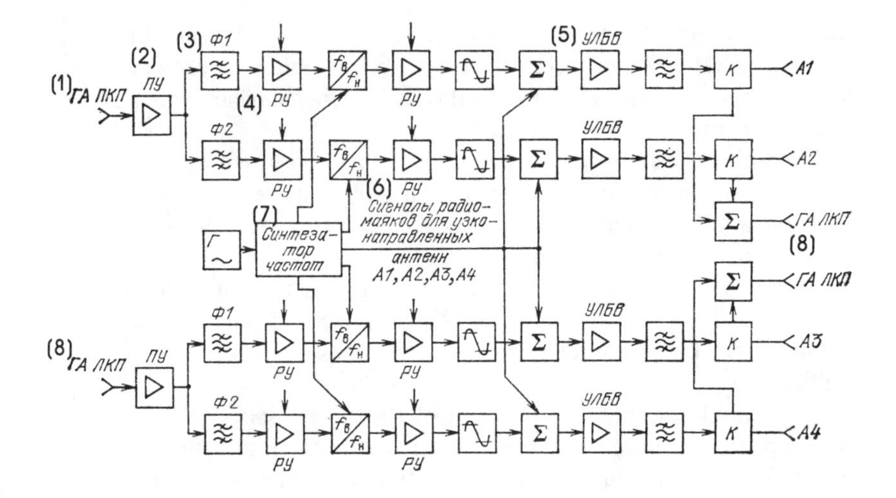

Key: **(1) Global antenna with right-hand circular polarization**
(2) Preamplifier
(3) Filter
(4) Gain control
(5) Traveling wave tube amplifier
(6) Radio beacon signals for spot beam antennas
(7) Frequency synthesizer
(8) Global antenna with left-hand circular polarization

FIGURE 9.6. Schematic diagram of typical multitransponder space repeater with FDM transponders and with two beams in the down link.

the signal levels can be regulated by using precision power control in the up link instead of hard limiting in the space repeater. From there, the radio signals are transposed to the frequency of the down line by shifting the frequency in a converter, for example by 750 MHz. For this, a heterodyne is used in the converter, which is synchronized with the aid of a phase-locked loop using a highly stable standard frequency. Then the traveling wave tube amplifier amplifies one or several of the signals to be transmitted in a given transponder. If there are several signals at the input of the traveling wave tube (for example in the case of FDMA), then nonlinear products can appear in the frequency band (due to a nonlinear amplitude response, and also due to the AM-PM conversion effect). To avoid this, the power level is reduced to the necessary level (power is stepped down). The output signal of the traveling wave tube amplifier passes through a mechanical SHF switch either to a spot-beam antenna, or to a global antenna with two kinds of polarization. The mechanical switch is controlled by commands from the earth.

Satellite beacon signals also can be inserted in the output signals of each transponder of the space repeater, usually outside of the band of the communications channel. The radio beacon signals are readily distinguishable and provide an opportunity to conveniently track a satellite with earth station antennas. In addition, the beacon signal is coherent with the stable heterodyne signal on-board the satellite, and can be used by an earth station for tracking long-term frequency changes of the signals which the satellite relays.

A simplified schematic diagram of the 12-transponder Intelsat-IVa space repeater, used in the Intelsat international satellite communications system, is shown in Figure 9.7(a), and its frequency plan is shown in Figure 9.7(b) [9.2.]. Each of four receivers consists of a tunnel diode amplifier in the 6-GHz band and a low-noise preamplifier, series-connected to which is is a band trap of the 4-GHz band, which provides attenuation outside of the working frequency band. The radio signals received in the 500-MHz band are transposed from the 6-GHz band to the 4-GHz band. The even and odd frequency division inputs each contain six bandpass filters with a 36-MHz bandwidth. Adjacent frequency transponders of the space repeater are separated by 40 MHz, and therefore there is a 4-MHz safety space between them. The responses of each of the 12 dividing networks of the even and odd transponders are corrected independently of each other. The main and standby traveling wave tube power amplifiers, harmonic filters, and switches for supplying the signals to the appropriate transmitting antenna are series-connected to these filters. The overwhelming majority of the components are redundant for the purpose of increasing the reliability of the space repeater.

9.5 SIMPLIFIED STRUCTURAL DESIGN AND BASIC CHARACTERISTICS OF THE SPACE REPEATER OF THE EHKRAN SATELLITE

A simplified schematic diagram of the space repeater of the Ehkran satellite is shown in Figure 9.8. The distinguishing feature of this kind of space repeater is the use of a high-power transmitter, which makes it possible to use simple receivers in the earth network. The space repeater consists of three main parts: a low-power receiver-transmitter, an output

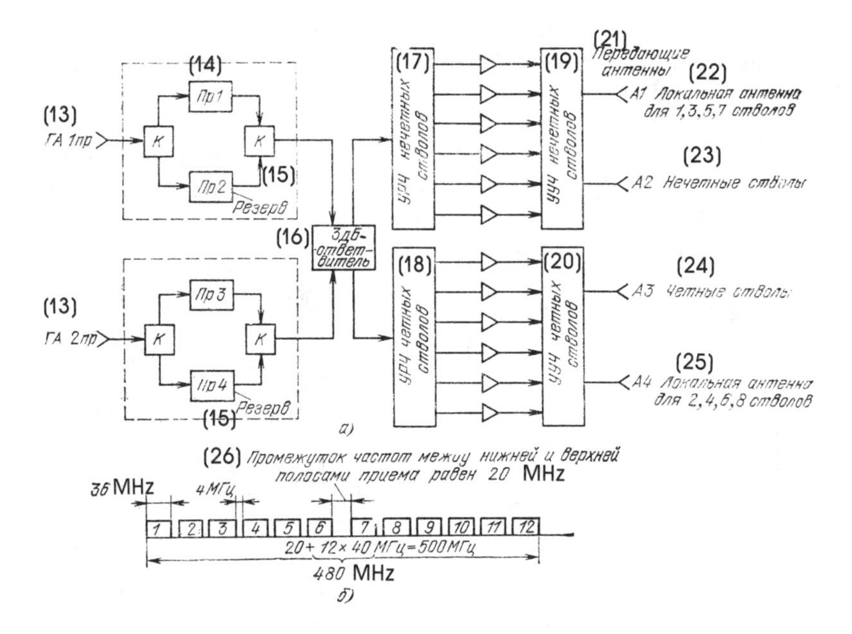

Key: (13) Global antenna
(14) Receiver
(15) Standby
(16) 3 dB dividing network
(17) FDM of odd transponders
(18) FDM of even transponders
(19) Frequency multiplexer of odd transponders
(20) Frequency multiplexer of even transponders
(21) Transmitting antennas
(22) Local antenna for transponders 1, 3, 5, 7
(23) Odd transponders
(24) Even transponders
(25) Local antenna for transponders 2, 4, 6, 8
(26) Frequency separation between lower and upper receiving bands is 20 MHz

FIGURE 9.7. Simplified schematic diagram of the Intelsat-IV space repeater.

power amplifier, and a power unit, not shown in Figure 9.8 for simplicity. All these instruments are connected to each other together and to the antennas by coaxial waveguide connectors, bunched connectors, and high-tension and HF cables.

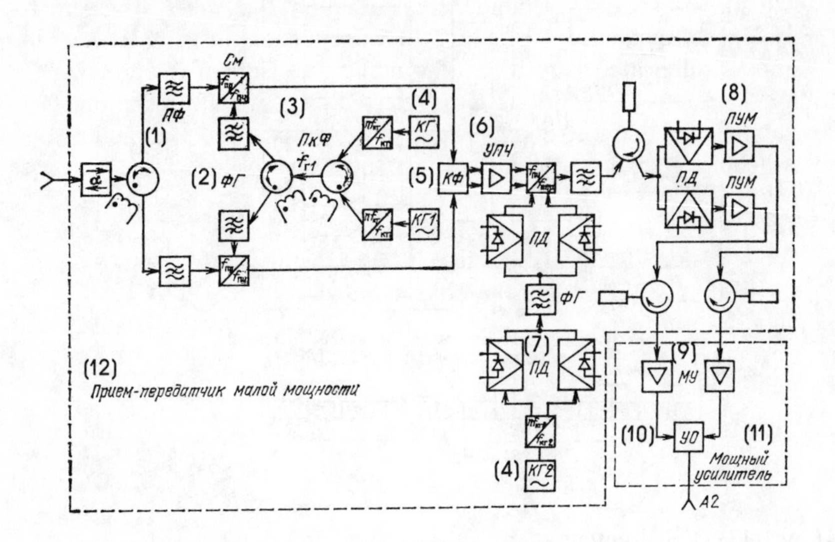

Key: **(1) Bandpass filter** **(7) Diode switch**
(2) Harmonic filter **(8) Power preamplifier**
(3) Ferrite switch **(9) Power amplifier**
(4) Crystal oscillator **(10) Combiner**
(5) Phase corrector **(11) Power amplifier**
(6) IF amplifier **(12) Low-power receiver-transmitter**

FIGURE 9.8. Simplified schematic diagram of space repeater of Ehkran satellite.

The SHF input signal in the (6200 ± 12)-MHz band, picked up by antenna A1, passes through a waveguide, isolating valve and ferrite switch, which selects one of two receiver sets as a working switch, to the input of a mixer (through a bandpass filter). There the received SHF signal is stepped down to an IF signal with f_{IF} = 70 MHz by means of heterodyne signal $f_{h\ 1}$, generated by multiplying frequency $f_{c.o}$ of a signal from a crystal oscillator. The IF signal passes through a phase corrector, intended for improving the group delay time response of the signal path, through an AGC IF amplifier for stabilizing the signal level, to the input of the mixer of the transmitter, where the IF signal is transferred to the transmitted frequency band of 714 ± 12 MHz, and then it is amplified in a power preamplifier to the level necessary for driving the output power amplifier. The main amplification to 200–300 W takes place in a transit klystron output amplifier [9.4].

Unit-wise "cold" redundance is used in the space repeater for improving its reliability. Virtually every independent instrument or a unit thereof is duplicated, and when one unit or instrument is operating the other is a standby. A working set is selected by commands from the earth by supplying to that instrument or unit power voltage and by simultaneously switching the necessary SHF or IF circuits with the aid of ferrite or diode switches (PF and PD).

Basic Technical Characteristics of the Space Repeater of the Ehkran Satellite

Number of sets	2
working	1
standby	1
Frequency band, MHz:	
transmitting[1]	714 ± 12
receiving[2]	6200 ± 12
Operating mode[3]	Television, radio broadcasting
Kind of modulation	FM
Passband[4], MHz	24
Signal level, dBW:	
input	−98
output[5]	25

[1]In satellite-earth link.
[2]In earth-satellite link.
[3]Simplex.
[4]At 2-dB level.
[5]At antenna input.

9.6 SPACE RADIO TRANSMITTERS

The transmitter output power is the main parameter of a space repeater, which determines the capabilities and performance characteristics of a communications system. The maximum transmitter power is limited by several factors: the maximum power of the primary satellite power sources; the possibility of removing dissipated heat from the satellite; degradation and a loss of reliability of electronic instruments when their power is increased; the size and weight of the satellite itself; the regulated power flux density, created at the earth's surface by the radiation of a satellite, which should not exceed -152 dBW/m^2 in a 4-kHz band [1].

The transmitters of most space repeaters are of the conventional design (Figure 9.9), consisting of a power frequency converter (mixer, heterodyne) and a power amplifier with the necessary assortment of filtering and matching elements. The transmitters most often are adapted for passing the signal band in the band of one transponder, and sometimes they are also used for simultaneously amplifying several signals. The main element, the heart of the transmitter, is the power output amplifier, because specifically it receives the greatest fraction of the input power of an entire space repeater, and it is the heaviest and largest component. Various SHF instruments—traveling wave tubes, klystrons, solid state instruments (transistors, tunnel and field effect amplifiers, avalanche transit diodes, *et cetera*)—are used as the amplifying element itself, depending on the purpose, necessary power, frequency band, weight and size, efficiency, service life, *et cetera*.

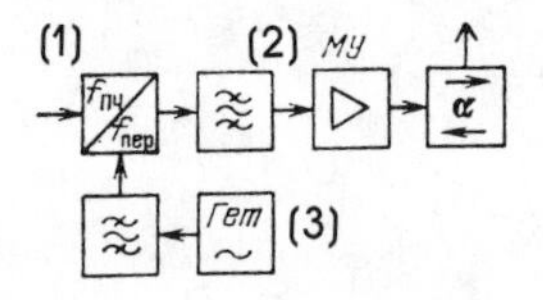

Key: **(1) f_{IF}/f_{tr}**
(2) Power amplifier
(3) Heterodyne

FIGURE 9.9. Schematic diagram of the transmitter of a space repeater.

Traveling wave tubes comprise the most numerous and most rapidly developing class of electronic SHF instruments for space technology [9.6], the extensive use of which in this field is explained by the fact that they have a multitude of advantages over other SHF instruments: high gain, wideband, capability of operating in the pulse, and continuous modes in a wide range of output powers.

The traveling wave tubes that are used in space repeaters have a wideband, high gain, high efficiency, they are exceptionally compact and lightweight, have a service life of up to 20,000 hours [9.6], and are highly reliable. These instruments operate at comparatively low voltages (below 5000 V); they are quite rugged and are capable of withstanding the strongest vibrations and impact stresses. In most cases, the dimensions and weight of a traveling wave tube are perfectly acceptable, but most often it is desirable to improve the attainable efficiency, because the traveling wave tube receives most of the power that is used by a space repeater.

In fact, all traveling wave tubes that are used in space repeaters have the identical construction, with the exception of minor modifications, connected with specific functions of repeaters [9.6]. The spiral usually is fastened by round or wedge-shaped tie rods in a metal cylinder with coaxial ports in the input and output ends. During the development of electron guns, an effort is usually made to reduce the electron flux density from the cathode surface (for the purpose of improving the endurance). Lightweight magnets, for example based on platinum-cobalt or samarium-cobalt, usually are employed for achieving the maximum focusing precision. The tube with the focusing system is placed in a lightweight metal can, which gives the entire traveling wave tube the necessary mechanical strength and also removes heat to the chassis of the spacecraft. Because of the interaction of electron beams through the length of the retardation system with the electromagnetic field of the traveling wave, considerable gain (up to 60 dB) is achieved in these instruments with a comparatively small beam current.

By using retardation systems with weak resonance properties, it is possible to amplify signals in a very wide frequency band, reaching two and more octaves. Traveling wave tubes are built for operating both in the pulse, and in continuous modes, and also for alternate operation in both modes, which makes it possible to build general purpose space repeater transponders, suitable for the retransmission of any kind of signal. Shown in Figure 9.10 [9.6] are curves that illustrate the highest output powers and efficiencies, achieved in different types of traveling wave tubes. Medium-power traveling wave tubes are of the greatest importance for space repeaters, and special economical and compact traveling wave tubes are built for these purposes. Exceedingly rigid requirements on the efficiency, durability, reliability, size and weight, and also on mechanical strength and climatic stability, are imposed on traveling wave tubes because of use in unattended space equipment.

Traveling wave tubes for space equipment are manufactured for separate, specifically assigned frequency bands in the 1.8–12.7-GHz range. Traveling wave tubes with $P_{out} \approx 20$ W were typical of the first models of space repeaters; traveling wave tubes with powers of 30 W and 70 W in the continuous mode were used in individual satellites. The preferred output power level in the continuous mode continues to be in the 10–20-W range. This should not be considered the limit, determined by mechanical strength. Even higher powers are obtained from this very same design. The most important factor in the choice of the output power of a traveling wave tube is the capacity of the satellite power source. Even a slight increase of efficiency produces a considerable gain for a satellite as a whole; fewer solar batteries and storage batteries are needed, the power sources are smaller and the cooling problem is simplified. For nearly all satellites, the main parameter of a traveling wave tube, which determines all other structural components, is the necessary input power from the space power source. Therefore, much attention is devoted to methods of increasing the efficiency of traveling wave tubes. These methods include the following: changing the phase rate of the retarded wave through the length of a tube, or correcting the rate synchronization, creating a discrete response, multistage recovery in the collector (step-down of the potential of a collector or of a series of collectors to a level below the potential of the retarding structure, which makes it possible to return some of the unutilized energy of the spent electron beam); the insertion of a harmonic (this method is based on a change of the relative level of the second harmonic in the RF input signal).

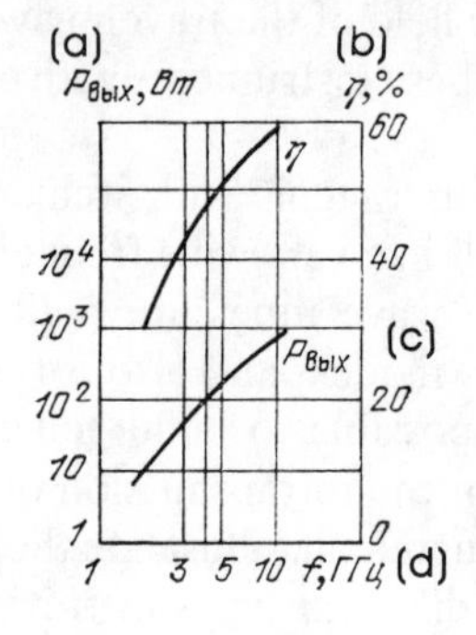

Key: (a) P_{out}, W
(b) η (%)
(c) P_{out}
(d) f, GHz

FIGURE 9.10. Maximum output power and efficiency as functions of frequency.

The parameters of traveling wave tubes used in the transmitters of space repeaters are: gain, output power, amplitude modulation to phase modulation (AM-PM) conversion factor, bandwidth, and efficiency. The gain of a traveling wave tube usually ranges up to 30-40 dB, and the electronic efficiency is not less than 25-30%. As a rule, traveling wave tubes are the main energy consumer on a satellite. The traveling wave tube is the main amplifier element of a space repeater and the major source of transmission nonlinearity. The transfer functions of traveling wave tubes, i.e, the instantaneous output voltages as functions of the instantaneous input voltages, are nonlinear.

Typical amplitude responses of traveling wave tubes are shown in Figure 9.11. When a traveling wave tube is in operation the necessary output power mode P_{out} is set by selecting the appropriate input power level P_{in}. There are three characteristic parts of the amplitude response of a traveling wave tube, which correspond to three modes of operation. In the linear mode (the weak signal mode) gain K_g is constant, and the output power is proportional to the input power. The output power in this mode, as a rule, is 3–6 dB less than the nominal power.

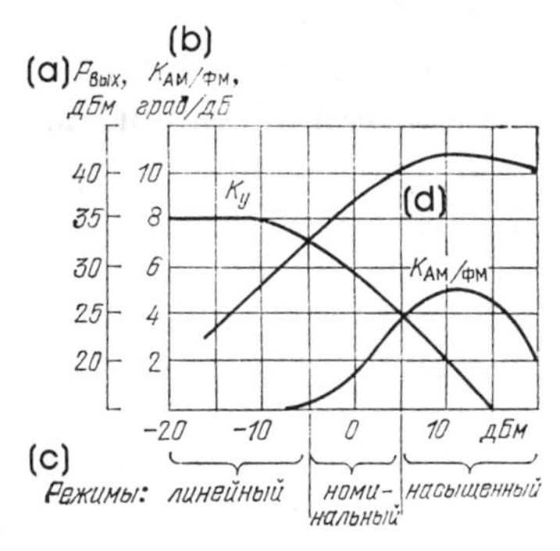

Key: **(a) P_{out}, dBp**
(b) $K_{AM/PM}$, deg/dB
(c) Modes: linear, nominal, saturation
(d) P_{out}

FIGURE 9.11. Typical responses of traveling wave tubes.

In the saturation mode (the strong signal mode) the output power is no longer a linear function of the input power, and an increase of the input signal does not produce a further increase of the output power due to the limited power of the electron beam. In the saturation mode, there is a sharp increase of the steepness of the phase-amplitude response of a traveling wave tube, which represents the phase shift ϕ as a function of the amplitude of the input signal, which causes amplitude modulation to be converted to phase modulation (amplitude-phase conversion). Thus, parasitic amplitude modulation, for example of the FM input signal of a traveling wave tube, will cause parasitic phase modulation and consequently nonlinear distortions of the signals.

In traveling wave tube output power amplifiers, for considerations of economy, it is desirable to use a tube in the close to saturated mode (the nominal mode). In the case of the transmission of wideband information signals in the saturation or near-saturation mode it is necessary to consider the tolerable level of distortions. Tolerable distortions depend on the kind of modulation, for example single carrier frequency modulation, multiple carrier frequency modulation with time or frequency multiple access.

Two effects that occur in the traveling wave tubes of space repeaters, i.e., amplitude nonlinearity and AM to PM conversion, are reasons why crosstalk interference appears when several signals are amplified. This problem arises when several FM signals are amplified in the same amplifier, for example when the Gruppa (see Chapter 7), Gradient-N and other channeling equipment is used.

Odd-order products of nonlinear distortions, which occur in an amplifier, get into the useful signal band and interfere with the useful signals, even when the spectra do not overlap. As a result, crosstalk interference occurs on telephone channels during the transmission of telephone signals. The calculation of interference and the analysis of methods of reducing it by choosing the right working region on the response of a traveling wave tube are examined in Chapter 7 [9.3] and are of great practical importance. Some parameters of several traveling wave tubes, used in the output amplifiers of communications space repeaters, are given in Table 9.1. Klystrons sometimes are used as a final amplifier of a space repeater. For instance, the Ehkran satellite TV system, developed in the USSR, and operating in the 702–726 MHz band, uses a direct transit klystron [9.4] with 200–300-W output power and a passband of about 24 MHz at the 2 dB level. The application of klystrons in space equipment is limited, because a klystron, as a rule, is a narrow-band instrument. The advantages of a klystron are its great simplicity, fewer power voltage ratings than a traveling wave tube, and somewhat higher efficiency. Otherwise the characteristics of klystron amplifiers are similar to those of traveling wave tube amplifiers, considering all their advantages and disadvantages (nonlinear transfer function, amplitude-phase conversion, *et cetera*).

Solid state instruments began to be used recently as output power amplifiers of space repeaters in connection with progress in semiconductor electronics, which has made it possible to significantly increase the power of SHF transmitters [9.7, 9.8]. Semiconductor SHF transmitters are being developed for space repeaters along two main lines:

- the development of new, more powerful SHF transistors and diodes;
- combining of the powers of semiconductor generators with the aid of networks and a phased array antenna.

TABLE 9.1

Parameter, unit of measurement	**RF parameters of traveling wave tubes of different satellite space repeaters**					
	Syncom	*ATS*	*Intelsat III*	*Intelsat IV*	*Molniya I*	*Intelsat V*
Power, W	2	4	12	6	40	4.5–8.5
Gain, dB	33	40	43	56	30–40	—
Bandwidth, MHz	10	100	250	30	12	36/72
Range, GHz	2–4	4–6	4–6	4–6	1	4–6 11–14
Efficiency	20	33	34	31	30	—

The advantages of solid state transmitters for space repeaters in comparison with electronic vacuum instruments are:

- as a rule they are substantially more durable;
- lower direct current (dc) source voltage (the former require not more than a few tens of volts for power and, as a rule, have only one nominal rating, and the latter require a whole collection of different voltages for power, the highest of which amount to several kilovolts, even for a comparatively weak SHF output signal);
- the use of semiconductor instruments makes it possible to use microelectronic techniques for manufacturing different assemblies and units of a space repeater transmitter, which makes it possible, in turn, to significantly reduce the weight and size of the latter;
- as a rule, solid state transmitters have significantly better industrial efficiency, which improves the utilization of available energy in space;
- large semiconductor instruments have a virtually instantaneous readiness for operation in comparison with electronic instruments, in which the heater circuit has to warm up. This makes a communications system more versatile and responsive.

The consequences of the above-listed advantages are the following: a significant reduction of weight and size, better economy, durability, and reliability of solid state transmitters of space repeaters in comparison with electronic vacuum instruments under otherwise identical conditions. Along with advantages, these transmitters also have some drawbacks:

- semiconductor instruments are sensitive to deviations, even brief ones, from the tolerable operating mode, which can cause the *p-n* junction to break down and an instrument to fail completely; therefore, it is necessary to take special measures in a transmitter to protect it from accidents;
- semiconductor instruments have limited power, and for most of them as frequency f increases it decreases in proportion to f^2; the output power of the most powerful types of bipolar SHF transistors is plotted in Figure 9.12 as a function of frequency.

The main electrical parameters of certain types of semiconductor instruments, used in the transmitters of space repeaters, in the 1–10-GHz range are given in Table 9.2 [9.8]. As can be seen in the table, the power of SHF instruments is limited. At the same time, the necessary output power of the transmitters of space repeaters can be many times greater than the power of one transistor. Therefore, special means of combining (summing) many amplifiers of a given type are used for significantly increasing the output power of a transmitter.

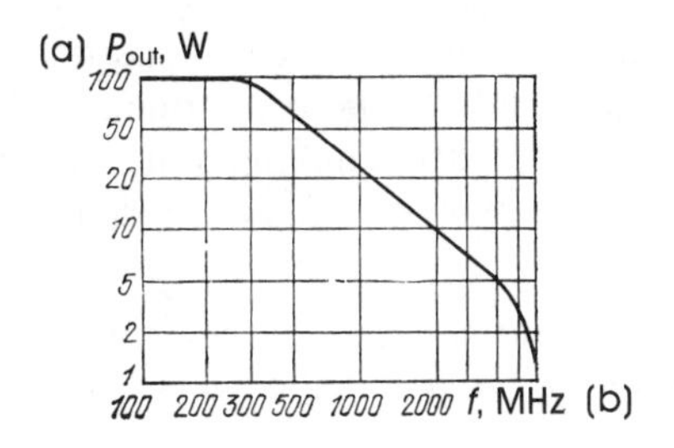

Key: (a) p_{out}, W
(b) *f*, MHz

FIGURE 9.12. Output power as function of frequency for most powerful types of bipolar SHF transistors.

TABLE 9.2

Frequency, GHz	*Type of instrument*	*Output power, W*	*Efficiency, %*	K_r, *dB*
		continuous operating mode		
1–2	Transistor bipolar	20–40	40–55	7–8.5
3–4	Transistor bipolar	5–8	30–35	7
	Transistor field effect	4–15	44	5.5–6
5–7	Transistor bipolar	1.5	23	4.4
	Transistor field effect	2.7–5.1	31	5
5–7	Avalanche transit diode in impact avalanche and transmit time (IMPATT) mode	8	13.3	

There are three main methods of combining powers: using multiterminal networks; using a multicomponent phased array antenna; and in a common resonator.

In the first method, many like-amplifiers, loaded into a common output, are connected to a special device; in the second method, the power of the signals are combined in space using a phased array antenna, containing many appropriately oriented irradiators, each of which is driven by an independent amplifier. In both cases, the powers both of transistor, and of diode SHF amplifiers are combined. The third method is used only for combining SHF generator diodes, placed in a common resonator.

In practice, the first method makes it possible to increase the power of a transmitter in relation to the power of one transistor by 15–20 dB, the second method by 30–40 dB, and the third by 10–13 dB. The main requirements that the above-described methods must satisfy are the following:

1. The power of the output signal of the combiner (in dB) is equal to or close to the sum of the nominal powers P_{nom}, i.e., of the individual n amplifiers:

   $$P_{tot} = nP_{nom}$$

2. All the amplifiers, the powers of which are combined, must be mutually independent, i.e., isolated from each other. The failure of any amplifier (idling or short-circuit) must not have an effect on the operating mode and the power of all the other amplifiers.
3. If m amplifiers of a total of n amplifiers fail, the power into the load should fall as little as possible, preferably by not more than nP_{nom}.

The powers of SHF amplifiers most often are combined with the aid of so-called bridges, which pair signals. Chiefly bridges of the directional coupler type, i.e., eight-terminal networks, are used. They are intended for the directional branching of energy, and their distinguished feature consists in the following: when one of four channels of a directional coupler is excited the energy goes only into two, and not into the fourth. This kind of directional coupler can be used for the reverse procedure—for dividing the power by two (i.e., for reducing the power by 3 dB).

There are various design versions of transistor transmitters in which the powers of the amplifiers are combined on the basis of bridges. Here the number of combined power amplifiers should equal 2^n, which is accomplished by using $(2^n - 1)$ bridges. Combiners for combining the powers of many SHF amplifiers are designed by using various versions of multiterminal adder-dividers. These devices consist of three main parts: a signal power divider, n identical SHF amplifiers, and a power adder. Shown in Figure 9.13 as an example is a network that combines the powers of four amplifiers, using quadrature bridges with external ballast loads. This design, which combines signals of rather great power, is not hard to apply to a large number of paired identical amplifiers or amplifier units.

It is important during the design of networks for combining the powers of individual amplifiers to meet the requirement on the phasing of the signals being combined. Structurally identical dividers D and adders Σ, connected conjugately as shown in Figure 9.13, are used for this purpose. In this case, when the amplifiers are of the identical design no additional phase inverters are needed. The diagram in Figure 9.13 is typical of transistor modules, based on hybrid integrated technology. The powers of up to 50-100 semiconductor instruments can be combined with the aid of multiterminal adders, and usually four transistors are first combined into a module, and then the powers of 8–16 of these modules are combined, depending on the necessary amplifier output power.

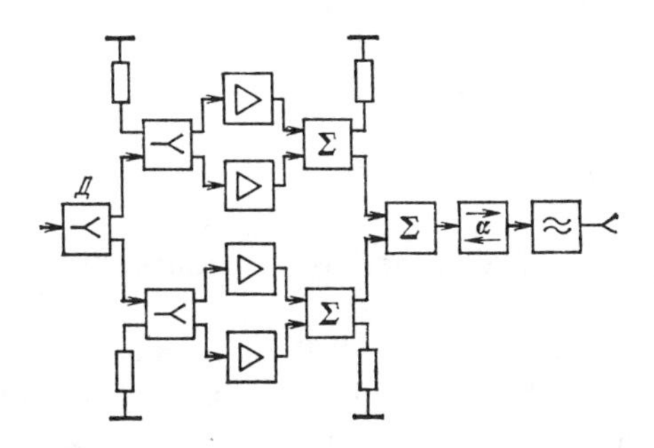

FIGURE 9.13. Diagram of a network that combines powers of four amplifiers.

Under actual conditions the combining of the powers of signals inevitably involves losses, related to the scattering of the parameters of individual amplifiers, to deviations of the *S*-parameters (scattering parameters) of adders and dividers [9.4] from the optimum in the frequency band, and to reflections in signal paths. The summary power of all amplifiers into a common load in these cases [9.4] is

$$P_{\mathrm{lo}} = \frac{P_{\mathrm{g.mean.nom}}}{n} \left|\sum_{k=1}^{n} a_k P_k s_k g_k \exp\left(-\mathrm{i}\phi_k\right)\right| \left(1 - |\Gamma_{\mathrm{lo}}|^2\right)$$

where a_k is a coefficient that takes into account the reflection in the *k*th channel, and when matching is good it is close to 1; $P_{\mathrm{g.mean.nom}}$ is the mean nominal power, identical for all *n* amplifiers; $p_k = \sqrt{P_{gk\mathrm{nom}}/P_{\mathrm{g.mean.nom}}}$ is a coefficient that takes into account the change of the amplitude of the input signal; $P_{gk\mathrm{nom}}$ is the nominal power of the *k*th generator; s_k is a coefficient that takes into account the change of the modulus of the transfer function of an individual amplifier relative to the nominal for identical amplifiers, close to 1; g_k is a coefficient, close to 1, which takes into account the change of the amplitude of the voltage of the incident wave in a common channel during the operation of the *k*th generator due to reflection in the ballast loads; ϕ_k is the resulting phase of the signal that goes into the common channel from the *k*th amplifier, and when phasing is good $\phi_k = 0$; Γ_{lo} is the return loss of the common load, and when matching is good $\Gamma_{\mathrm{lo}} = 0$; and *n* is the number of amplifiers, connected to the adder.

The efficiency of a transmitter, tuned by adding the powers of individual amplifiers, is determined by the transfer function

$$k_{\text{loss}} = \frac{P_{\text{lo}}}{\sum_{k=1}^{n} \text{P}_{gk\,\text{nom}}}$$

Power losses (in dB) in an adder are

$$b_{\text{loss}} = 10 \log (1/k_{\text{loss}})$$

When all the power of all generators is transmitted into the load, $k_{\text{loss}} = 1$.

Given in Figures 9.14 and 9.15 are functions that illustrate the influence of different factors on losses when the powers of signals are combined with the aid of multiterminal adders: in Figure 9.14—b_{loss} is a function of the change (imbalance) of amplitude of combined signals, on the condition that $\Gamma = 0$, $\phi_k = 0$, $a_k = 1$, $s_k = 1$, and all nominal amplifier powers are $P_{gk\text{nom}} = P_{\text{g.mean.nom}}$ for two particular cases:

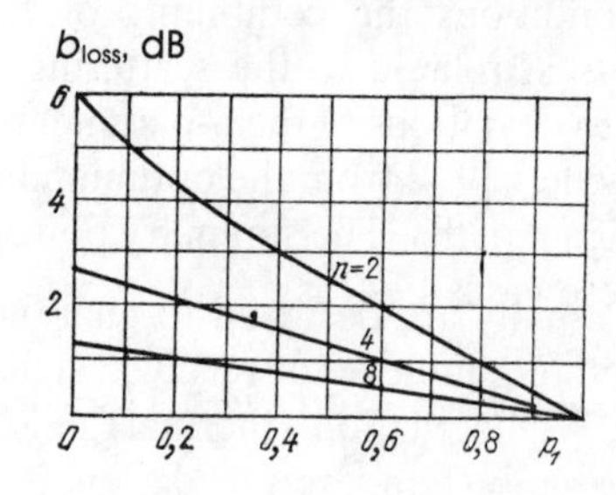

FIGURE 9.14. Power loss of an adder as a function of the amplitude imbalance.

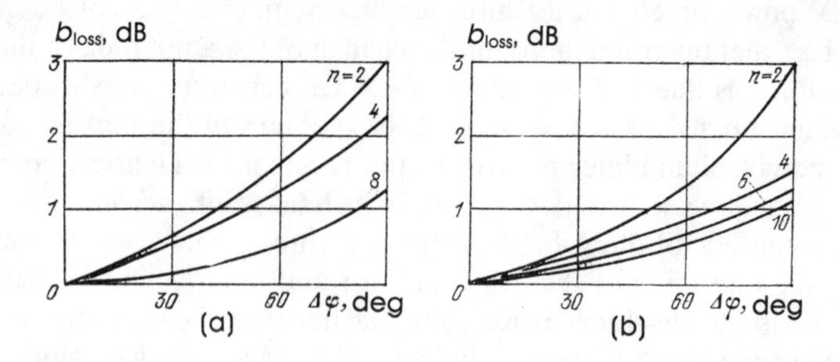

FIGURE 9.15. Power losses of an adder as a function of phase imbalance.

1. the powers of all generators are identical and constant, with the exception of one of them, for example the 1st;
2. the amplitudes of all combined signals of the amplifiers are uniformly distributed from P_{min} to P_{max}.

The curves in Figure 9.15 show b_{loss} as a function of the phase disbalance of combined signals for two cases:

1. the phases of the signals of all the amplifiers, with the exception of one, are constant and equal to the nominal phases (Figure 9.15(a));
2. the phases of all the signals from all the amplifiers are uniformly distributed from ϕ_{min} to ϕ_{max} (Figure 9.15(b)).

An analysis [9.4] shows that in order to achieve a power transfer function K_t of not less than 90–95% in a transmitter in which powers are combined, it is necessary to guarantee that the scattering of the powers of the individual amplifiers will fall within the 20–30%-range and that the phase difference of the combined signals will not exceed 20°–30°, and the changes of the scattering parameters (S-paramters) here for a given frequency band must be kept within 20–30%, and the standing wave ratio of each of the inputs should be not worse than 1.4–1.5. Parameters of versions of power adding amplifiers that have been developed, based on foreign data [9.8], are presented in Table 9.3.

9.7 INPUT RECEIVERS OF SPACE REPEATERS

Input receivers usually are designed to guarantee reliable reception of weak signals. The minimum level of received signals, or the sensitivity, is determined by the set fluctuation (thermal) noises. In practice, when selecting the effective noise temperature designers proceed, on the one hand, from the requirement that the contribution of the noises of the earth-satellite link be five to ten times less than the noise of the satellite-earth link and, on the other hand, the minimum effective noise temperature of a satellite receiver be not less than T_e, which is the equivalent temperature of the earth, because satellite receivers usually are aimed toward the earth.

The noise temperature of the input receiver of a space repeater is

$$T_{sat} = T_e + T_{atm} + bT_{cosm} + T_{rec}$$

where T_{atm} is the equivalent temperature of atmospheric noises, and for antennas of stationary satellites in the 1–20 GHz range, it varies from 2°–25°; T_{cosm} is the equivalent temperature of cosmic noises, which depends

TABLE 9.3

Type of base transistor	Number of channels in amplifier		Parameter			
	Parallel output	*Common*	*Frequency, MHz*	*Power, W*	*Gain, dB*	*Efficiency, %*
Bipolar	8	14	1 550	40	55	41
	8	10	1 550	110	16	39
	8	14	860	110	62	44
Field effect	3	5	4 000	7.3	17	33
	2	2	7 800	0.5	7	15
	8	22	7 500	3	35	5
	1	1	12 000	0.2	5	30

on the area of the sky for which the antenna is aimed, and it can be determined using special maps of the heavens [9.11]. The maximum values on 1 GHz do not exceed 30° and drop sharply as frequency increases; b is a coefficient, considerably smaller than unity, which determines the fact that cosmic noises are received only through the sidelobes; and T_{rec} is the noise temperature of the receiver of a space repeater.

It was shown in [9.1] that the following practical conclusion relative to the choice of T_{sat} follows from the first condition:

$$T_{sat} = (10\text{–}20)T_{rec} \tag{9.1}$$

where T_{rec} is the noise temperature of the receiver of an earth station, operating with a given satellite. It was shown in Chapter 8 that input receivers of modern earth stations, using low-noise amplifiers of different types in the satellite communications bands, have a cumulative noise temperature within 300°. In these cases, the cumulative noise temperature T_{sat} accordingly [see (9.1)] can range from 800 to 6000.

The set noise temperature of a reciever is

$$T_{rec} = [T_{sat} - T_a - (L_f - 1)T_f]/L_f \tag{9.2}$$

where T_a is the noise temperature of the antenna ($T_a \approx 300°$ K); T_f is the equivalent temperature of the SHF feed ($T_f \div 300°$ K); and L_f are losses in the SHF feed (L_f = 1–2 dB). When L_f = 1 dB (1.26), the set noise temperature of a space receiver, computed by (9.2), can range from 500° to 4500°. These noise temperatures of space repeater receivers can be achieved in several ways:

- by using crystal mixers; this is the easiest, most technological, compact and reliable method and guarantees a minimum noise temperature of 2500°–3000° on frequencies up to 6 GHz;
- by using different types of preamplifiers, primarily the same as for earth stations, parametric amplifiers, transistor amplifiers, and tunnel diode amplifiers; in addition, in contrast to earth stations, traveling wave tubes often are used in space repeaters as input preamplifiers.

The ranges of the frequency dependence of the noise characteristics of different types of input amplifiers are shown in Figure 9.16 [9.12], where it can be seen that the noise temperatures of different types of input receivers often are practically comparable. Therefore, the noise temperature not always conclusively determines the desirability of using a given type of input in different kinds of space repeaters. The choice of the necessary type of input is also determined by other energy parameters, in addition to noise characteristics: the passband, dynamic range, stability, energy consumption, and also by design simplicity, manufacturability, size, weight, cost, *et cetera*.

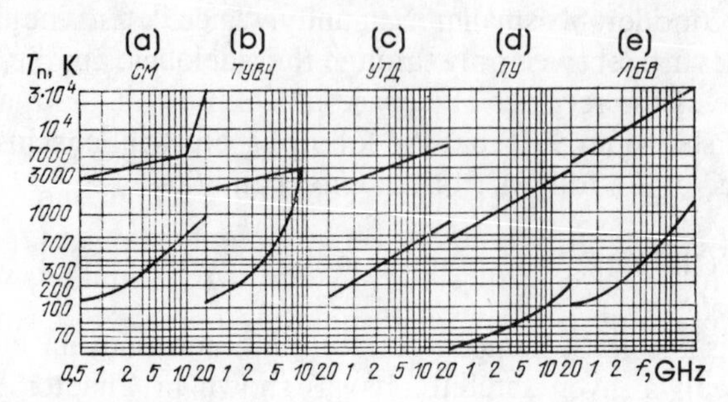

Key: **(a) Mixer**
(b) Transistor amplifier
(c) Tunnel diode amplifier
(d) Parametric amplifier
(e) Traveling wave tube

FIGURE 9.16. Ranges of noise characteristics of different types of space repeater input receivers as functions of frequency.

Main Requirements on the Inputs of Space Repeaters

1. The passband should be as wide as possible.
2. The noise temperature of an input, characterizing the sensitivity of the receiver, should be minimal. However, the noise temperature should be lowered only to the point that satisfies the following inequality:

$$T_{rec} < T_A/3$$

where T_{rec} is the noise temperature of the receiver and T_A is the noise temperature of the antenna. The noise temperature of the antenna of a space repeater may not be lower than 300°, because the main beam of the antenna is aimed toward the earth. Therefore, it is sometimes possible in principle to set the noise temperature of the input of the space repeater at about 120°–150°, if other circumstances so permit.

3. The transfer function of the input circuit should be sufficient for effectively suppressing the noises of the subsequent amplifier-converters. If a low-noise input amplifier is used the noise coefficient is

$$n_{rec} = n_{q.a} + (n_{2s} - 1)/K_g$$

where $n_{q.a}$ is the noise coefficient of the low-noise amplifier; k_g is the gain of the low-noise amplifier; and n_{2s} is the noise coefficient of the second stage of the input circuit. If K_g is large enough the second term of this equation may be ignored, and then n_{rec} of an input receiver is determined only by the noise level of the quiet amplifier.

4. The inputs of space repeaters should have a high enough saturation level, otherwise undesirable distortions of the received signals can occur in it. This is especially risky during the simultaneous reception of several signals, for example, during operation in an [expansion unknown; MD is multiple access] system.
5. The requirements on the amplitude and phase responses of an input receiver, albeit different from one system to another, nevertheless are quite rigid. For instance, in communications systems, the phase response (or the derivative of the group delay time response) must have little nonlinearity in order to reduce distortions, and the amplitude-frequency response must be highly stable in the received signal band.
6. The line-on time of an input receiver to the operating mode should be minimal.
7. Inputs should perform stably in a wide temperature range and withstand severe mechanical stresses.
8. Inputs should be simple of design and be as compact, lightweight, and use as little power as possible.

Parameters of different low-noise amplifiers of space repeaters are given in Table 9.4 [9.12]. Crystal mixers are used most widely as the inputs of space repeaters due to their simplicity and performance reliability. Transistors, Schottky diodes, ordinary point contact silicon diodes, and sometimes tunnel and inverted diodes are used at the present time in these SHF mixers. There are mixers with one diode (single-cycle), and with two diodes (balanced). Schottky diodes, which provide a set noise temperature that does not exceed 2,500 K in satellite communications bands up to 20 GHz, are used in most cases in balanced mixers.

9.8 MONITOR AND TEST EQUIPMENT OF SPACE REPEATERS

It was shown in the article [9.10] that monitor and test equipment is used for qualitatively assessing the technical characteristics of space repeaters, both independently, and as a part of a satellite. The number of prescribed technical requirements on space repeaters that are to be tested determines the number of characteristics which must be checked with the aid of monitor-test equipment and the necessary measurement precision. In addition, monitor-test equipment usually contains auxiliary technological equipment to enable a space repeater that is being checked to function independently.

Shown in Figure 9.17 as an example is a schematic diagram of monitor-test equipment for testing the space repeater of the Ehkran satellite TV broadcasting system. The main functional elements of the diagram are the test transmitter; test receiver; receiver-transmitter (test retransmitter); space repeater and monitor-test equipment control panel, monitor and meter panel, including monitor modulator and demodulator; technological frame for the installation, fastening and connection of a space repeater; heat control system; technological power source. Monitor-test equipment also usually includes a set of standard test instruments.

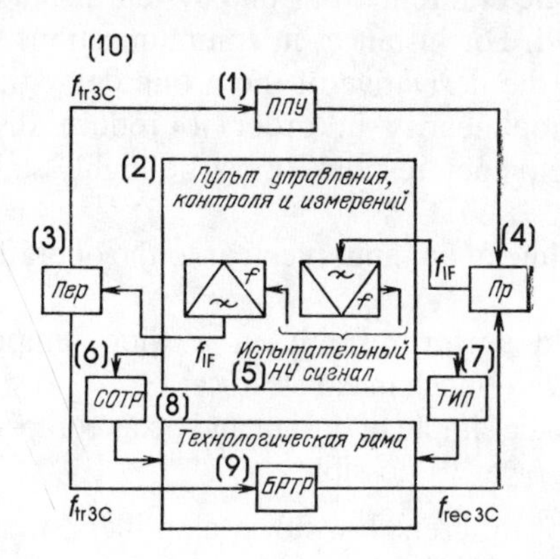

Key: **(1) Receiver-transmitter**
(2) Control, monitor and test panel
(3) Transmitter
(4) Receiver
(5) LF test signal
(6) Heat control system
(7) Technological power source
(8) Technological frame
(9) Space repeater
(10) 3C=earth station

FIGURE 9.17. Schematic diagram of monitor-test equipment of space repeater.

TABLE 9.4

Parameter, unit of measurement	**For different quiet amplifiers**			
	Transistor amplifier	*Tunnel diode amplifier*	*Traveling wave tube*	*Parametric amplifier*
Working band, GHz	0.5–8	0.5–19	0.5–20	0.5–20
Amplification band, %	Up to 50	3.5–67	36–67	0.5–7
Typical gain K_g, dB		10–20	25–35	8–45
per stage	5–8			
per amplifier	15–30			
Nonuniformity of K_g in band, dB	±(0.1–1.5)	±(0.25–5)	0.5–1.5	0.5–1.5
Noise coefficient (in dB)		3–7	11–20	(30–50)°
in frequency band, GHz:				
0.5–1	3–10.5			
1–2	3.5–7			
2–4	4.5–25			
4–8	5.5–25			

Chapter 10
Waveguides and Multiplexers

10.1 WAVEGUIDE DESIGNS

Different waveguide designs and multiplexers can be used, depending on the equipment of a station, the type of antennas, and the number of transponders. A most general diagram of a waveguide is shown in figure 10.1 [10.1]*. The entire guide can be divided into three sections: combiner (used in receiving and transmitting simultaneously); receiving guide (connects the inputs of the receivers to the outputs of the combiner), transmitting guide (connects the outputs of the transmitters to the input of the combiner).

The combiner consists of a sealing section, which separates the airtight pipe from the antenna; round waveguide sections; a polarization unit, which converts one kind of field polarization to another (linear to rotary and rotary to linear) and separates the received and transmitted signals. All elements of this part of the guide operate in the receiving and transmitting frequency bands at the same time. The receiving guide is connected to the output of the polarization unit and contains the following: elevation and azimuth swivels; a sealing section, which separates the outer part of the guide from the unsealed guide, located in the room; a trap, which protects the input of the receiver from signals from the transmitters; and a switch for connecting the main or standby set. All elements of the receiving guide are connected by rectangular waveguide sections and elbows in the *E* and *H* planes.

The transmitting guide is connected to the input of the polarization unit. The main elements of the transmitting guide are: elevation and

---

*Lines, built of this design, are used at stations of the Intersatellite communication system.

azimuth swivels; sealing section; signal adder for several transmitters; rectangular waveguide sections; and elbows, in some of which are mounted sensors that react to light from SHF breakdown. The signals from these sensors are used for turning off the transmitters. In addition, the transmitting guide also contains such components as a harmonic filter, a ferrite valve, and a switch, designed for passing high transmitter power levels.

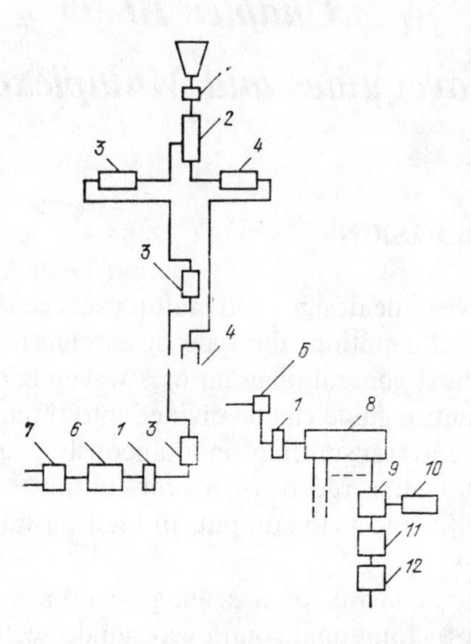

FIGURE 10.1. Diagram of waveguide 1=airtight section; 2=polarization unit; 3 and 4=swivels of the receiving and transmitting guides; 5=elbow fitting with photoelectric cell; 6=safety filter; 7=switch in receiving guide; 8=signal combiner for several transmitters; 9=switch in transmitting guide; 10=ballast load; 11=harmonic filter; 12=ferrite valve.

At some stations the receiving equipment is installed either in special booths, which rotate along with the antenna around the vertical axis, or are connected straight to the output of the polarization unit in order to reduce losses in the receiving part of the guide. In these cases, the length of the receiving guide is considerably shorter, and one or both of the swivels are eliminated*. There are also waveguide designs and multiplexing systems for antennas with light pipes. These guides contain neither receiving, nor

*These guides are used at some of the stations of the Orbita system.

transmitting swivels, and the lengths of guides can be made comparatively short with the minimum number of bends. This kind of guide has low attenuation, both on receiving and on transmitting frequencies, has improved linear phase responses, and is more reliable.

It is suggested that both directions of rotation of the polarization plane of the electromagnetic field be used for receiving and transmitting simultaneously to further increase the number of transponders in satellite communications systems. In this case, the polarization unit has two outputs (for the received signals with the polarization plane rotating in the opposite direction) and two inputs for connecting the transmitters, the signals of which are transmitted as fields, the polarization planes of which rotate in different directions. The design and construction of this kind of guide are considerably more complicated, because, in addition to a combiner section, there must be two receiving and transmitting guides with a complete set of all the necessary components. These guides obviously are best in antennas with light pipes or in cases when the receiving equipment is connected directly to the outputs of the polarization unit. Descriptions of designs, parameters, and methods of calculating individual parts of waveguides and multiplexing equipment are described below.

10.2 WAVEGUIDE SECTIONS

10.2.1 Round Waveguides

The section of a guide that connects the antenna to the polarization unit is made of waveguide sections with a circular cross section. An electromagnetic field propagates in this kind of waveguide as an H_{11} wave. The running attenuation for this wave in a copper waveguide is

$$\alpha = \frac{0.019\sqrt{f}}{a} \frac{0.416 + (\lambda/\mathrm{a})^2}{\sqrt{1 - (\lambda/3.41\mathrm{a})^2}} \qquad (10.1)$$

where f is frequency, GHz; λ is wavelength, cm; a is the cross section radius of the waveguide, cm; and α is attenuation, dB/m [10.2]. The calculated linear attenuations of power in a 70-mm diameter copper waveguide are given in Table 10.1 for different frequencies. Also given in the table are the real (maximum) attenuations, which are 30–50% higher than the calculated figures due to the roughness of the walls, the presence of oxide, and varnish films on the inner surfaces, flanges, *et cetera*. A round waveguide is connected to the antenna and polarization unit by means of flanged connectors, shown in Figure 10.2.

TABLE 10.1

f, GHz	3.4	3.9	4.2	5.6	6.0	6.4
α, dB/m (calculated)	0.0145	0.012	0.011	0.009	0.0088	0.0086
α, dB/m, actual values (maximum)	0.022	0.018	0.016	0.014	0.013	0.012

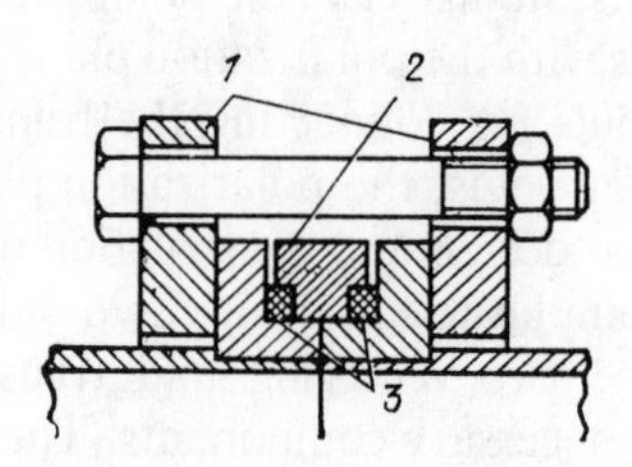

FIGURE 10.2. Flanged connector for round waveguides: 1 = adjustable flanges; 2 = centering coupling; 3 = gasket.

10.2.2 Rectangular Waveguides

The linear attenuation (in dB/m) in a rectangular waveguide for an H_{10} wave is

$$\alpha = \frac{0.019\sqrt{f}}{b}\left[1 + \frac{2b}{a}\left(\frac{\lambda}{2a}\right)^2\right] \Big/ \sqrt{1 - \left(\frac{\lambda}{2a}\right)^2} \tag{10.2}$$

where f is frequency, GHz; λ is wavelength, cm; and a and b are the dimensions of the wide and narrow walls of the waveguide, respectively, cm [10.2]. The calculated linear attenuations for rectangular copper waveguides with different cross sections are given in Table 10.2. The actual attenuations are approximately 1.3–1.5 times greater than the calculated values. Flanged connections of rectangular waveguides are shown in Figure 10.3.

TABLE 10.2

$a \times b$, mm	72×34	58×25	48×24	40×20
Frequency range, MHz	3400–3900	3600–4200	5600–6200	5600–6400
α, dB/m	0.017	0.03	0.03	0.04

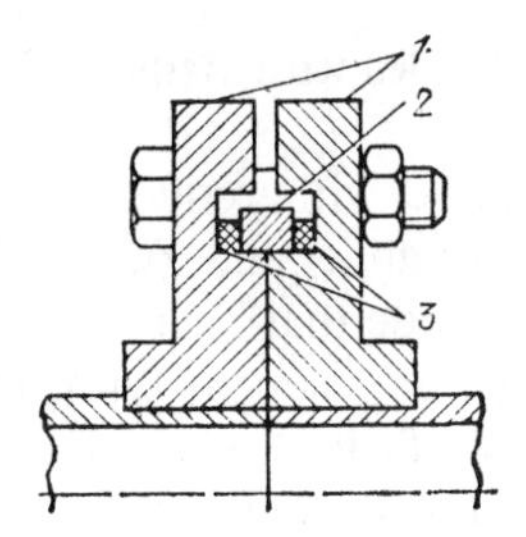

FIGURE 10.3. Flanged connector for rectangular waveguides: 1 = flange; 2 = centering coupling; 3 = gasket.

10.2.3 Elliptical Corrugated Waveguides

The features of these power guides are, first, that they have a certain amount of flexibility, by virtue of which a rather complicated configuration can be imparted to them on the spot during installation, and second, that they are manufactured in very long sections (up to 100 m and longer), wound on cable drums, which requires the minimum number of flanged connectors during installation. The cross section of corrugated waveguides is elliptical, and the dimensions (the major and minor axes) are selected such that only the main type of wave can exist in the waveguide within the working frequency range. Special end fittings are used for connecting elliptical flexible waveguides with rectangular rigid waveguides. The types and parameters of elliptical waveguides, manufactured by the industry, are given in Table 10.3.

TABLE 10.3

Type of waveguide	*WG*-2	*WG*-4	*WG*-6	*WG*-8
Frequency range, MHz	3650	5950	7900	10 800
α, dB/m	0.04	0.047	0.08	0.16
Standing wave ratio	1.16	1.16	1.16	1.16

When flexible waveguide sections are used in transmitting guides, through which high power levels pass, it is important to consider the possibility of their becoming overheated due to the poor thermal conductivity of the outer protective layer. More detailed data on elliptical corrugated waveguides are given in the book [19].

10.2.4 Rectangular Waveguides with Reduced Losses

The section of a guide with reduced losses [10.3] consists of an oversized rectangular waveguide and two fittings, which connect this waveguide to parts of the guide that have the standard cross section. One of the fittings is a waveguide segment with a gradually changing cross section. Because the tapered fitting ends with the standard cross section, it represents a short circuit for all higher types of waves that occur in the oversized waveguide. The second fitting is made in such a way that higher types of waves are completely absorbed. A diagram of it is shown in Figure 10.4. It consists of center waveguide segment 1 with an increased cross section, connected through waveguide coupler system 4 to two lateral waveguides 2 with smaller cross section. Both lateral waveguides are combined through a T-connector into one common standard waveguide 3. There is ballast load 5 on one side of the center waveguide, and the main oversized waveguide is connected to it on the other side. The dimensions of the lateral waveguides and of the coupling waveguides, and also the number of them are selected such that, on the one hand, all the power of the H_{10} wave from the larger waveguide will pass into the standard waveguide and, on the other hand, all higher types of waves that occur in the larger waveguide will be absorbed in the ballast load. The above-described waveguide with reduced losses is designed for the 5600–6400-MHz band. A 90 × 45-mm cross section waveguide is used in it. The linear attenuation for this kind of waveguide, calculated by (10.2), is 0.011 dB/m due. To the losses in the waveguide also must be added 0.1 dB due to losses in fittings.

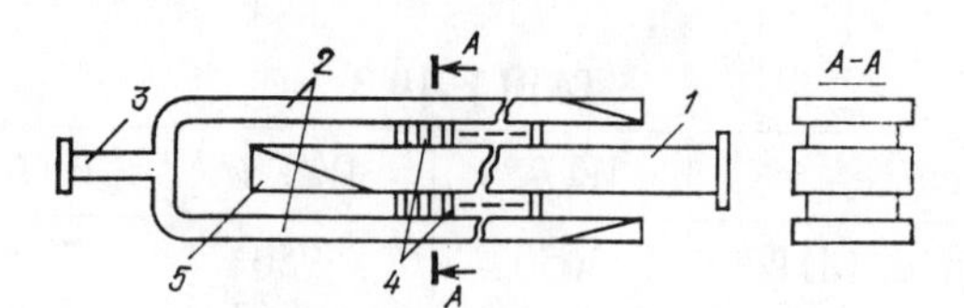

FIGURE 10.4. Waveguide fitting that absorbs higher types of waves.

Thus, the maximum total losses in a 20-m oversized waveguide are $0.011 \cdot 20 \cdot 1.5 + 0.1 = 0.43$ dB. The coefficient 1.5 takes into account the increase of actual losses in relation to the calculated losses. The losses in a guide with the 40 × 20-mm standard cross section of the same length are $0.044 \cdot 20 \cdot 1.5 = 1.35$ dB. So, the gain produced by this kind of section of guide was approximately 1 dB, or 20% of the power.

During the assembly of an oversized waveguide, it is important to pay special attention to the precision with which all the elements of the guide are joined and to lay it in a straight line. Failure to observe these requirements will result in an increase of losses due to the increased excitation of higher types of waves and the absorption of them in the ballast load.

10.3 WAVEGUIDE FITTINGS

10.3.1 Corner Fittings (Elbows)

Elbows, in which SHF breakdown sensors are mounted, are used in the transmitting part of a waveguide, in addition to conventional wave-guide elbows. The heart of this sensor is a photoresistor, the resistance of which changes in response to a light beam, which occurs when an arc appears in the waveguide. A drawing of a waveguide elbow with a sensor is shown in Figure 10.5. In the wall of the bent part of the waveguide, there are two small holes, which are the entrance ports of the channels and through which a light beam strikes the photoresistor. The axes of the channels are aligned with the longitudinal axes of the straight sections of the waveguide. Elbows are manufactured with sensors both in the *E* plane, and in the *H* plane. When the light of an arc appears in a section of a waveguide, the resistance of the corresponding photoresistor changes sharply. As a result, a signal appears which turns off the transmitter, and indicates in which part of the guide a breakdown occurred. In the absence of a light flux, the resistance of the photoresistor is higher than 100 MΩ, and when an arc appears, it decreases by a factor of more than 100 MΩ. Photodiodes may be used instead of a photoresistor.

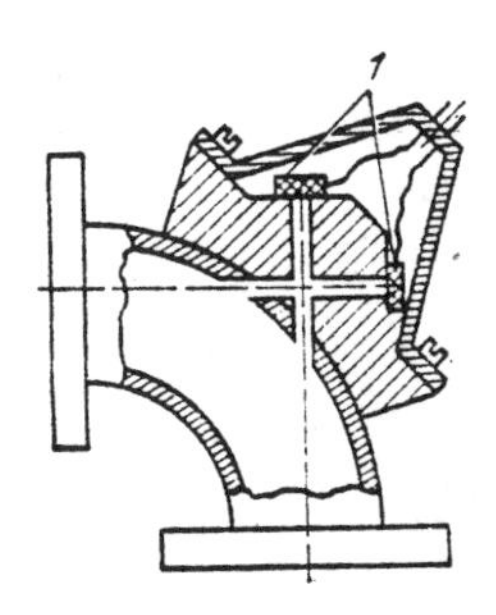

FIGURE 10.5. Elbow with photoelectric sensors: 1 = photoresistor.

10.3.2 Seal Sections

A seal section, a drawing of which is shown in Figure 10.6, is installed in a combiner. It is a 270-mm waveguide section, into which are inserted two 0.2-mm thick Teflon plates. The distance between the plates is set close to a quarter wavelength, corresponding to the middle frequency of the entire working range of the combiner, which is 18.8 mm. At this distance, a reflection from one of the plates is offset by a reflection from the other plate, and the total coffecient of reflection from the seal section does not exceed 1.5-2.5%, both on receiving and transmitting frequencies. Because the seal section is located in the outer part of the line and separates the unsealed part from the sealed part, moisture can build up in half of the section that is connected to the unsealed part, which in the winter can freeze. A seal section is equipped with a heater and a cock for draining accumulated water to prevent this. A seal section is equipped with a pipe with a valve, with which the sealing plates can be bypassed if the irradiator of the antenna is sealed. A seal section is intended for an excess pressure of not more than 0.5 atm.

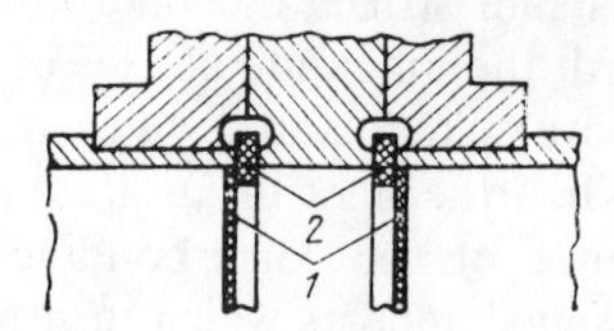

FIGURE 10.6. Seal section in round waveguide: 1 = Teflon film, 2 = frame.

Rectangular seal sections are used in receiving and transmitting guides. They consist of two rectangular flanges, between which are held a 0.2-mm thick teflon plate and a 3-mm thick teflon frame. There is a tuning element in the form of a small inductive membrane in the cross section of the plate. The coefficient of reflection from rectangular seal sections does not exceed 1-2%. Their mechanical strength is not less than 0.5 atm. They are manufactured with cross sections of 58 × 25, 48 × 24, and 40 × 20 mm. Seal sections used in combiners and transmitting guides are intended for passing an average power of up to 10 kW.

10.4 POLARIZATION UNITS

10.4.1 Designs of Polarization Units

A polarization unit in a waveguide performs two functions: it converts one kind of field polarization to another (linear to rotary or *vice* *versa*), and it combines receiving and transmitting guides into a common combiner. Fields with a left-hand* polarization vector (**L**) in the 6000-MHz range are being used at the present time in Soviet satellite communications systems of the centimeter band for transmitting signals from earth to space; fields with a right-hand polarization vector (**R**) in the 4000-MHz range are used for transmitting signals from space to earth. A diagram of a polarization unit, used in waveguides of earth stations for received and radiated signals with different polarizations, is shown in Figure 10.7. In this diagram the polarizer converts polarization in a wide frequency range (3400–6400 MHz).

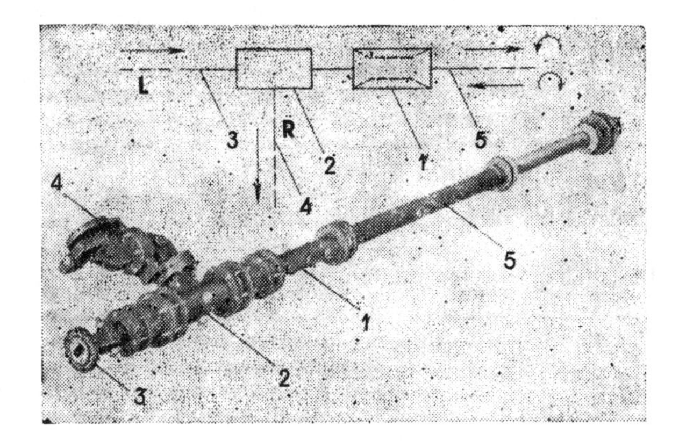

FIGURE 10.7. Diagram of polarization unit for received and radiated signals with different polarizations: 1 = wideband polarizer; 2 = polarization filter; 3 = transmitting guide; 4 = receiving guide; 5 = combiner.

Both directions of rotation of polarization may be used for receiving and for transmitting in satellite communications systems to increase the volume of information that can be transmitted. Diagrams of a polarization unit for waveguides of these kinds of communications systems are shown in Figure 10.8**. The polarization unit consists of wideband polarizer 1, which converts one kind of polarization to another, both on receiving frequencies and on transmitting frequencies; section 2 that separates received and transmitted signals, and which also separates orthogonally

---

*Left hand means that the polarization vector rotates counter-clockwise.

**Polarization units of this kind are used, for example, at stations in L'vov and Dubna.

polarized received signals, and polarization filter 5, which combines orthogonally polarized signals from the transmitter in a round waveguide. Between this section that separates received and transmitted signals and the polarization filter there are two narrow-band polarizers, which operate only in the transmitting frequency band. One of these polarizers 3, called a $\pi/2$ polarizer, converts linear polarizations to rotary polarizations; the second 4 is a π polarizer, which does not alter the field polarization, but makes it possible to change the direction of the field intensity vectors by any given angle. The purpose of these narrow-band polarizers, located in the transmitting guide, is to eliminate distortions of the field polarization of the signals radiated by the antenna.

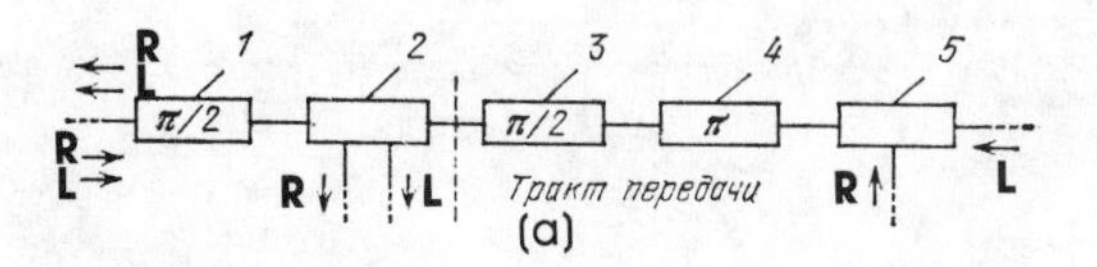

Key: **(a) Transmitting guide**

FIGURE 10.8. Diagram of polarization unit for received and radiated signals with identical polarizations, with a wideband polarizer: 1 = wide-band (π) polarizer; 2 = receiving and transmitting guide junction; 3 = narrow-band ($\pi/2$) polarizer; 4 = narrow-band (π) polarizer; 5 = polarization filter.

Shown in Figure 10.9 is a diagram of a polarization unit, which corrects polarization distortions not only in the signals radiated by the antenna, but also in the signals it receives. This kind of polarization unit consists of a section that separates received and transmitted signals and diverts them into two branches, which are round waveguides of the appropriate sizes.

Narrow-band polarizers, operating in either the receiving band or the transmitting band, are inserted in each branch. After the polarizers, polarization filters, which separate or combine linearly orthogonally polarized signals, are installed in each branch. The design of the polarization filter is complicated, but it offers the best polarization characteristics for an earth station in both the receiving and the transmitting bands.

10.4.2 Polarizers

A device that converts one kind of field polarization in a circular waveguide to another is a section of circular waveguide, in which there are

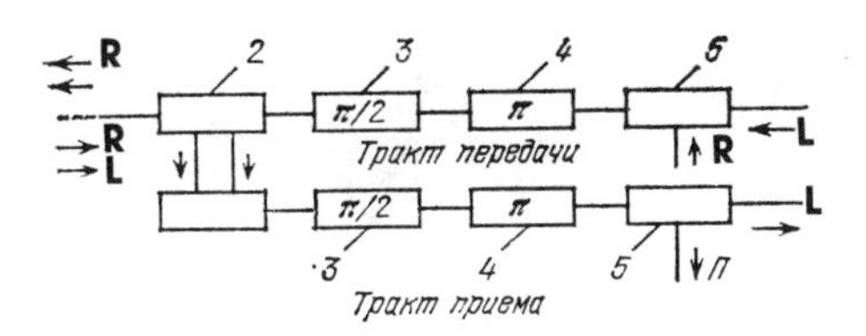

Key: **(a) Transmitting guide**
(b) Receiving guide

FIGURE 10.9. Diagram of polarization unit for received and transmitted signals with identical polarizations, with narrow-band polarizers (the symbols are the same as in Figure 10.7).

longitudinal inhomogeneities in the form of dielectric and metal plates, or a row of metal rods. The phase velocities of waves, electric field intensity vectors **E** of which are parallel or perpendicular to the plates or rods, are obviously different. In a circular waveguide with longitudinal inhomogeneities, there propagates a linearly polarized wave, whose vector **E** makes a 45°-angle with the plane of the inhomogeneities (Figure 10.10). We will expand this vector into two components: the parallel and the perpendicular planes of an inhomogeneity. At the polarizer input, both field components are identical and have the identical phases. If the length, the parameters, and shape of the plates or rod are such that the phase difference between the parallel and perpendicular components of vector **E** at the output is $\pi/2$, then at the output of the polarizer we will have a circularly polarized field instead of a linearly polarized field. (That is what a $\pi/2$ polarizer does.) Obviously, if a circularly polarized field enters this kind of polarizer, it will be converted to a linearly polarized field. If vector **E** of the linearly polarized field makes with the plane of the inhomogeneities an angle that is different from 45° by Δ, then not a circularly polarized, but an elliptically polarized field will appear at the output of the polarizer, i.e., at the output we will have not only a circularly polarized field, rotating in the necessary direction, but also a parasitic field, rotating in the opposite direction. The cross polarization of the field will be $b = 20 \log \tan \Delta \approx 20 \log \Delta$ dB. A difference of the phase shift, produced by the longitudinal inhomogeneities, from $\pi/2$ also will produce a parasitic field, the polarization of which rotates in the opposite direction. If the phase shift is different from $\pi/2$ by δ, then the amount of cross polarization will be $b = 20 \log \sin(\delta/2) \approx 20 \log(\delta/2)$. The ellipticity factors in the former and latter cases are, respectively,

$$\eta \approx 1 + 2\Delta \quad \text{and} \quad \eta \approx 1 + \delta$$

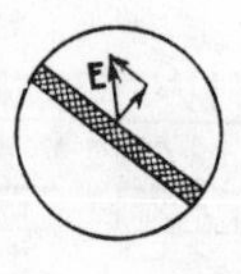

FIGURE 10.10. Cross section of polarizer.

If the length, parameters, and shape of the longitudinal inhomogeneities are such that the phase difference for orthogonal field components at the polarizer output is equal to π, then this kind of polarizer will not change the polarization, but it will rotate vector **E** by angle 2ϕ, where ϕ is the angle made by vector **E** with a line perpendicular to the plane of the inhomogeneities [10.4]. The standing wave ratio and ellipticity factor (η) of a wideband polarizer in the receiving and transmitting bands are given in Figure 10.11. Losses in a polarizer do not exceed 0.02 dB. Metal and dielectric plates or metal plates and a row of metal rods are used as longitudinal inhomogeneities for achieving a wide band.

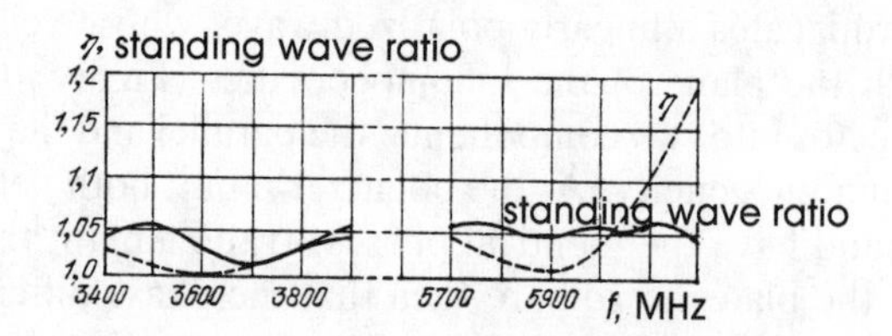

FIGURE 10.11. Electrical parameters of wideband polarizer.

A diagram and the dimension of a wideband polarizer are given in Figure 10.12. The polarizer can have a very wide band if dielectric plates or a row of resonator cavities, coupled with the field in a round waveguide through coupling slots, are used as longitudinal inhomogeneities. Resonators are detuned relative to the working frequency bands [10.13]. The parameters of this kind of polarizer are given in Table 10.4.

10.4.3 Receiving and Transmitting Guide Junctions

Polarization filters, described in detail in the book [10.4], are used in polarization units, used in communications systems, in which the received and transmitted signals are orthogonally polarized, for combining these signals. A drawing of the filter is shown in Figure 10.13. Signals received by the antenna in the 4-GHz band are selected through rectangular side

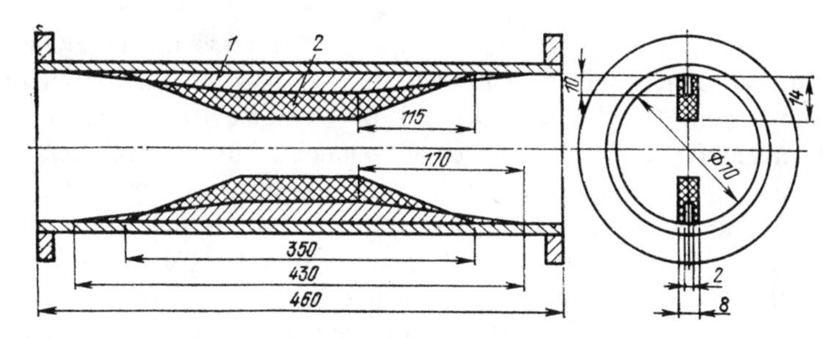

FIGURE 10.12. Wideband polarizer: 1 = metal plate; 2 = dielectric plate.

TABLE 10.4

Parameter	For frequency band	
	Receiving (4 GHz), not less than	*Transmitting (6 GHz), not less than*
Insertion losses	0.04 dB	0.04 dB
Standing wave ratio	1.1	1.06
η	1.035	1.035

branch II. The transmitted signals pass through the corresponding waveguide fitting into circular branch III.

Structural and Electrical Parameters of Polarization Filter

Diameter of circular waveguide, mm	70
Cross section of rectangular branch	58 × 25
Length of filter, mm	460
Coeffiecient of reflection from polarization filter, %, not more than	2–3
Cross attenuation between receiving and transmitting guides, dB, not more than	30
Losses in polarization filter, dB, not more than	0.02

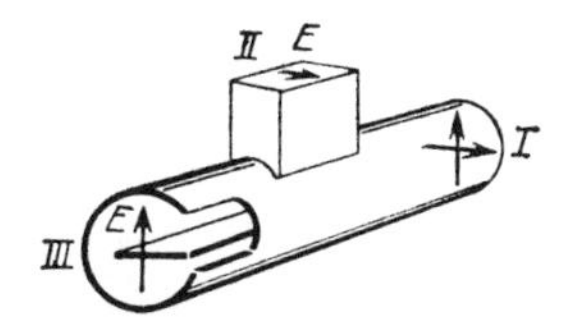

FIGURE 10.13. Diagram of polarization filter.

Shown in Figure 10.14 is a coupling that joins the receiving and transmitting branches in a polarization unit, used in dual-polarization multiplex systems (see Figures 10.8 and 10.9). The device consists of center round waveguide section 1, to which are connected four lateral rectangular waveguides 2. One end of the round waveguide section is connected to the combiner, and the other is connected to a waveguide fitting, through which the transmitting guide is connected. Because the fitting ends have a cross section that is smaller than the critical cross section for the received frequencies, the received signals will not enter the transmitting guide, but are diverted into the side branches, which are connected in pairs through double waveguide *T*-joints. Signals with just one of the orthogonal polarizations enter each pair of rectangular waveguides. A trap is inserted in each of the rectangular waveguides where it is connected to the central circular waveguide to prevent some of the power of the transmitters from entering the receiving guide. The characteristics of this device are listed in Table 10.5.

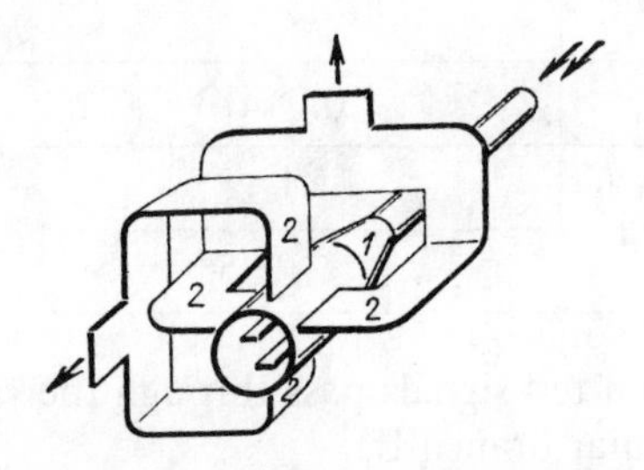

FIGURE 10.14. Diagram of device for combining reception and transmission with wideband (a) and narrow-band (b) polarizers.

TABLE 10.5

Parameter	**For frequency band**	
	Receiving (4 GHz)	*Transmitting (6 GHz)*
Insertion losses	0.07 dB	0.04 dB
Standing wave ratio	Less than 1.06	Less than 1.05
Crosstalk attenuation	—	More than 45 dB
Polarization isolation	More than 45 dB	More than 45 dB

The device described also is used in a polarization unit, in which only narrow-band polarizers are used (see Figure 10.9). In this case, the side waveguides are not connected in pairs through *T*-pipes, but are connected symmetrically to another round waveguide, into which the received signals pass. Then the receiving and transmitting round waveguide branches, to which the narrow-band polarizers are connected, end with polarization filters, described in detail in the book [19].

10.4.4 The Parameters of Polarization Units

(a) A polarization unit for guides in which the received and transmitted signals are differently polarized (see Figure 10.7)*:

	Receive	*Transmit*
Working band, MHz	3400–3900	5600–6200
Voltage ellipticity factor	<1.1	<1.25
Standing wave ratio	<1.1	<1.1
Insertion losses	<0.03	<0.05
Transmitted power, kW	–	10
Crosstalk attenuation between receiving and transmitting guides, dB	>30	>30

(b) A polarization unit for guides in which the received and transmitted signals are identically polarized (see Figures 10.8 and 10.9)**:

	Receive	*Transmit*
Working band, MHz	3700–4200	5925–6425
Voltage ellipticity factor	<1.06	<1.06
Standing wave ratio	<1.25	<1.25
Insertion losses	<0.2 (see Figure 10.8)	<0.27
	<0.3 (see Figure 10.9)	–
Transmitted power, kW	–	10
Crosstalk attenuation between receiving and transmitting guides, dB	>85	>85
Crosstalk attenuation between inputs (outputs) for differently polarized signals, dB	30	30

*Polarization units of this kind are installed in the guides of stations of Soviet satellite communications and TV systems and of the intersatellite system.

**The polarization unit described above (see Figure 10.8) is used at the L'vov station in direct communications between the USSR and the US and at the station in Dubna.

10.5 ROTARY JOINTS

10.5.1 Rotary Joints of a Receiving Guide

A diagram of a rotary joint is shown in Figure 10.15. The joint consists of two coaxial waveguide end fittings, joined by a contactless coaxial connector, so that one fitting can be turned relative to the other without breaking the electrical contact between them [10.5]. The contactless coaxial connector is outlined in the drawing by the two vertical dashed lines. One-half of the joint rotates relative to the other around the longitudinal axis, shown in the figure by the dot-dashed line. Electrical contact is made without mechanically connected conductors because the general dimensions of the bores in the outer and inner coaxial line conductors are made equal to one-half wavelength (see Figure 10.15). In this case the distribution of the longitudinal currents in the bores is such that they are equal to zero at points *a* and *a*′. This makes it possible to mechanically break the guides at these points without breaking the electrical contact.

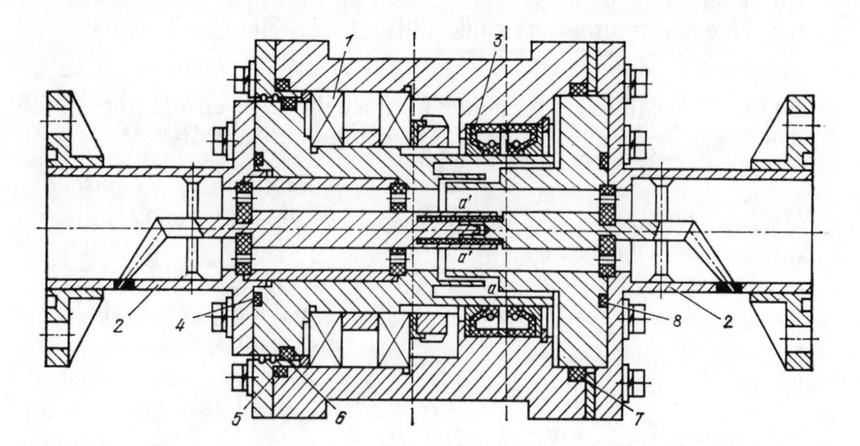

FIGURE 10.15. Diagram of a rotary joint of a receiving guide: 1 = bearing; 2 = coaxial waveguide fitting; 3 = seal; 4–8 = gaskets. General dimensions: length = 260 mm; diameter = 112 mm.

The fitting has the following dimensions: the diameter of the inside conductor of the coaxial part is 10 mm, the diameter of the outer conductor is 28 mm, and the waveguide cross section is 72 × 34 mm. The dimensions and shape of the couplings between the coaxial and waveguide parts are such that the rotary joint is well matched with the waveguide. Tuning pins, located near the couplings, are also used for this purpose. The electrical parameters are not changed when one part of the joint is turned relative to the other by any angle. The rotary joint is sealed and can be installed outside of rooms.

Characteristics of the Rotary Hinge of a Receiving Guide

Working range, MHz	3400–3900
Standing wave ratio, less than	1.05
Losses, dB, less than	0.1

10.5.2 Rotary Joint of a Transmitting Guide

Rotary joints with greater electrical strength than coaxial rotary joints are used in transmitting guides. A diagram of a joint is shown in Figure 10.16(a). It consists of two polarizers, a rotary section with a contactless waveguide connector, and fittings between round and rectangular waveguides. A polarization filter with a load is used as one of these fittings, which simultaneously functions as an elbow and a ballast load for parasitic waves. Linearly polarized waves are converted by the first polarizer to circularly polarized waves with an axially symmetric field. This field passes through the rotary section and enters the input of the second polarizer, which reconverts this field to a linearly polarized field. Then the signals pass through the corresponding fitting into the rectangular waveguide of the transmitting guide. The rotary section is a contactless connector of two round waveguides and is made in the same way as the connector of the outer conductor of the coaxial line in the rotary hinge of the receiving guide. The fittings described in Section 10.4 are used as polarizers. Because the phase shift in the polarizers is a little different from $\pi/2$ in the frequency band, a field that is not circularly, but elliptically polarized appears in the rotary section. Because of this an orthogonally polarized parasitic field appears in the waveguide after the second polarizer, in addition to the main field, linearly polarized in the necessary direction. In addition, E_{01} waves can be excited in all the assemblies of the rotary hinge due to a certain amount of structural asymmetry. A circular ballast load is connected to the free branch of the polarization filter to prevent resonant effects on orthogonally polarized waves and on the E_{01} wave. Because the level of excitation of parasitic waves depends on the relative position of the polarizers, the parameters of the rotary hinge depend somewhat on the angle of rotation ϕ of one part of the hinge relative to the other. The return losses p of the input of the hinge are given in Figure 10.16(b) for different rotation angles ϕ. All the assemblies of the rotary hinge are sealed and are connected by flanges with rubber gaskets, which make the entire assembly airtight.

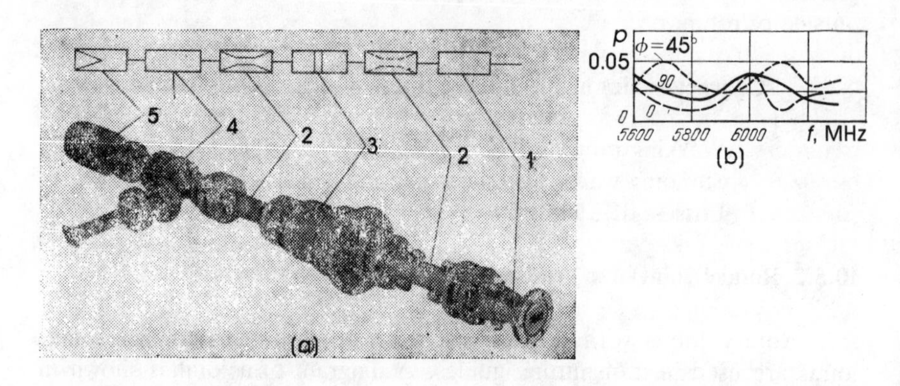

FIGURE 10.16. Rotary hinge of transmitting guide (a) and return loss as function of frequency (b): 1 = fitting; 2 = polarizer; 3 = rotary section; 4 = polarization filter; 5 = ballast load.

Characteristics of the Rotary Hinge of a Transmitting Guide

Maximum power losses in rotary hinge, dB, not more than	0.15
Working range, MHz	5600–6250
Transmitted power, kW	10
General dimensions, mm:	
length	1405
maximum diameter	165

10.6 FITTINGS FOR COMBINING THE SIGNALS OF SEVERAL TRANSMITTERS

Resonatorless devices for combining the signals of different frequencies [10.6] are used in Soviet satellite communications systems for frequency multiplexing waveguides by signals from several large transmitters. These fittings have high electrical strength, low losses, minor distortions of the phase response, and the waveguide and transmitter outputs are well matched. Depending on the number of signals to be combined, the difference between their carrier frequencies and the bandwidth of each signal, either an amplitude differentiator [10.7], a phase differentiator [10.12], or a combination of them is used for combining them. Different versions of these fittings are described below.

10.6.1 Phase Differentiators for Combining the Signals of Two and Three Transmitters

A diagram of a phase differentiator for combining two signals on different frequencies is shown in Figure 10.17. It consists of two waveguide bridges 1, connected by two transmitting guide sections of different length. The lengths of these sections are such that the phase difference on the frequency of one of the signals being combined is equal to 0, and on the frequency of the second signal it is equal to π. The transmitters are connected to the two inputs of the first bridge. One of the outputs of the second bridge is connected to the common guide. Ballast load 2 is connected to the other output. One of the properties of the bridge is that signals on frequencies f_1 and f_2, entering the different inputs of the bridge, are separated by its outputs into two components with different amplitudes. If the phase shift between the components of signal f_1, entering from one of the inputs, is equal to Φ, then the phase shift between the components of signal f_2, entering from the second input, is $\Phi + \pi$ at the same points. This phase shift is offset by the difference of the lengths of the transmitting guide segments that connect both bridges. Thus, at the inputs of the second bridge, the phase ratio of the components of both signals is identical and they both cross over into the same arm of the bridge, to which the common waveguide is connected. However, the difference of the lengths of the transmitting guides that connect both bridges produces a phase shift of π only on the middle frequencies of the signals that are being combined. When a deviation from these frequencies occurs the phase shift becomes different from π. Therefore, some of the power of the combined signals goes into the ballast load in the working bands. As is known [10.1], power losses due to a deviation of the phase shift from π are $A = -20 \log[\cos(\theta/2)]$, where θ is the deviation of the phase shift from π. In actual transmitting guides with comparatively minor frequency differences, the following expression may be used:

$$A = -20 \log \left[\cos \frac{\pi}{2} \frac{\Delta f}{f_2 - f_1}\right],$$

where f_1 and f_2 are the middle frequencies of the combined signals, and Δf is the deviation from the middle frequency. When the frequency separation is 100 MHz and the working band of signal $2\Delta f_0$ is 34 MHz, losses on the edge of the band are 0.35 dB, and when the frequency separation is 200 MHz, losses are 0.09 dB. To these losses must be added losses due to the attenuation of power in the waveguide walls. Slotted bridges, polarized filters, double waveguide bridges, *et cetera,* may be used as waveguide bridges for specific developments.

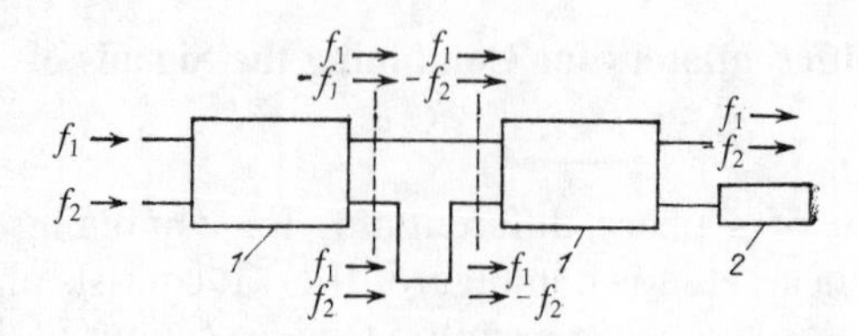

FIGURE 10.17. Diagram of phase differentiator for combining two signals: 1 = bridge; 2 = ballast load.

Shown in Figure 10.18 is a phase differentiator for combining the signals of two transmitters, in which polarization filters are used as bridges [10.12]. The device consists of five series-connected polarization filters. The first and second filters are arranged such that their side arms make an angle of 90°. The third and fourth filters are arranged so that their side arms lie in the same plane, making an angle of 45° with the side arms of the first and second filters, and they are connected so that the flat plates in them are aligned. The fifth filter is arranged in the same way as the first. Ballast loads are connected to both ends of the chain of polarization filters. The side arms of the third and fourth filters are connected by a waveguide loop.

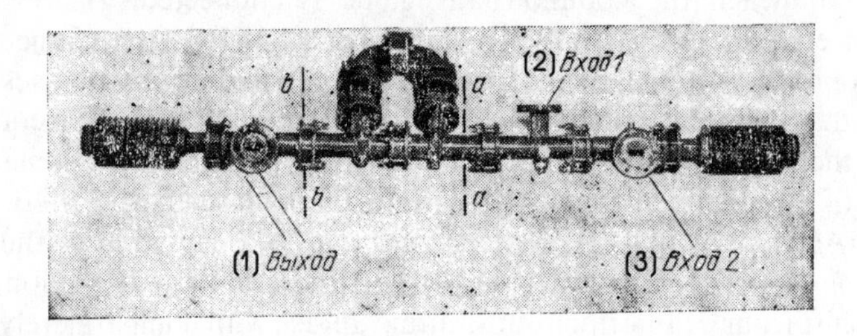

Key: (1) Output
(2) Input 1
(3) Input 2

FIGURE 10.18. Phase differentiator for combining signals of two transmitters.

The combiner works in the following way. The signals of two transmitters enter the side arms of the first and second polarization filters. These signals in cross section *aa* produce fields, the field intensity vectors of which are orthogonal to each other (Figure 10.19). Each of these vectors

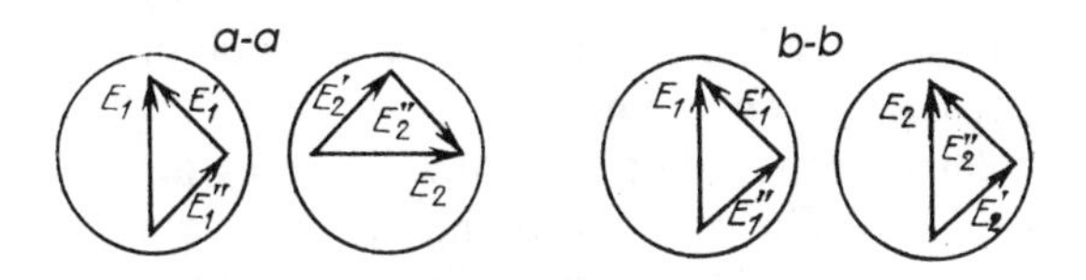

FIGURE 10.19. Distribution of field vectors in signal combiner.

can be expanded into two components with equal amplitudes (E'_1, E''_1 and E'_2, E''_2). Components E''_1 and E'_2 of the signals pass through the third and fourth polarization filters through a round waveguide, and components E'_1 and E''_2, through the rectangular waveguide loop. The length of the loop is such that in consideration of the length of the round waveguide segment between sections *aa* and *bb*, a phase shift of π will appear between components E'_2 and E''_2, and a phase shift equal to 0 between components E'_1 and E''_1. For this reason, in section *bb* the field intensity vector of both signals points in the same direction (Figure 10.19(b)) and both enter the same arm of the fifth filter, to which the waveguide of the transmitting guide is connected. Field components that are orthogonal to the main signals appear in the frequency band of each transmitter due to the dispersion of the phase shift. These signals are absorbed in the ballast load, connected to the fifth polarization filter. The ballast load, connected to the first filter, prevents resonant effects from occurring due to the excitation of parasitic waves. The length of the rectangular loop is determined by the considerations set forth above. This length is found from the relation:

$$l_1 - l_2 \approx \frac{\Lambda_0}{2} \frac{f_0}{|f_1 - f_2|} \left(\frac{\lambda_0}{\Lambda_0}\right)^2$$

where $f_0 = (f_1 + f_2)/2$; f_1 and f_2 are the frequencies of the combined signals; Λ_0 is the wavelength in the transmitted guide on frequency f_0; and λ_0 is the wavelength in free space on frequency f_0. The length found by the preceding formula is corrected a little so as to satisfy the condition

$$2\pi\left(\frac{l_1}{\Lambda_1} - \frac{l_2}{\Lambda_1}\right) = K\pi,$$

where Λ_1 is the wavelength in the transmitting guide on frequency f_1; and K is an integer. The change (longer or shorter) of length should not exceed $\Lambda_1/4$. The amplitude-frequency response of the combiner is determined by the expression $A = 20 \log[\cos(\pi\Delta f/|f_1 - f_2|)]$, where Δf is the deviation of the middle frequency of the transmitted signal. The instrument shown in Figure 10.18 is used for combining the signals of two transmitters in the 6000-MHz range, the frequencies of which differ by 100 or 200 MHz. In the former case, the length of the rectangular waveguide branch is 1300 mm, and in the latter case it is 650 mm.

The calculated amplitude-frequency responses for both devices are shown in Figure 10.20. The continuous line shows the amplitude-frequency responses at one input, and the dashed line at the other input. The circles in the same figure show experimental responses. The working bands of the transmitters in these figures are confined to the vertical dashed lines (34 MHz). The standing wave ratio at either input does not exceed 1.15. The crosstalk attenuation between the inputs to which the different transmitters are connected is not less than 35 dB.

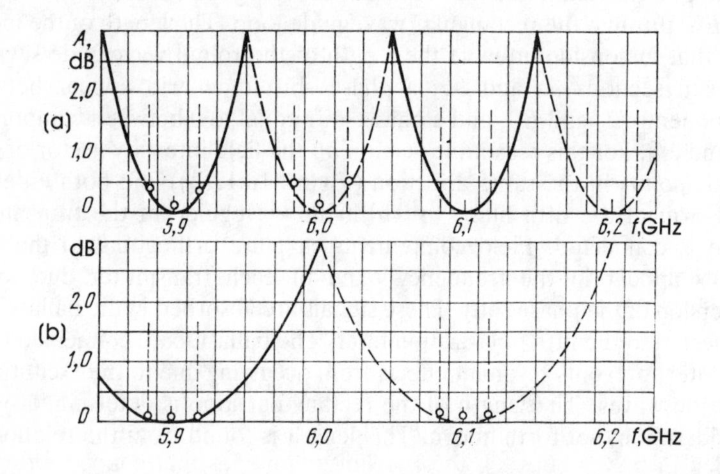

FIGURE 10.20. Amplitude-frequency responses of signal combiners (phase differentiator): (a) 100 MHz frequency separation; (b) 200 MHz frequency separation.

A two-stage combiner is used for combining the signals of three or four transmitters in a common guide, for which the separation between adjacent frequencies is 100 MHz. The above-described combiner with a short loop, in which the signals of transmitters, separated by 200 MHz, are combined, is used as the first stage. The outputs of the first stage are connected to the inputs of the second stage, which has a long loop. If the signals from three transmitters need to be combined, then the output of the third transmitter can be connected directly to one of the inputs of the second stage. The amplitude-frequency response of a two-stage signal combiner is determined by the expression (in dB)

$$A = -20[\log \cos \pi(\Delta f/\Delta f_1) + \log \cos \pi(\Delta f/\Delta f_2)],$$

where Δf_1 and Δf_2 are the frequency separations between adjacent transmitters, the signals of which are combined in the first and in the second stages, respectively.

10.6.2 Amplitude Differentiators for Combining the Signals of Six Transmitters

As can be seen by the material presented above, phase differentiators have amplitude-frequency responses that are acceptable for combining signals with a 100-MHz separation. In the case of the combining of signals with a smaller separation, for example 50 MHz, losses on the edges of the working bands of the transmitters (± 17 MHz) are unacceptably high (1.4 dB). For this reason, amplitude differentiators of the design shown in Figure 10.21 are used in these cases for combining signals. The device consists of two waveguide bridges (I and II), connected by two identical transmitting guides, in which there is an odd number of pairs of planar inhomogeneities. Waveguide bridges are used, in which, when one of the input arms is driven, signals of different amplitudes, phase-shifted by $\pi/2$, are excited in the output arms. One of the arms of each bridge is used for connecting the transmitters. The two remaining arms are connected by transmitting guide sections and by a third bridge (III). A common transmitting guide is connected to one of the outputs of bridge III, and a ballast load is connected to the other. The operating principle of an amplitude differentiator is explained as follows [10.9]. A signal from one of the transmitters on frequency f_1 enters arm a of bridge I. This signal is divided in halves between the transmitting guides that connect bridges I and II. Because there are planar inhomogeneities in the transmitting guides, the signals that enter the guides are divided into passing signals and reflected signals. The passing signals enter arms b and d of bridge II, and the reflected signals return to bridge I through its arms b and d. Since the signals that pass into arms b and d are identical, they cross over in each bridge to arm c. The signals pass from these arms into input arms a and c of bridge III. It has been shown [10.9] that by choosing planar inhomogeneities with the right parameters and the right distances between them, it is possible to make the amplitudes of the reflected and passing signals identical or close to each other in a given working frequency band of a transmitter, and the phases of these signals will differ by $\pi/2$. These signals, entering the input arms of bridge III, are combined in one of its output arms, connected to the transmitting guide. Because the amplitudes of the signals that enter the input arms of bridge III are a little different from each other a small fraction of the power will go into the ballast load. If the signal from the second transmitter, operating on frequency f_2, enters arm a of bridge I, it also will be divided in the connecting guides into a reflected signal and a passing signal. And the amplitudes of these signals in the working frequency band of the second transmitter, just as for the first transmitter, will be close to each other, but their phases will differ by 180°. Because of this, the signals of the second transmitter go into the ballast load. However, if the signals of the first and second transmitters enter arm a of bridge II, then everything will be reversed—the signal of the first transmitter will enter the ballast load, and the signal of the second transmitter will enter the common guide. Thus, to combine the signals of both transmitters, operating on frequencies f_1 and f_2, it is necessary in the common waveguide to connect the output of one of the transmitters to the input of bridge I, and the output of the second transmitter to the input of bridge II. The amplitude-frequency response of the described system, when there are two inhomogeneities in the connecting guides, will be of the form shown in Figure 10.22. The amplitude-frequency response is calculated by a procedure set forth in the article [10.8]. The continuous curve is the amplitude-frequency response at one input, and the dashed curve at the other input; A_{max} are the maximum losses in working bands $2\Delta f_0$ of the transmitters, confined to the vertical dashed lines. Maximum losses A_{max} are plotted in Figure 10.23 as functions of the ratio $2\Delta f_0/(f_1 - f_2)$. The curve in Figure 10.23 shows that an amplitude differentiator can combine signals on different frequencies with exceedingly small losses. For instance, when the separation between the transmitters is 50 MHz and the working bandwidth of each transmitter is 35 MHz [$2\Delta f_0/(f_1 - f_2) = 0.7$], losses in the working bands do not exceed 0.16 dB. These losses do not account for the attenuation of power in the walls of the waveguides of which an amplitude-differentiating signal combiner is made. Therefore, the actual losses will be a little greater than indicated in Figure 10.23.

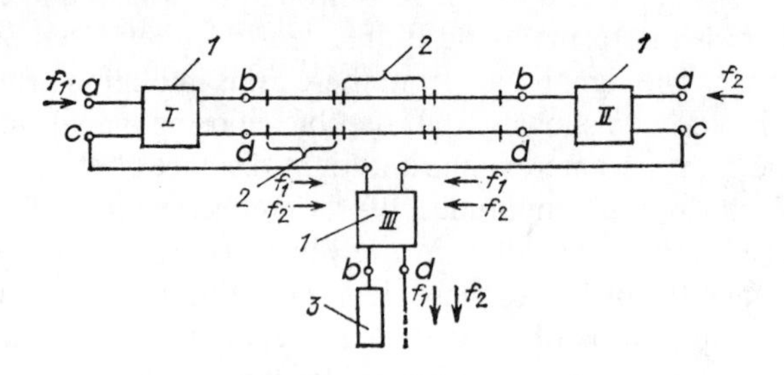

FIGURE 10.21. Diagram of amplitude differentiator for combining two signals: 1=bridge; 2=a pair of planar inhomogeneities; 3=ballast load.

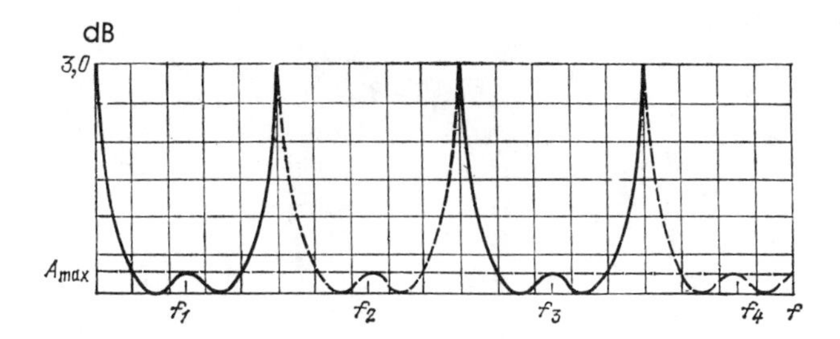

FIGURE 10.22. Amplitude-frequency response of an amplitude differentiator for combining two signals.

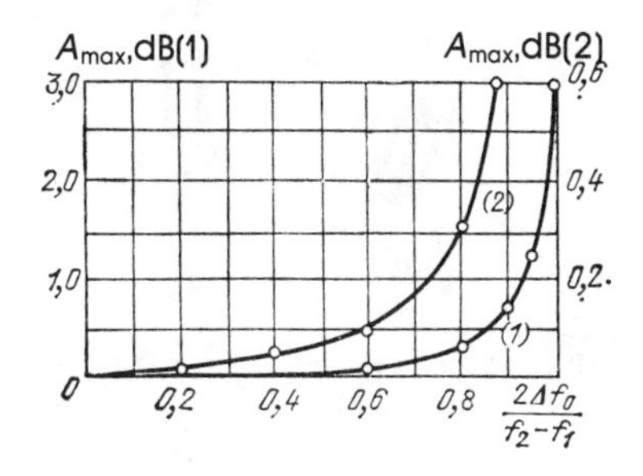

FIGURE 10.23. Maximum losses in amplitude differentiator with two inhomogeneities.

Shown in Figure 10.24 is a general view of an amplitude differentiator that combines the signals of two transmitters, operating on 6000 MHz with a 50-MHz separation. The working band of each transmitter is 34 MHz. Slotted waveguide bridges and 40 × 20-mm waveguide sections are used in the instrument. Short ($\lambda/8$) waveguide segments with a narrower wall are used as inhomogeneities. A three-stage combiner (Figure 10.25) is used for combining the signals of six transmitters with a 50-MHz separation. First the signals of transmitters with a frequency separation of 200 and 100 MHz are combined with the aid of the units described in the previous section (phase differentiators), and then the signals of two groups of transmitters are combined with the aid of an amplitude differentiator in a common guide*. The experimental amplitude-frequency responses of each of the six inputs of this instrument are shown in Figure 10.26.

---

*The system described above for combining the signals of six transmitters is used at large Soviet satellite communications stations, in particular at stations of the Moscow Space Communications Center.

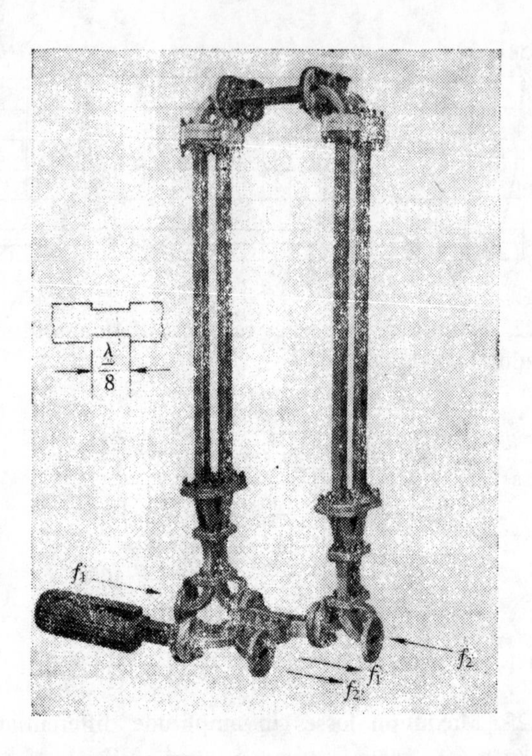

FIGURE 10.24. Amplitude differentiator for combining signals of transmitters with 50-MHz separation.

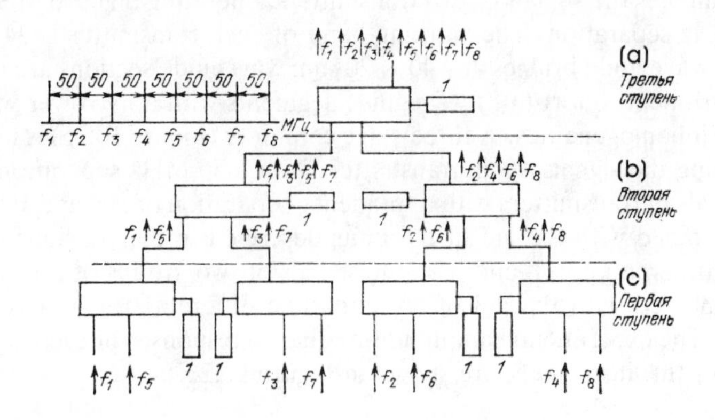

Key: (a) Third stage
(b) Second stage
(c) First stage

FIGURE 10.25. Three-step signal combiner: 1=ballast load.

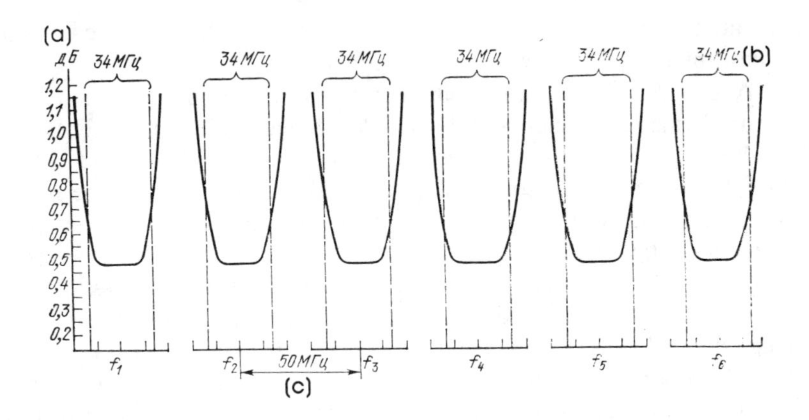

Key: **(a) dB**
(b) MHz
(c) 50 MHz

FIGURE 10.26. Amplitude-frequency response of six-transmitter signal combiner.

Electrical Parameters of the Device

Working frequency band, MHz	5600–6250
Losses for signals of any transmitter in ±17-MHz working band, dB, not more than:	
in middle of band	0.5
on edges of band	0.9
Standing wave ratio at any input in ±17-MHz band, not more than	1.15
Crosstalk attenuation between any two inputs, dB, not less than	23
Total transmitted power, kW	15

10.6.3 Amplitude Differentiators for Combining the Signals of Wideband Transmitters

An examination of the curve in Figure 10.23 shows that when there is one pair of inhomogeneities in the connecting guides, the combiner can have satisfactory responses if the ratio $2\Delta f_0/(f_1-f_2)$ does not exceed 0.9. However, in the practice of the development of satellite communications systems, cases are encountered when the signals of transmitters must be combined, for which $2\Delta f_0/(f_1-f_2) = 0.95$. As can be seen by the curve, losses in this case will be 1.5 dB, to which must be added another 0.2–0.3 dB due to losses of power in the waveguide walls. Therefore, an amplitude differentiator, in the connecting guides in which are inserted three pairs of inhomogeneities [10.10], was used for solving this problem.

A general view of the combiner is shown in Figure 10.27. It is structurally made of round and rectangular waveguide segments. Polarization filters 1, together with polarizers 2, are used as bridges I and II (see Figure 10.21), and both connecting guides are combined into one round waveguide 4, in which propagate two orthogonally polarized waves, phase-shifted by $\pi/2$. The side arms of the polarization filters are connected through rectangular waveguides to double waveguide T-fitting 3. Asymmetric arm E of the double T-fitting is connected to the common waveguide. A ballast load is connected to arm H. The signals of the two transmitters are fed into polarization filters I and II through the side arms of two other polarization filters, series-connected to them. The side arms of two adjacent polarization filters must make a 90° angle. The polarization filters, polarizers and double waveguide T-fitting are described at length in [10.4]. Teflon washers, inserted in a 46-mm diameter round waveguide, are used as inhomogeneities. The arrangement of the inhomogeneities and their dimensions and electrical parameters are given in Figure 10.28.

FIGURE 10.27. Device for combining two wideband signals: 1=polarization filter; 2=polarizer; 3=double waveguide *T*-fitting; 4=waveguide with inhomogeneities.

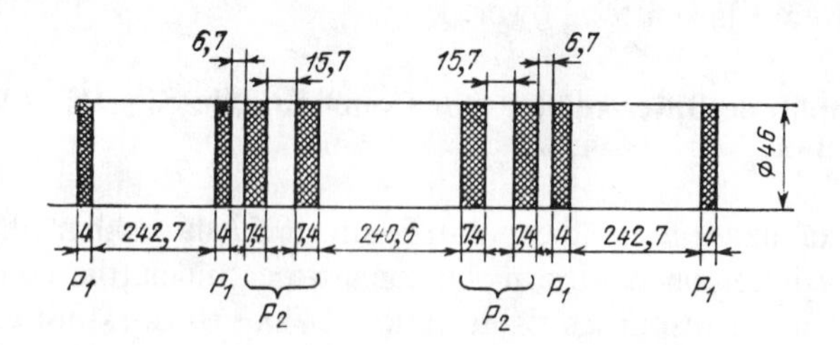

FIGURE 10.28. Diagram of inhomogeneities, forming a semireflective structure: P_1 = 0.357; P_2 = 0.753; P_1 and P_2=return losses of inhomogeneities.

The described combiner operates in the 6000-MHz range. Its amplitude-frequency responses at both inputs are shown in Figure 10.29(a). The continuous curve corresponds to one output, and the dashed curve to the other. In the same figure, the little circles indicate experimental data. As can be seen, the actual losses are 0.2 dB higher than the calculated losses due to additional losses in the waveguide walls. The standing wave ratio and crosstalk attenuation T between the inputs of the signal combiner are shown in Figure 10.29(b). The described instrument was tested at a high power level. The maximum temperature to which the combiner was heated with a power of 10 kW passing through it did not exceed 60° C without forced cooling, and breakdown and arcing did not occur*.

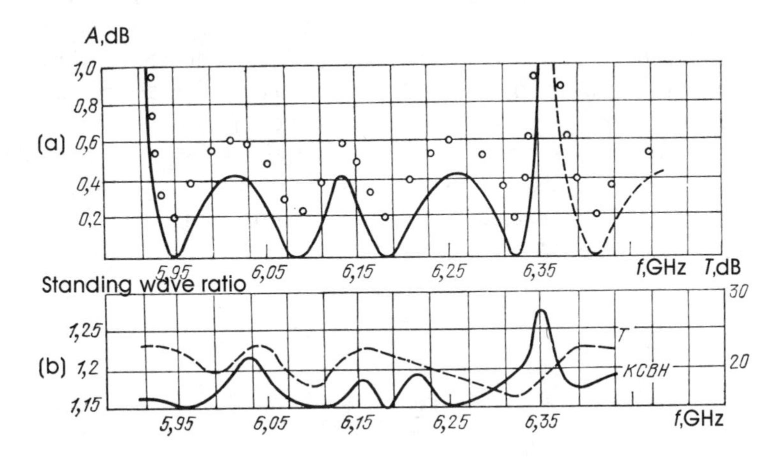

FIGURE 10.29. Electrical characteristics of wideband signal combiner.

10.7 SAFETY FILTERS

A safety filter is installed in the receiving guide. Its purpose is to protect the receiver input from signals from the transmitters. A safety filter is installed near the receiver input in some guides (see Figure 10.1), and in other guides it goes into the polarization unit (see Figure 10.14) and is installed in places where received and transmitted signals are separated. A drawing of a safety filter [10.11] is shown in Figure 10.30. It is made of a 58

*The described combiner is used at the L'vov station in direct communications between the USSR and the US.

× 25-mm waveguide section. To keep higher waves from propagating in this kind of waveguide on a frequency of 6000 MHz, the waveguide section is partitioned by a longitudinal plate into two 58 × 12-mm waveguides. Four rows of rods are placed in each waveguide. The top end of each rod is soldered to the wide waveguide wall. The total length of a rod is 14 mm, and the diameter is 1 mm. Rods of this size are tuned to resonance in the 6000-MHz range and reflect practically all signals from the transmitters. Their length is considerably different from $\lambda/4$ on receiving frequencies, and therefore the return loss is small. Four rows of rods are placed in a safety filter for increasing the return loss on transmitted frequencies and for reducing the return loss on received frequencies. The distance between the first and second, and also between the third and fourth rows is made equal to $\Lambda_{tr}/4$ (Λ_{tr} is the wavelength in the waveguide on transmitted frequencies). This increases the return loss on transmitted frequencies. The distance between two pairs of rows is such that reflections from them are compensated in the receiving frequency band. The described safety filter produces not less than 40 dB attenuation on transmitted frequencies. The return loss of this kind of filter in the receiving frequency band does not exceed 2.5%; losses in the same band do not exceed 0.05 dB. Two of these filters, series-connected, are installed for producing more than 80 dB attenuation on transmitted frequencies.

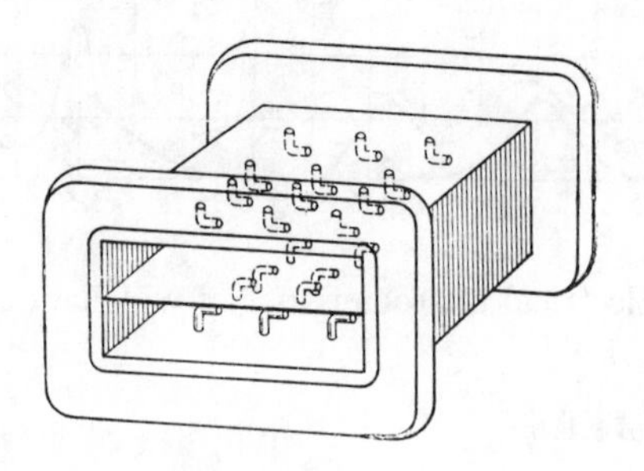

FIGURE 10.30. A trap.

10.8 THE PARAMETERS OF WAVEGUIDES AND MULTIPLEXERS

The total losses (in dB) in a waveguide and in a multiplexer are determined by adding up the losses inserted by all of their elements, including waveguide sections:

$$b = \sum b_m + b_1 L_1 + b_2 L_2$$

where b_m are the losses in the mth element; b_1 and L_1 are the linear attenuation and the length of the waveguide of the transmitting or receiving section of a guide; and b_2 and L_2 are the linear attenuation and the length of the waveguide of the combined section of a guide.

Losses on transmitted frequencies depend on the size and construction of the antenna and of its base and turntable, because they determine the length of the transmitting guide. For instance, losses on 6 GHz for a 25-m antenna (KTNA-200) do not exceed 3 dB, and for a 12-m antenna (TNA-57) do not exceed 2 dB. Losses in the receiving guide depend on the deployment of the receiver equipment. When the low-noise amplifier is installed directly at the output of the polarization unit, the losses are minimal and do not exceed 0.4 dB. When the low-noise amplifier is installed in the shack, rotating along with the antenna around the azimuthal axis, losses increase slightly and do not exceed 0.6 dB. Losses in the receiving guide not only weaken the received signal, but they also increase the input noise of the low-noise amplifier. This increment is determined by the expression:

$$T = 300\,(1 - \eta)\ K$$

where η is the efficiency of the waveguide.

When the losses in a guide are small, the approximate formula $T = b \cdot 70$ may be used, where b are the losses in the guide, in dB. The match between waveguides and transmitting and receiving equipment is determined by the standing wave ratio. In turn, the standing wave ratio is determined through the return loss of a guide:

$$K = (1 + |P|)/(1 - |P|)$$

where $|P|$ is the modulus of the return loss of a waveguide. Shown in Figure 10.31(a) are the standing wave ratios of several transponders in the transmission frequency bands at the input of a combiner, connected to a transmitting guide. As can be seen in the figure, the standing wave ratios do not exceed 1.4. To improve matching further, it is necessary to reduce the return loss of each of the components and elements that comprise a waveguide and multiplex system. A receiving guide is a little better matched, because it is considerably shorter and has fewer elements. The standing wave ratio of a receiving guide does not exceed 1.2.

10.8.1 Phase Responses (Nonuniformity of the Group Delay Time)

The nonlinearity of the phase response of a waveguide and of a multiplex system, characterized by the nonuniformity of the group delay time, is determined by the return losses of the components that comprise a waveguide and multiplex system, on the antenna irradiator and on the receiving-transmitting equipment, and also on the length of a waveguide. The nonuniformity of the group delay time as a function of frequency represents the sum of different components: linear, parabolic, and a few sinusoidal components. The linear and parabolic components are corrected with the aid of special phase correctors, located in the receiving-transmitting equipment. Sinusoidal components are virtually impossible to correct, because it requires exceedingly sophisticated phase correctors. In order to reduce sinusoidal components of the nonuniformity of the group delay time, it is necessary to achieve good matching between the guide and antenna and receiving-transmitting equipment, to shorten the guides and to reduce reflections from its elements. The transmitting guide makes the greatest contribution to the nonuniformity of the group delay time, because it is substantially longer than the receiving guide. The nonuniformities of the group delay time in the frequency bands of a few transponders are given in Figure 10.31(b).

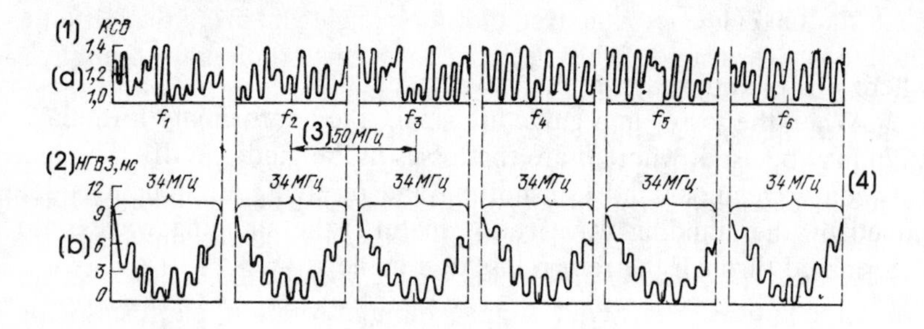

Key: **(1) Standing wave ratio**
(2) Nonuniformity of group delay time, ns
(3) 50 MHz
(4) MHz

FIGURE 10.31. **Electrical parameters of (a) waveguide and of (b) multiplex system.**

Chapter 11
Antennas of Satellite Communications Systems

11.1 GENERAL INFORMATION AND REQUIREMENTS OF ANTENNA SYSTEMS

As was mentioned earlier, satellites in a geostationary orbit at an altitude of about 36,000 km, and in an elliptical orbit of the Molniya type with an apogee of about 40,000 km, are used most widely for satellite communications. Ideally, a satellite in a geostationary orbit, relative to an earth station, is located at a fixed elevation angle and azimuth. Actually, a satellite moves around slightly, relative to an earth station due to imprecise insertion into orbit and other destabilizing factors, with the result that the elevation angle and azimuth vary from fractions to units of a degree. A satellite in an elliptical orbit moves relative to an earth station in wide ranges of elevation angle and azimuth. Therefore, in order to maintain communications, it is necessary in both cases that the antenna beam track the satellite. The tracking problem is solved with the aid of 360°-platforms (for an elliptical orbit) and simpler limited-rotation platforms (for a geostationary orbit), on which the reflector is erected, and automatic tracking and programmable steering systems.

The main factor that determines the requirements on the antenna of an earth station is the distance of several tens of thousands of kilometers between the corresponding earth station and satellite repeater, both in a geostationary orbit and in an elliptical orbit. The necessary receiver input signal-to-noise ratio at such great distances can be achieved only by using an earth station antenna with a high gain (G_a), and consequently with the necessary geometric dimensions, and a receiver with high sensitivity, achieved with the aid of low-noise input systems. Reception conditions cannot be improved by increasing the level of the power flux that a space antenna transmits toward an earth station because the transmission toward the earth of powerful signals (that can interfere with other electronic systems) cannot be tolerated. Moreover, a space power source has limited power.

Earth station antennas are expensive installations because of their great geometric size. Therefore, it is important for a given geometric area to achieve the highest possible gain. The maximum gain is achieved with an antenna that transmits a wave that has a uniform amplitude and a perfectly flat leading edge. Actually, the field amplitude falls off toward the edges of the antenna aperture. In addition to that, the leading edge of a wave becomes different from flat due to the imprecision of the reflector surface, errors in the positioning of antennas, imperfect transmitting properties of a reflector, deformations caused by gravitational and wind stresses, unidirectional solar heating, *et cetera*. The ratio of the actual gain to the maximum attainable is called the effective area. The effective area of modern antennas is about 0.7.

The signal-to-noise ratio is proportional to the ratio of G_a to the sum of the noise temperatures (T_a) of an antenna of the antenna-waveguide (T_{line}) and quiet input system (T_{rec}) (the parameter G/T). T_{line} is reduced by shortening the antenna-waveguide or by using a light guide, and T_a and T_{rec} are commensurate with each other. Therefore T_a, like G_a, is an important parameter that influences the power potential of a space-earth link. The factor that determines T_a is the power level of the noises created by the environment and received by an antenna. This level depends on the shape of the radiation pattern, particularly around its sidelobes. The earth, the noise temperature of which is about 300 K, is one of the main sources of noises. These noises during communications at small angles of elevation arrive from directions that are adjacent to the peak of the radiation pattern, i.e., in the range where the sidelobe levels are relatively high. Therefore, in this case the contribution of noises of the earth to T_a is substantial. In addition, at small elevation angles a wave travels the longest distance through the atmosphere, and the noises created by absorption in the atmosphere make a considerable contribution. In view of what was said above, the minimum working elevation angle is limited to 5° and the ratio G/T is standardized for this angle: for example, in the Intelsat system for class I stations (the antenna diameter D_a is 28–32 m) this ratio is 40.7 dB, and for class II stations ($D_a \approx 12$ m) it is about 32 dB. Depending on the depth of the reflector and the effective area, the T_a of modern antennas for elevation angles of 5° and 90°, is 40–60 K and 10–20 K, respectively (on 4 GHz).

Weather conditions, especially the working and limit wind velocities, have an important influence on the construction of an antenna. The main requirements on satellite communications earth station antennas are:

- to have a high gain with an adequate effective area (0.6–0.7) and, if possible, low noise temperature and sidelobe levels;
- to point the beam at the satellite using a turntable, programmable and manual steering, and automatic tracking systems;

- the dynamic characteristics and the construction of an antenna must be such as to meet electrical specifications and to guarantee reliable operation in given weather conditions.

Reflector antennas, which are used in satellite communications earth stations, are best suited for meeting the above requirements.

11.2 METHODS OF MEETING ELECTRICAL SPECIFICATIONS

We will examine the physical picture of the operation of a single-reflector axially symmetric antenna, for simplicity in the transmitting mode. A section of a parabola of revolution is used as a reflector. An irradiator is placed at the focus, which emits a spherical wave toward the reflector. Being reflected from the latter by virtue of the familiar property of a parabola, the wave is transformed into a plane wave. In the aperture (a plane, passing through the edge of the reflector), the phase of the field is constant and the amplitude changes in accordance with the frontal distribution of the wave that strikes the reflector. To achieve a high effective area, it is necessary to have an amplitude distribution in the aperture that is as close as possible to uniform. The distance from the focal point to the vertex of a parabolic reflector is shorter than to its edges. Therefore, in order to offset the weakening of the field in a long path, the radiation pattern of the irradiator must have elevations toward the edges of the reflector. This elevation can be achieved by means of additional excitation in the irradiator of higher types of waves with the appropriate amplitude and phase, by dephasing the field in the aperture of the irradiator, *et cetera* [11.1]. However, beyond the range of elevation the radiation pattern falls off gradually and the energy "spills over" the edge of the reflector, and this results in a loss of gain. Therefore, the levels of irradiation of the edges of a reflector are selected by making a compromise between the effective area of the aperture (the "aperture" effective area) and spillover losses.

This compromise can be reached more effectively with a double-reflector antenna (rather than a single-reflector), consisting of a horn irradiator and main and auxiliary reflectors of a subreflector. In the simplest case, a spherical wave, emitted by the irradiator, is aimed at the subreflector, which has the shape of a hyperboloid of revolution. One of its foci is aligned with the phase center of the irradiator, and the other with the focus of the main parabolic reflector. Therefore, the reflected wave has a spherical leading edge, originating from the focus of the main reflector, and the main reflector is irradiated in exactly the same way as in a single-reflector antenna, the irradiator of which has a decaying radiation pattern. In order to elevate the radiation toward the edges of the main reflector, the shape of the subreflector is modified so that some of the power that strikes the center of the main reflector will be directed toward the perimeter, and the amplitude distribution of the aperture field will be close to uniform. Now, if the subreflector is large enough in comparison with wavelength, the field can be made to fall off steeply beyond the edges of the main reflector, and energy spillover can be reduced significantly. This provides a compromise for a high effective area. However, for a modified subreflector, the surface shape of which is different from a hyperboloid, the incident spherical wave no longer is transformed into a spherical wave and phase distortions occur. These distortions are corrected by modifying the shape of the main reflector, which alters the paths traveled by the beams in such a way that the aperture field of the main reflector becomes cophasal.

The above-described two-reflector antenna with a modified main reflector and a subreflector is the main type of antenna used at an earth station. Along with a high effective area, the antenna of an earth station must also have a low noise temperature and low sidelobe levels, which do not exceed certain international standards. In order to meet this requirement, it is necessary that the amplitude distributions fall off toward the edges of the main reflector and subreflector and, to the extent possible, to reduce shading of the aperture. Then the levels of the first few ("aperture") sidelobes decrease, and so do the levels of spurious emissions due to spillage over the edges of the subreflector and main reflector. However, the effective area gets smaller when the distribution decays. Therefore, in this case a compromise is made first, and then the loss of the effective area is offset by increasing the area of the main reflector (for example, at the latest class I stations of the Intelsat system, the reflector diameter is 32 m instead of 25 m of the first generation of antennas). The spurious emissions levels can be reduced considerably by using an asymmetric antenna with an extended irradiator, which completely eliminates shading of the aperture by the irradiating system and platforms of the subreflector. These antennas are beginning to be used at earth stations [11.2].

11.3 CALCULATION OF THE RADIATION PATTERN AND GAIN OF A TWO-REFLECTOR ANTENNA

Calculations of the directive properties of antennas are based on the use of complicated methods of diffraction theory and are done with the aid of an electronic computer [11.3, 11.4]. The explanation of these methods extends beyond the scope of this work. The ensuing presentation will be confined to a clear approximate calculation of the classical Cassegrain antenna [11.5], simplified for the practical axisymmetric case* (see Figure 11.1). The meaning of the symbols that are used is explained in the figure.

Radiation pattern G_F of the horn feed and radiation pattern G_s of the horn-subreflector feed system are connected by the relation:

$$G_s(\psi) = \left(\frac{\sin\theta}{\sin\psi}\right)^2 G_F(\theta) \quad (11.1)$$

where

$$\tan(\psi/2) = M \tan(\theta/2) \quad (11.2)$$

where M is the magnification factor of the hyperboloid. In consideration of the physical picture of the operation of a double-reflector antenna (see Section 2), the resulting effective area may be written as

$$\eta = \eta_1 \eta_2 \eta_3 \quad (11.3)$$

where η_1 is the aperture effective area; η_2 is a coefficient that takes into account the energy spillover; and η_3 is a coefficient that takes into account working losses (the shading of the aperture, phase errors due to flaws, *et cetera*).

Considering that the gain of an aperture with a cophasal and uniform amplitude field distribution is $G_0 = (\pi D/\lambda)^2$, and the geometric parameters of a paraboloid are connected by the relation

$$F = \frac{D}{4} \operatorname{cotan} \frac{\psi_{max}}{2}$$

it is easy to show [11.5] that

$$\eta_1 = \frac{2 \operatorname{cotan}^2\left(\frac{\psi_{max}}{2}\right) M^2 \left| \int_0^{\theta_{max}} [G_F(\theta)]^{1/2} \tan\frac{\theta}{2}\, d\theta \right|^2}{\int_0^{\theta_{max}} G_F(\theta) \sin\theta\, d\theta} \quad (11.4)$$

Writing the ratio of the power that strikes the subreflector to the power radiated by the horn irradiator, we obtain

---

*The symbols that are used in the cited works are abbreviated here and below for the reader's convenience.

$$\eta_2 = \frac{\left[\int_0^{\theta_{max}} G(\theta)\sin\theta\, d\theta\right]}{\left[\int_0^{\pi} G(\theta)\sin\theta\, d\theta\right]} \tag{11.5}$$

The effective area of a Cassegrain antenna (see Table 11.1) should be calculated for an irradiator with a given radiation pattern and for reflectors with a given geometry by using (11.4) and (11.5) for a given η_3. Approximation of the radiation pattern is based on its representation as the sum: (a) of the main radiation, generated by the currents on the paraboloid; (b) the radiation generated by the shading of the aperture by the subreflector (shading by supports is not considered); and (c) of the radiation generated by spillage over the edges of the subreflector.

The calculation for (a) is done for a perfect paraboloid. The symbols that are used are given in Figures 11.1 and 11.2. The irradiating system is replaced with the equivalent point radiator with a radiation pattern, placed at the focus of the paraboloid. For an axially symmetric field distribution, normalized to unity, in the aperture of the paraboloid with radius ρ the radiation pattern in the far area is

$$G(\theta') = 2\pi^2_a \int_0^1 F(\rho)J_0\left[\frac{(2\pi a}{\lambda)}\rho\sin\theta'\right]\rho\, d\rho \tag{11.6}$$

where J_0 is a zero-order Bessel function and $a = D/2$.

In consideration of the fact that $\rho = \mathrm{cotan}(\psi_{max}/2)\tan(\psi/2)$:

$$F(\rho) = [G_s(\psi)]^{1/2}\cos^2(\psi/2) \tag{11.7}$$

and the expression for the normal radiation pattern is written as

$$G_{norm}(\theta') = \frac{\int_0^{\psi_{max}} [G_s(\psi)]^{1/2} J_0\left[\left(\frac{2\pi a}{\lambda}\right)(\sin\theta')\mathrm{cotan}\,\frac{\psi_{max}}{2}\tan\frac{\psi}{2}\right]\tan\frac{\psi}{2}\, d\psi}{\int_0^{\psi_{max}} [G_s(\psi)]^{1/2}\tan\frac{\psi}{2}\, d\psi} \tag{11.8}$$

To simplify the calculation of the component for (b), we will assume that the amplitude of the field of the incident wave is constant in the region occupied by the subreflector. Then from (11.6), we derive an expression for the normal radiation pattern of the shaded part of the aperture:

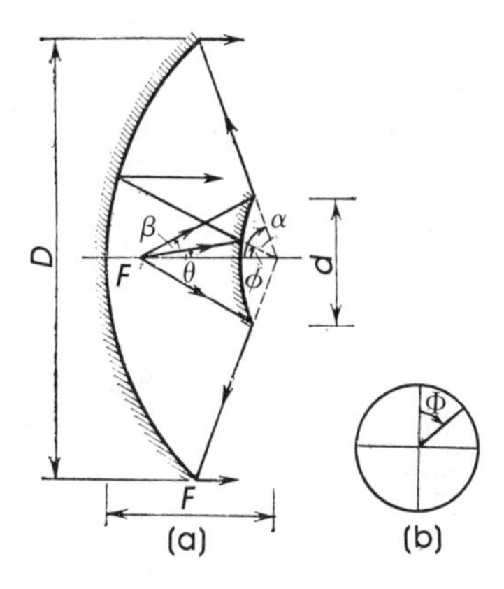

FIGURE 11.1. Geometry of classical Cassegrain antenna (a) and the aperture plane of the main reflector (b): F' = the phase center of the irradiator.

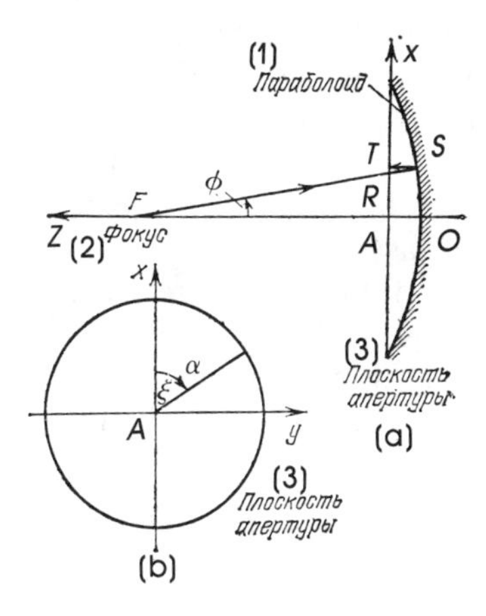

Key: (1) Paraboloid
(2) Focus
(3) Aperture plane

FIGURE 11.2. Geometry of the main reflector (a) and XY, the aperture plane (b).

$$A_{\text{norm}}(\theta') = \frac{\int_0^1 \rho J_0 \frac{2\pi b}{\lambda} \rho \sin\theta' \, d\rho}{\int_0^1 \rho \, d\rho} = 2\int_0^1 \rho \, J_0 \frac{2\pi b}{\lambda} \rho \sin\theta' \, d\rho \qquad \textbf{(11.9)}$$

where $b = d/2$.

The components for (c) are determined as the product of the experimentally found gain $G_{\text{irr.r}}(\theta')$ and its radiation pattern E_F. Thus, for the sector of angles near the peak the radiation pattern may be expressed as the sum of the components for (a)-(c):

$$D_r(\theta') = E_m G_{\text{norm}}(\theta') - E_b\, A_{\text{norm}}(\theta') + E_F G_{\text{irr.r}} \qquad \textbf{(11.10)}$$

The gain is found in consideration of (11.3):

$$G_0 = (\pi d/\lambda)^2, \quad \text{and } E_b = 2E_m(d/D)^2$$

We mention in conclusion that the CCIR recommends that the envelope of the peaks of the sidelobes of the antennas of satellite communications stations, in the sector of angles $\theta = 1$–$48°$, not exceed the levels determined by the expression $(32 - 25 \log \theta)$ dB, and in the sector $\theta = 48$–$180°$, the level of 10 dB relative to the isotropic level. Angle θ is read relative to the peak of the radiation pattern. These levels may be exceeded in 10% of the cases.

It is intended in the future to realize recommended tolerable levels determined by the expression $29 - 25 \log \theta$ in the sector $\theta = 1$–$36°$, and later on to 180° to the level of 10 dB relative to isotropic. Formulas for calculating the main parameters are given in Table 11.1.

11.4 THE PROCEDURE FOR CALCULATING REFLECTOR SURFACES

The purposes of and the physical prerequisites for modifying the shape of the surfaces of reflectors were examined in Section 11.2. Presented here is a procedure for increasing the effective area of an axially symmetric antenna for an irradiator with a given radiation pattern [11.5, 11.6].

Two differential equations and one integral equation are composed on the basis of the laws of reflection and conservation of energy. The symbols and the coordinate systems that are used are explained in Figure

TABLE 11.1

Description	*Formula*
Gain (G)	1) $\eta \frac{4\pi S}{\lambda^2}$, where S is the aperture area; λ is wavelength; η is the effective area; 2) (Formula 3.3) $\eta \approx 0.6$–0.7
The radiation pattern of a circular aperture with a uniform amplitude distribution* [11.1]:	2 arcsin $(1.22\lambda/D)$, where D is the diameter of the aperture
width of the radiation pattern at zero is $2\theta_0$ at one-half power	$1.02\lambda/D$
level of the first sidelobe relative to the peak radiation pattern	-17.6 dB
directive gain	$4\pi S/\lambda^2$
formula of the radiation pattern	$J_1(u)/u$; $u = (\pi D/\lambda)\sin\theta$; θ is the angle relative to the peak axis

11.3, which shows the upper half of an axially symmetric double-reflector antenna:

$$\frac{dr}{d\theta} = r\tan\frac{(\theta+\nu)}{2} \tag{11.11}$$

$$\frac{dy}{dx} - \tan\frac{\nu}{2} \tag{11.12}$$

$$x^2 = x_{\max}^2 \frac{\left[\int_0^\theta G_F(\theta)\sin\theta\, d\theta\right]}{\left[\int_\eta^{\theta_{\max}} G_F(\theta)\sin\theta\, d\theta\right]} \tag{11.13}$$

$$r + \frac{(x-r\sin\theta)}{\sin\nu} + y = c \tag{11.14}$$

*The amplitude distribution of the antennas of satellite communications systems is close to uniform.

where $G_F(\theta)$ is the radiation pattern of the irradiator; $(\theta + \nu)/2$ and $\nu/2$ are the angles of incidence of the current beams on the subreflector and main reflector; and c is a constant.

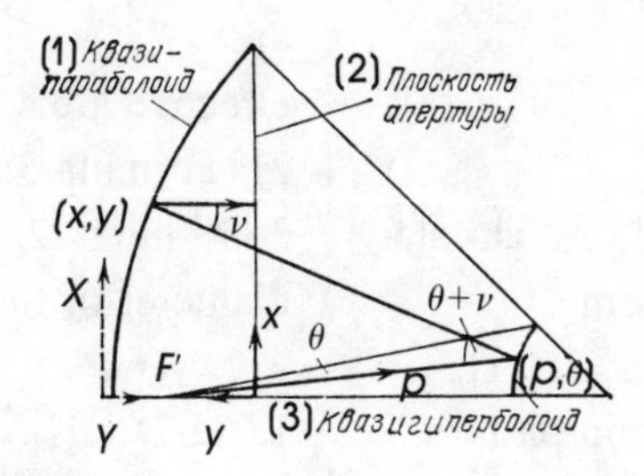

Key: **(1) Quasiparaboloid**
(2) Aperture plane
(3) Quasihyperboloid

FIGURE 11.3. Calculation of reflectors of a special shape.

The equations of a quasiparaboloid and of a quasihyperboloid are found as a result of the joint solution of these four equations by numerical methods with the aid of an electronic computer. The problem can be reduced by convergence to the solution of three differential equations:

$$\frac{dx}{d\theta} = \frac{x_{\max}^2 \, G_F(\theta)(\sin\theta)}{2A\,x} \tag{11.15}$$

$$\frac{dy}{d\theta} = -\frac{x_{\max}^2 \, G_F(\theta)(\sin\theta)}{2\,Ax} \tan\frac{\nu}{2} \tag{11.16}$$

$$\frac{dr}{d\theta} = r\tan\frac{(\theta+\nu)}{2} \tag{11.17}$$

where

$$A = \int_0^{\theta_{\max}} G_F(\theta)\sin\theta \, d\theta$$

If the main reflector has the shape of a paraboloid, which cannot be corrected, then the shape of the surface of the subreflector is corrected in consideration of the additional requirement that the difference between the resulting quasiparabolic surface and the existing parabolic surface be minimal. This problem was solved for the antenna network of the Orbita station [11.6]. The problem of calculating modified surface shapes of nonaxially symmetric antennas is exceedingly difficult and the reader is referred here to the article [11.7], in which are given one of the calculation procedures and a bibliography.

11.5 CALCULATION OF THE NOISE TEMPERATURE

We will examine a method of calculating the noise temperature in consideration of the constraints given in Section 11.1 [18]. The cumulative noise temperature (the noises of the antenna, signal path, and receiver) is conveniently reduced to the receiver input:

$$T_a = \alpha T'_a + T_0(1-\alpha) \ldots \qquad \textbf{(11.18)}$$

where α takes into account all impedance losses in the antenna and between the antenna and receiver input; T'_a is the ambient temperature; T_0 is the physical temperature of the elements responsible for losses (usually 290 K); losses usually are expressed in decibels:

$$\alpha = 10 \log(1/\alpha), \qquad \alpha < 1 \qquad \textbf{(11.19)}$$

T'_a is the cumulative weighted actual temperature of the noises that are picked up by the antenna from the environment:

$$T'_a = \frac{1}{4\pi} \int_0^{2\pi} \int_0^{\pi} T(\theta, \Phi) G(\theta, \Phi) \sin\theta \, d\theta \, d\Phi \ldots \qquad \textbf{(11.20)}$$

where $T(\theta, \phi)$ is the actual temperature of the sky and of the earth; and $G(\theta, \phi)$ is the radiation pattern of the antenna.

Graphs for determining $T(\theta)$ for frequencies of 4, 6 and 10 GHz are given in Figure 11.4, where the elevation angle, the supplement of angle of inclination θ, is plotted on the abscissa axis. (It is assumed here that $T(\theta_t, \Phi) = T(\theta)$, i.e., the distribution is symmetric with respect to the axis $\Phi = 0$). By using the curves in Figure 11.4 for a given radiation pattern it is possible, using (11.20), to calculate T'_a, and for known losses, using

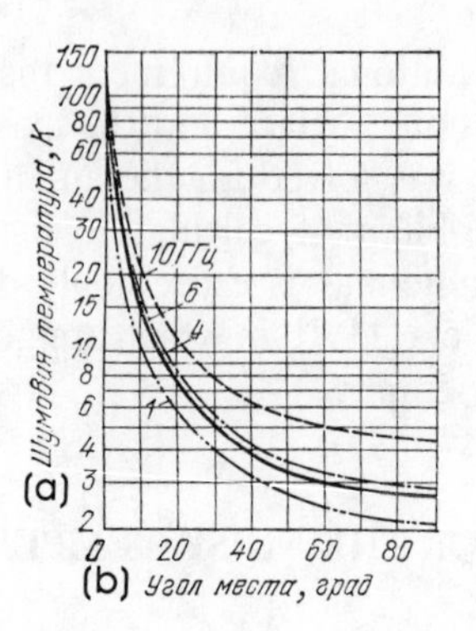

Key: **(a) Noise temperature, K**
(b) Elevation angle, deg

FIGURE 11.4. ***T*(*θ*) as a function of the elevation angle for the frequencies of 1, 4, 6 and 10 GHz.**

(11.18), to determine the cumulative noise temperature T_a. An idea of the orders of magnitudes can be obtained by way of example of an antenna with the fractional power distribution indicated in Table 11.2. The partial noise temperatures were determined graphically (Figure 11.5) [18], and the symbols in the figure are explained in Table 11.2.

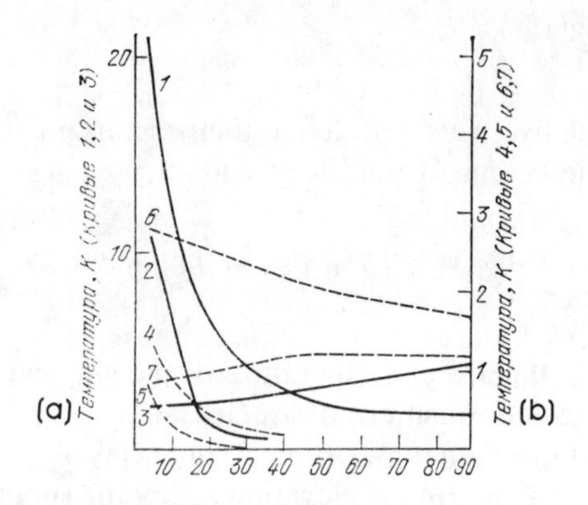

Key: **(a) Temperature, K (curves 1, 2 and 3)**
(b) Temperature, K (curves 4, 5 and 6, 7)

FIGURE 11.5. Graphs for determining partial noise temperatures at 4 GHz.

TABLE 11.2

Region of radiation pattern	*Fraction of power, %*	T_n for three elevation angles (degrees)			*Curve in Figure 11.5*
		5°	20°	45°	
Main lobe	80	23.8	6.2	2.7	1
First side lobe	5	1.4	0.4	0.2	4
Near side lobe	1	0.7	0.1	—	5
Spillage over subreflector	10	10.5	1.2	0.3	2
Far side lobes	2	2.8	2.6	2.0	6
Spillage over main reflector	1.5	2.2	2.4	3.0	3
Rear lobes	0.5	0.7	1.0	1.3	7
T'_a cumulative	—	42.1	13.9	9.5	—
T_a cumulative (for 0.4 dB losses)	—	63.9	38.2	34.2	—

11.6 ROTARY BASE STRUCTURES [18, 11.8]

Modern satellite communications are conducted through a repeater, installed on a satellite, which rotates in a stationary circular equatorial orbit (the altitude is about 36,000 km) or in an elliptical orbit (for example of the Molniya type with a 40,000 km apogee). In the former case, angular displacements of the satellite relative to the earth station are small (in the Intelsat system, for example, a satellite is held to an accuracy of 0.1°). This facilitates the problem of pointing the antenna beam of the earth station at the satellite and makes it possible to simplify the rotary base of the antenna. However, for high-latitude regions of the globe, when the angles of elevation to a stationary satellite become smaller than 5°, the reception of noises of the earth and of the atmosphere increases sharply on 4 GHz. This necessitates the use for communicating with the indicated regions of a satellite in an elliptical orbit, the apogee of which is located in the necessary hemisphere. Low-altitude circular and elliptical orbits are used in special systems. In this case, the lower the orbit, the shorter the period of rotation, and thus the higher the angular velocities of displacement of the satellite relative to the earth station and the more complicated the rotary base and the systems that steer the beam, which is accomplished by elevation angles in the range of 0–90° and by azimuth of ±180°. The angular velocities and accelerations as the antenna rotates depend, in addition to the altitude of the orbit, on the design of the rotary base of the antenna. It is customary to characterize these designs in accordance with the axes that are used (Figure 11.6). The structurally simple azimuth-elevation angle design with X- and Z-axes is most popular (see Figure 11.6). The primary axis of a rotary base of this design (fixed in space) is the vertical Z-axis, called the azimuth axis. The horizontal X-axis is the secondary axis, and it rotates azimuthally relative to the primary axis, and the antenna rotates by elevation angle relative to the secondary axis.

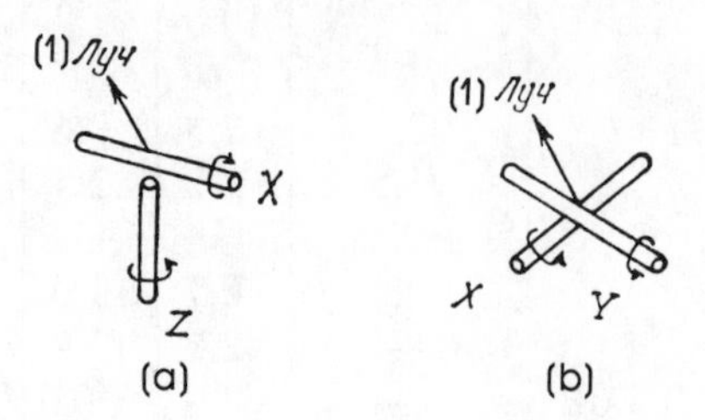

Key: **(1) Beam**

FIGURE 11.6. Diagram of antenna suspension in axes *X* (a); *X* and *Y* (b).

There are two kinds of rotary bases of radically different designs for the azimuth-elevation angle suspension of reflector antennas. Rotary bases with a central tower are usually used for antennas of average sizes (Figure 11.7). The antenna is mounted on a structure, located in the center of the tower. The azimuth drive is accomplished with a gear transmission, also located in the tower. Elevation angle drive is accomplished with a gear transmission located above the tower. The other version, which as a rule is used for large antennas, is based on the use of a rotary base of the merry-go-round type, in which carriages are installed on a large-diameter race. In the azimuth drive, there is usually a gear wheel of approximately the same diameter as the race. A merry-go-round rotary base is also used in the antennas of satellite communications stations that are equipped with a light guide. It is important for these antennas that the central part of the rotary base be free of structural elements and that a light guide can be installed in that part (see below, Figure 11.17). Examples of antennas with an azimuth-elevation angle rotary base are described in Sections 11.8 and 11.9.

It is important to mention that all azimuth-elevation rotary bases share an important drawback—the appearance of a "dead funnel" at elevation angles close to the zenith. Let us explain this. An analysis showed [11.8] that the angular rotation velocity relative to the Z-axis is propor-

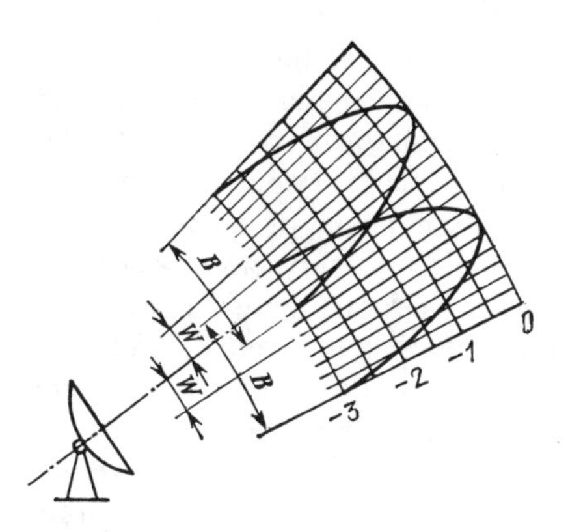

FIGURE 11.7. Two opposed beams in conical scanning.

tional to the secant of the elevation angle. Therefore, starting at some elevation angle, called the critical angle, it becomes impossible to maintain the necessary angular rotation velocity of the antenna relative to the azimuth axis due to an increase of the secant. As a result the antenna beam stops the satellite from moving and its tracking is interrupted. Tracking can be restored only after the satellite, passing through the zenith, drops to a smaller than critical elevation angle, and the required angular velocity of azimuthal rotation becomes possible. The aperture angle of the "dead funnel" depends on the orbital altitude of the satellite and on the maximum angular velocity of which the azimuth drive is capable. In cases when the "funnel" is too big, it becomes necessary to use a structurally more complicated and unwieldy suspension in *X*- and *Y*-axes. In such a design, the primary stationary axis is horizontal, "dead funnels" are located in the range of nonworking small elevation angles (Figure 11.6 (b)), and problems in the tracking of the satellite near the zenith do not occur.

In certain cases, for example when an antenna is installed on a ship, the vertical is normal to the deck and the horizontal lies in its plane and their direction changes as the ship rolls and turns. As a result, problems can occur in tracking in directions close to the primary axes in both X-Z and X-Y systems. This problem can be surmounted, for example, by turning the primary axis away from the satellite. This can be done at a slow speed with the aid of a simple rotary base that works out an angle with low precision. Rotary bases of this kind are called a suspension in *X*-, *Y*-, and *Z*- axes.

In conclusion, we will discuss a simplified version of a rotary base with a tripod suspension. Motions of the beam in a small range are sufficient for operation with a satellite in a stationary orbit. This makes it possible to simplify the rotary base and makes it cheaper. Although studies have shown that, for example, the cost of a standard azimuth-elevation angle rotary base cannot be cut very much by limiting the range of rotation by azimuth and elevation angle, it is nevertheless feasible to use a structurally entirely different rotary base, which is considerably simpler and cheaper. This kind of rotary base has a tripod suspension. In this rotary base design, the antenna is fastened to a frame, which is connected to a fixed suspension by three rods, similar to the turntable of a theodolite. One of them is of fixed length, and the other two can be adjusted. By using the adjustable rods it is possible to change the direction of the beam within certain limits. The range in which the rods can be adjusted is selected such that a large enough region of the sky can be covered, and the stationary rod is fixed in the position where the satellite will be in the middle of the range of adjustment. A detailed analysis showed that it is easy to build a rotary base with a ±5°-range of coverage, and the range can be expanded to ±10°. Rotations of the beam in these ranges are sufficient for covering all the positions of two satellites in adjacent stationary orbits, separated by approximately 5° by longitude. If it becomes necessary for the station to work with a different satellite in a stationary orbit, separated from the first by a larger angle, for example by 20°, then the fixed support can be moved to that angle. However, this takes a certain amount of time. Therefore a tripod rotary base is better for stations that operate with a pair of satellites in a stationary orbit with an angular separation of about 5°. Another advantage of this kind of rotary base is the fact that a fixed polarization can be used, which in the case of an azimuth-elevation angle suspension is not guaranteed for points close to the subsatellite point in a geostationary orbit. Details on the designs of actual rotary bases, data on their structure and details of operation are presented below by way of examples of specific antenna systems.

11.7 POINTING THE BEAM OF AN ANTENNA AT A SATELLITE [11.8]

The antennas of satellite communications earth stations create exceedingly directional radiation. For example, the width of the radiation pattern at the 3 dB level of 12 and 25 m Soviet reflector antennas, operating in the 4 GHz (reception) and 6 GHz (transmission) ranges, are approximately 16 and 7 arc minutes in the higher-frequency transmission range. For stations operating on each working frequency with a polarization vector that rotates in one direction without polarization multiplexing, it is assumed that the deviation of the beam from the line to the satellite should not exceed a tenth of the width of the radiation pattern at the 3 dB level. For our example this gives 1.6 and 0.7 arc minutes. Antennas that operate in the 11/14- and 20/30-GHz ranges, which are beginning to be adopted, have even narrower radiation patterns. For stations operating in systems in which counterrotating polarizations are used, it is desirable to make the deviation of the beam even smaller, to reduce the levels of the cross-polarization components and to increase polarization isolation. In order to point a beam within these narrow sectors, it is obviously necessary to direction-find the satellite with significantly greater precision.

The operating principle of all direction-finding systems is based on a comparison of the actual beam axis with the direction of arrival of the signal from a satellite by receiving from two radiation patterns. The peaks of the radiation patterns are shifted relative to each other so that the resulting equal signal line will coincide with the electrical axis of the antenna. Signals arriving from the direction coinciding with this axis and received from both radiation patterns have the identical strengths and their difference is equal to zero. When the direction of arrival deviates from the signal line the signals become nonidentical and difference signals appear (error signals). Their voltages and phases correspond to the projections of the angular error onto the axes of the rotary base, relative to which the antenna rotates. In the case of automatic satellite tracking, the difference signals are supplied to final controls, which generate commands that start the electric drives of the rotary base of the antenna in the right direction. As a result the difference of the directions of the beam and of the arrival of the signal from a satellite decreases, and this means that the difference signals also get weaker. When these signals fall below some threshold, the final controls generate commands to stop the drives of the rotary base. After the steering error builds up again and the threshold is reached the entire cycle is repeated. (The axial drives obviously are started and stopped not necessarily at the same time.)

The most effective direction finding is based on the radiation pattern of the steered antenna itself, when there are no errors due to errors in the determination of the actual beam line, which can change under the influence of gravitational forces during rotations by elevation angle, wind stresses, uneven solar heating, *et cetera*. Errors caused by manufacturing and mounting tolerances, and also errors that occur during the generation and execution of commands by servomechanisms are also eliminated.

The equal signal line is found either on the basis of a comparison of signals, received one after another from different radiation patterns (for example in conical scanning systems with an extremal automative drive), or by utilizing the patterns simultaneously (the monopulse method). For conical scanning, rotation takes place relative to the equal signal line of the radiation pattern, the peak of which is deflected by some angle (see Figure 11.7). If the signal does not arrive from the equal signal line, then amplitude modulation of that signal occurs, which generates a voltage, an error signal. A reference voltage generator, relative to the signals of which the phases of the error signals are read, is used for determining the directions of displacement of the antenna, necessary for setting the beam in the equal signal position.

In terms of operating, principle monopulse direction finding is based on the use of sum-and-difference radiation patterns (Figure 11.8). The signal received through the sum radiation pattern goes into a sum channel and is used as a reference signal for measuring the phase. Signals received through difference radiation patterns go to difference channels and are used for generating error signals. After error signals appear everything happens just as above.

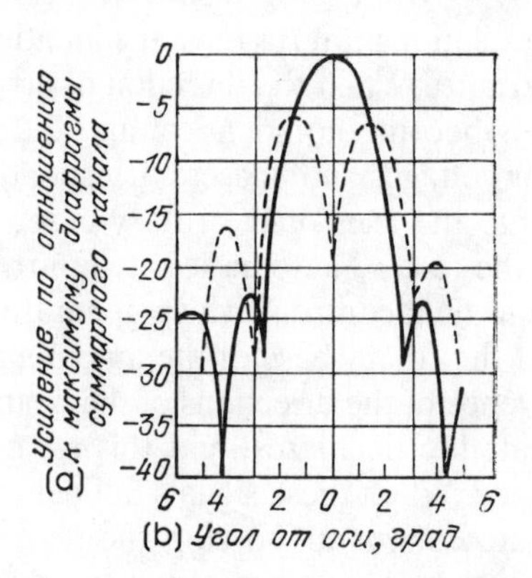

Key: **(a) Gain in relation to peak of pattern of sum channel**
(b) Angle from axis, deg

FIGURE 11.8. Sum-and-difference monopulse direction finding.

Automatic tracking is accomplished in Soviet satellite communications stations basically by conical scanning direction finding. Two types of scanners are used. In one, a scanning attachment, consisting of a bearing, rotary transmission drive, electric motor, reference voltage generator, and a seal with a radio transparent film and a phase shifting element, installed eccentrically, is installed on the inner wall in front of the aperture of a finned conical horn. The phase shifter consists of two crossed dipoles, which resonate on receiving frequencies. Because the dipoles are positioned eccentrically, the radiation pattern is deflected, which in rotation produces conical scanning on receiving frequencies. The radiation pattern stays undeflected on transmitting frequencies, by virtue of which parasitic modulation with the scanning frequency during transmission is excluded. The automatic tracking system is phased by the signal of the reference voltage generator.

The second type of scanner is the rotating center part of a subreflector, pressed onto the shaft of an electric motor, in which eccentrically crossed slits are cut for excluding modulation during transmission, and which resonate on receiving frequencies. A cavity resonator is placed on the inside behind the slits, and a reference voltage generator is attached to the other end of the motor shaft.

Still another version of automatic tracking system, which is very popular, is a system with an extremal automaton. In this system the radiation pattern of the antenna is shifted discretely in small "steps." The position corresponding to the peak signal can be determined by measuring the sign of the difference of the signal levels before and after a step. The advantages of this kind of system are its simplicity and low cost, but its drawback is its comparatively low speed, which renders a system with an extremal automaton suitable chiefly for operation through a satellite in a stationary orbit.

A programmable steering system that points the beam at a satellite in accordance with previously calculated target indications is also used at satellite communications stations, in addition to an automatic tracking system. Combined operation in the programmable steering mode, when target indications are corrected with the aid of an automatic tracking system, is also possible.

The monopulse method requires special auxiliary receivers, the equipment is more complicated, and it costs more. However, the monopulse method is used, as a rule, in systems that utilize polarization isolation, when greater satellite tracking precision is necessary, for example in Intelsat antennas. A detailed description of direction-finding methods and automatic satellite tracking systems is given in [11.8]. We will examine different versions of steering systems in greater detail.

11.7.1 Programmable Steering

The trajectory of a satellite is known, as a rule, and therefore the main programmable steering equipment is a machine which, by storing in its memory in some form or another data on this trajectory, generates signals that continuously control the antenna drive. Systems that perform this operation are called programmable steering systems. Two types of programmable steering systems exist and are used, with autonomous, and with centralized calculation of programmed steering angles.

As is known [11.9], a limited number of parameters is sufficient for the complete determination of the trajectory of a satellite, a knowledge of which, combined with the known geographic coordinates of the earth station, makes it possible to calculate steering angles on each axis of the antenna suspension at any given time. Thus, in programmable steering the very same orbital parameters and the individual geographic coordinates of the given earth station are entered in the memory of the first type of programmable steering system. A specialized electronic computer, which is a part of the programmable steering system, calculates the steering angles, as a rule in real time, on the basis of these data. A signal is generated at the output of the programmable steering system, which is proportional to the steering angles at a given moment of time, and which then goes to the antenna drive.

When a programmable steering system of the second type is used the earth station network control center is equipped with a general purpose high-speed electronic computer, which calculates programmed steering angles (target indications) on a centralized basis for each station of the network. These target indications then pass on service communications channels to each earth station, where they are entered in the programmable steering system. The target indications that are transmitted to each earth station are a discrete tabulated representation of the continuous trajectory of the satellite in azimuth-elevation angle coordinates (or in other coordinates, depending on the type of antenna suspension). The table of target indications for each axis is a list of pairs of numbers, one of which is the current time, and the other a steering angle, and digitization can be accomplished both by time, and by angle. Thus, steering angles are indicated in target indications only in a limited number of so-called reference points. Intermediate steering angles are generated in real time in the programmable steering system itself by means of interpolation. The interpolator is the main part of a programmable steering system of the second type. Two methods of interpolation are used in existing programmable steering systems (linear and nonlinear), and a steering angle as a real function of time is replaced in the space between reference points with straight line segments in the case of linear interpolation, or with segments of a second- or third-order curve in the case of nonlinear interpolation. The parameters of the substituting functions are worked out in the programmable steering system. The choice of interpolation method is determined by the tolerable interpolation precision and the complexity of the interpolator itself.

An analysis of the orbits of satellites, used for satellite communications systems, disclosed that the simpler linear interpolators accomplish interpolation with a precision of not less than 2′ with a limited number of reference points (not more than 50 for a satellite in a high elliptical orbit).

It should be pointed out that the volume of target indications that are transmitted to an earth station in the case of centralized calculation is considerably larger than the volume of data necessary for autonomous calculation, but then each earth station needs a rather sophisticated specialized electronic computer. Economic calculations and experience in the operation of the Orbita station have proved the advantages of centralized calculation.

11.7.2 Automatic Tracking Systems Based on Conical Scanning

A schematic diagram of a single-loop goniometric automatic tracking system is shown in Figure 11.9. The control signal for this system is mismatch signal α, the angle between the line to the satellite and the geometric axis of the antenna, information about which exists in the system at any moment of time. Any automatic tracking system obviously must contain two components: one that determines the deviation of the line to a satellite from the current peak line of the radiation pattern, and one that automatically scans the radiation pattern, trying to reduce the output signal of the first component to zero.

One way to determine mismatch angle α is to shift the peak line of the radiation pattern in a certain way. By comparing how the radiation pattern shifts with how the level of the received signal changes, it is obviously possible to determine the true direction to a satellite. The most popular method is conical scanning, when the antenna beam, the axis of which is deflected relative to the geometric axis by some angle (the angle of intersection of the partial patterns), rotates continuously.

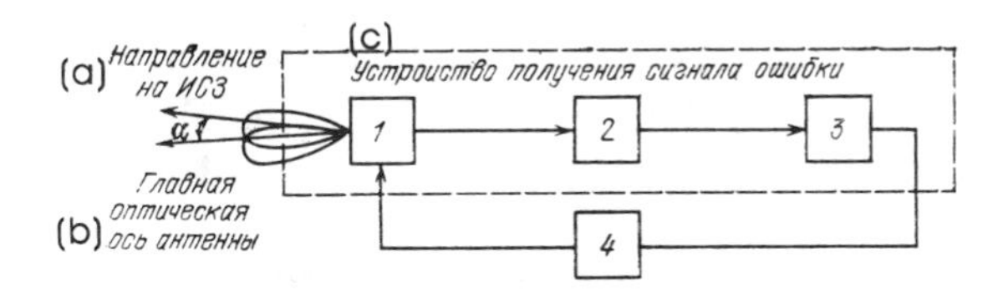

Key: (a) Direction to satellite
(b) Main optical axis of antenna
(c) Error signal generator

FIGURE 11.9. Schematic diagram of goniometric system.

The signal received from the satellite will be amplitude-modulated, the frequency of the envelope will be equal to the rotation frequency of the radiation pattern, and the modulation index will depend on the shape (steepness) of the radiation pattern, the angle of intersection of the partial patterns, and the mismatch angle (the angle between the line to the satellite and the axis of the antenna). Thus, the size of the mismatch angle can be determined on the basis of the amplitude modulation index of the received signal, and its sign can be determined on the basis of the phase of the envelope. A schematic diagram of an automatic tracking system that utilizes the conical scanning principle is shown in Figure 11.10.

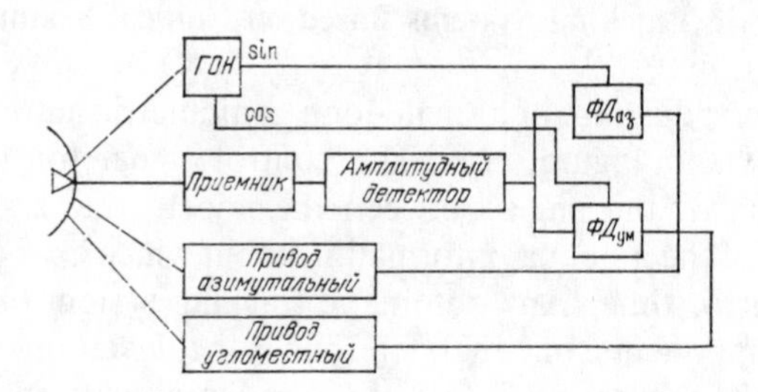

Key: **(1) Reference voltage generator (5) Elevation drive**
(2) Receiver (6) PD_{az} (phase detector)
(3) Amplitude detector (7) PD_{el}
(4) Azimuth drive

FIGURE 11.10. Schematic diagram of conical scanning automatic tracking system.

The signal, amplitude-modulated in the rotation of the radiation pattern, from the output of the antenna-waveguide, after being amplified and filtered in the receiver, passes through the amplitude detector to phase detectors, the second inputs of which receive mutually orthogonal harmonic reference signals, the frequency of which is the rotation frequency of the radiation pattern. The mismatch signal is expanded with their aid into two orthogonal signals, which determine the mismatch angles on each of the axes of the antenna suspension (for example, azimuth and elevation angle).

If the radiation pattern of the antenna in the main lobe is $F(\beta) = \exp[-1.4(\beta/\beta_{0.5})^2]$ (where $\beta_{0.5}$ is the width of the radiation pattern at one-half power, and β is the mismatch angle), the direction-finding characteristic of the automatic tracking system (the dependence of the output voltage of the phase detector on the mismatch angle) is written as

$$U(\alpha) = \kappa \exp[-1.4(\alpha_{0n}^2 + \alpha_n^2)(2.8\alpha_{0n}\alpha_n + \frac{3}{8\cdot 31}\, 2.8^3\alpha_{0n}^3\, \alpha_n^3 + \cdots)] \qquad \textbf{(11.21)}$$

where α_n is the mismatch angle, normal relative to $\beta_{0.5}$; α_{0n} is the normal angle of intersection of the partial patterns. The direction-finding characteristics, calculated in accordance with this formula, are shown in Figure 11.11.

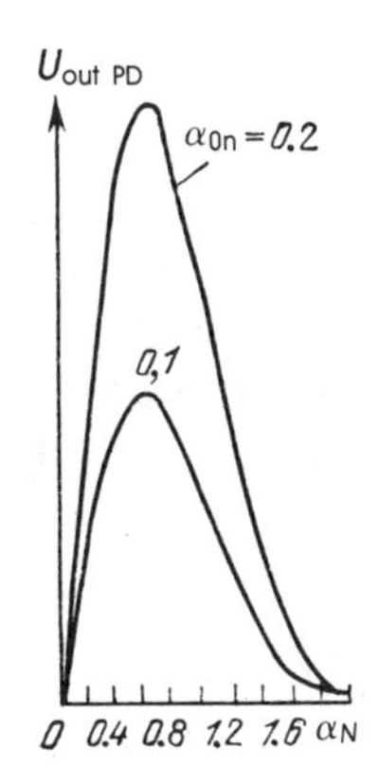

FIGURE 11.11. Direction-finding characteristic.

The angle of intersection of the partial patterns, corresponding to the maximum steepness of the direction-finding characteristic, can easily be determined by the familiar methods used for finding the extremum. The value of α_{0n}, corresponding to this point, is approximately 0.5–0.6. The level of the signal received here from the equal signal line (from the line corresponding to the point of intersection of the partial patterns) is approximately 3–4 dB lower than the level of the signal received from the direction corresponding to the peak of the radiation pattern.

The correct choice of angle of intersection of the partial patterns is determined by the criterion by which this choice is made. Thus, for instance, for earth stations that perform trajectory measurements, the main requirement imposed on the automatic tracking system is that the error in the determination of the angular coordinates of satellites be minimal, and in view of this the angle of intersection of the partial patterns should assure the steepest possible direction-finding characteristic. At the same time, the automatic tracking systems at the earth stations of satellite communications systems must minimize losses of the received signal, which corresponds to the minimum deviation of the peak axis of the radiation pattern from the line to the satellite, which represents the sum of the initial fixed deviation, which determines the angle of intersection of the partial patterns, and the fluctuation deviation, which is determined by the behavior of an automatic tracking system under the influence of fluctuation noise of the automatic tracking channel.

The fluctuation error of automatic tracking is determined by the expression [11.13]:

$$\sigma_\alpha = \frac{1}{\mu} \sqrt{2N_0\Delta F/P}\, \sqrt{1+N_0\Delta f/2P}$$

where σ_α is the mean square fluctuation error of automatic tracking; μ is a coefficient that depends on the parameters of the antenna and is proportional to the steepness of the direction-finding characteristic; N_0 is the spectral input noise density of the receiver; P is the power of the received input signal of the receiver; ΔF is the noise band of the tracking system (drive); and Δf is the noise band of the linear (in front of the detector) part of the receiver.

In this expression the first radical expresses the signal-to-noise ratio in the passband of the tracking system, and the second takes into account the increment of this ratio due to the effect of the suppression of the signal by noise in the amplitude detector. The coefficient is

$$\mu = 2.8\left(\frac{\alpha_0}{\beta_{0.5}}\right)\left(\frac{1}{\beta_{0.5}}\right)$$

Losses of the received signal as a function of the angle of intersection of the partial patterns obviously should have some minimum, because as this angle increases, on the one hand, losses increase due to an increase of the initial deviation of the equal signal line from the peak axis of the radiation pattern and, on the other hand, losses decrease, because in this case the direction-finding characteristic becomes steeper and the fluctuation component of the mismatch angle decreases. The angle of intersection of the partial patterns, corresponding to the minimum sum $(\alpha_{0n} + 3\sigma_{\alpha n})$, is

$$\alpha_{0nopt} = 1.034\ \sqrt[4]{2q + 1/q^2\,\kappa}$$

where

$$q = \frac{P}{N_0\Delta f} \text{ and } \kappa = \frac{\Delta f}{\Delta F}$$

Given in Figure 11.12 are graphs, with the aid of which the optimum angle of intersection of the partial patterns can be determined on the basis of the known parameters κ and q.

Automatic tracking systems based on extremal regulation, which

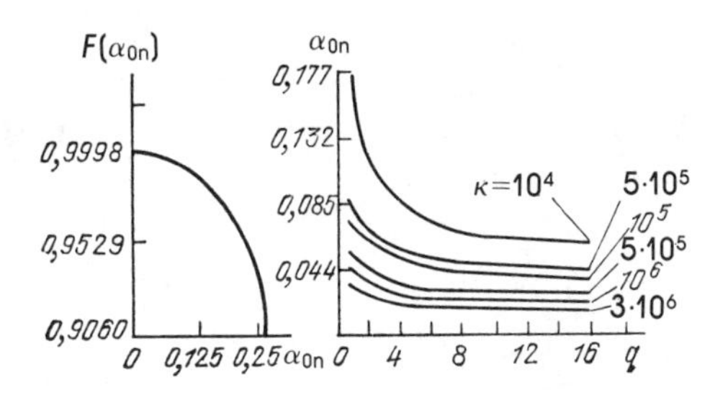

FIGURE 11.12. Determination of the optimum angle of intersection of partial patterns.

makes it possible to simplify the irradiating system, and in some cases even the antenna-waveguide significantly, have become more and more popular in recent years. Shown in Figure 11.13 is a generalized schematic diagram of an extremal system that continuously seeks the extremum, which by way of example makes it possible to explain the operating principle of systems of this kind. The system consists of a controlled process (antenna), a meter, which determines the sign of the derivative of the signal level, measured as the radiation pattern is scanned, a logic unit, which controls the controlled process by a certain algorithm, depending on the kind of sequence of signals that reach its input through final control motors.

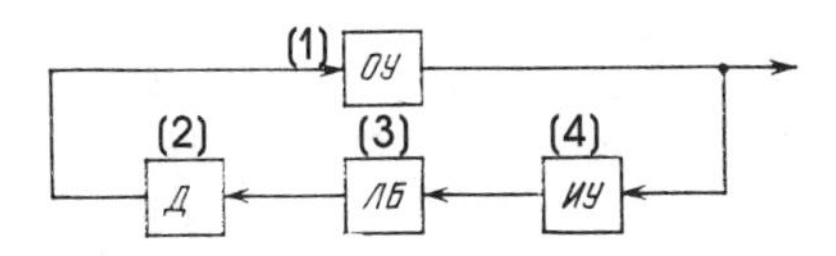

Key: **(1) Controlled process**
(2) Motor
(3) Logic unit
(4) Meter

FIGURE 11.13. Generalized schematic diagram of extremal continuous search system.

The operating algorithm of an automatic tracking system, based on the extremal automaton that is used at stations of the Moskva system, is illustrated in Figure 11.14 as a flow chart. The following symbols are used in this figure: $X1$ = a signal that appears at the output of the meter if the received signal falls below the set sensitivity of the meter, after the next step and during measurement in the position when the drives of both axes

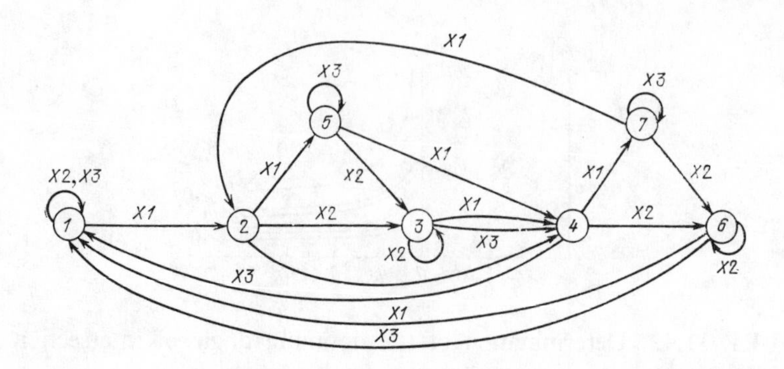

FIGURE 11.14 Operating algorithm of extremal regulator with stops.

are turned off; $X2$ = a signal, similar to $X1$, but which appears when the level of the received signal increases; $X3$ = a signal that corresponds to a constant received signal level in two successive measurements; 1 = a cycle in which the antenna does not move and only the continuous measurement of the received signal level is accomplished; 2 = activation of the azimuth drive and a step in the same direction; 3 = the continuation of motion of the antenna step by step by azimuth in the direction of a step of cycle 2; 4 = deactivation of the azimuth drive, activation of the elevation drive and the taking of one step in the direction of the last step on that axis; 5 = reversal of the azimuth axis and a step in the new direction; 6 = continuation of the motion of the antenna step by step by elevation angle in the direction of a step of cycle 4; 7 = reversal of the elevation axis drive and a step in the new direction. Calculated data on the steering precision when an extremal automaton is used as the automatic tracking system of 12-m antennas in the 4-GHz band are given in Table 11.3. Losses corresponding to the relative noise power σ_n = 0.023 in the steering signal path are given in the numerator, and to 0.012 in the denominator. More detailed information on the design of antenna steering systems can be found in the book [11.11].

11.8 SEVEN-METER 360° REFLECTOR ANTENNAS

A 360° antenna with a 7-m reflector is used on the Mars transportable station. The antenna has two reflectors, and t-e diameter of the subreflector is large enough (1100 mm) to give the horn-subreflector system a good radiation pattern. The surfaces of the subreflector and of the main reflector are modified [11.6], and the irradiator is a conical horn with a finned structure on the inside. All this gives the small antenna a high effective area of ~0.7 on receiving frequencies and ~0.6 on transmitting frequencies. The noise temperatures "to the zenith" and at an elevation angle of 5°

TABLE 11.3

Velocity of satellite, arc minutes per second	Signal losses, %, for step, arc minutes				
	0.5	1	2	4	8
0.1	2.14/1.17	1.2/0.7	1.26/0.8	2.54/2.1	7.3/7.3
0.2	7.79/2.49	1.49/0.9	1.23/0.84	2.64/2.25	7.45/7.35
0.4	Tracking interrupted	3.39/1.74	1.52/0.95	2.84/2.56	7.71/7.57
0.6	Tracking interrupted	17.1/11.54	1.88/1.47	2.97/2.61	7.97/7.67
0.8	Tracking interrupted	Tracking interrupted	2.78/1.73	3.01/2.67	8.21/7.71
1.2		Tracking interrupted	17.3/15.1	3.21/2.83	8.61/7.95

are 20 K and 50 K, respectively, which is usual for long focal reflectors (the angular size of the aperture here is 128°). On receiving and transmitting frequencies $\phi_{0.5} = 41'$ and $26'$.

Automatic tracking is based on conical scanning, which is accomplished with the aid of the scanning attachment described in Section 11.7 with a resonant phase shifter [11.9]. The main reflector is made of aluminum alloy. The foundation frame is riveted. The skin, made of polished sheets, is riveted to the frame and is a bearing member of the structure. This made it possible to reduce the weight of the reflector (it weighs about one ton). The subreflector is supported by four oval tube struts, the ends of which, for eliminating attenuation of the spherical wave, coming out of the subreflector, are fastened to the outer edge of the main reflector. The reflector is erected on a 360°-rotary azimuth-elevation angle base. The structure of the latter consists of a stationary foundation with the azimuth axis, parts that rotate by azimuth and elevation angle, and electrical auxiliary equipment. The stationary foundation, to which the vertical shaft is fastened, with bearing members and the gear wheel of the azimuth steering mechanism, is the main bearing structure, which receives all the stresses from the rotating part. The rotary part is a mount with housings, on which all the main drive mechanisms and the drive motor are installed, and it rotates on bearings of the bearing members. An equipment booth is assembled on the end of the mount.

The trunnion beam and sectors go into the rocking part of the rotary base. The electrical equipment of the rotary base consists of electric steering drives and lighting, heating, ventilation equipment, and cables. The fasteners of the alignment instruments, cables and SHF line, traps, and barriers are auxiliary equipment. The rotary base can function properly in an ambient temperature range of −50° to +50° C in up to 20 m/s winds. The limit wind head is up to 40 m/s.

11.9 A TWELVE-METER REFLECTOR ANTENNA

A 12-m reflector antenna (TNA-57M) usually is used for operation in the 4/6-GHz bands in the Orbita and Intersputnik satellite communications networks. The antenna is a short-focal double-reflector: the ratio of the focal length to the aperture diameter is 0.25, to which corresponds an angular aperture of 180°. Two versions of irradiating systems were developed for this reflector.

In the first version (Figure 11.15), the irradiating system is a small conical horn, the aperture diameter of which is equal to the wavelength in the receiving band, and a subreflector. The subreflector consists of a conical center part with a vertex angle of 145° and a base diameter of 440 mm, and a flat 640-mm diameter ring, fastened to it [11.2]. The subreflec-

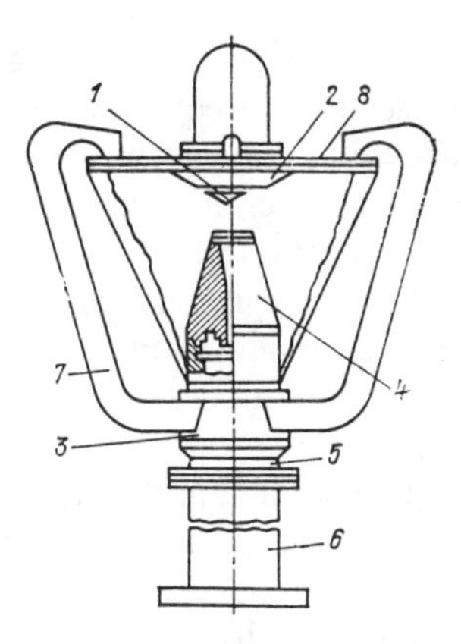

FIGURE 11.15. Diagram of an irradiator, the center part of the counter-reflector of which rotates eccentrically.

tor is mounted on three plastic brackets, installed in star fashion at an angle of 120° relative to each other. At the bottom, the brackets rest on a slide (3), which can be moved by means of a screw along a support sleeve (5), fastened to a pipe (6), installed in the center of the reflector. This is done for focusing the antenna by placing the subreflector in the optimum position. The optimum position usually is the one in which one-half of the height of the conical part coincides with the aperture plane of the main reflector.

For accomplishing scanning, the conical part of the subreflector is divided into two parts (see Figure 11.15): one of them is a truncated cone (2), which is stationary. The other is the center (1), which at receiving stations is installed eccentrically [11.8] with a 10-mm displacement on the shaft of a 30-W motor. The reference voltage generator is fastened to the other end of the shaft. To the nominal rotation speed of the motor, which is 3000 rpm, corresponds a scanning frequency of 25 Hz. The phase of the voltage of the reference voltage generator determines the position of the offset part as it rotates, and accordingly the instantaneous direction of deflection of the antenna beam. The voltage from the reference voltage generator goes to automatic tracking equipment.

The described eccentric scanner cannot be used at receiving-transmitting stations, because the beam is deflected not only on receiving frequencies, but also on transmitting frequencies, so that parasitic modulation occurs with the scanning frequency in the transmitting band. As a result, the signal that goes to the receiver of the satellite repeater is modulated with that frequency. The modulation lasts even after the signal passes through the equipment of the repeater and is radiated by the transmitting antenna. As a result the automatic tracking equipment receives a signal with the scanning frequency, both when the beam of the receiving antenna is deflected (the purely conical scanning mode), and when the beam is pointed precisely at the satellite. This makes it impossible for an earth station to select an error signal and for all the earth stations that receive signals from a given satellite to automatically track the satellite. To suppress parasitic modulation, the center part of the cone is pressed onto the motor shaft without eccentricity, two crossed slots are cut in its surface, and a cavity resonator is installed behind the cone. The slots resonate on receiving frequencies and are virtually not excited on transmitting frequencies. As a result, the beam is scanned only in the receiving frequency band.

In the second version, the irradiating system consists of a horn irradiator, unified with the irradiator of a 7-m antenna, the inner surface of which is finned, with a scanning attachment [11.8] and a modified 1.2-m diameter quasihyperbolic subreflector [11.6]. The subreflector is installed in the reflector on three oval tube struts made of aluminum alloy.

Both irradiating systems have effective areas of about 0.6 and 0.7, respectively, in the transmitting and receiving frequency bands. The envelope of the sidelobes of the radiation pattern with the first irradiating system meets the CCIR recommendation of $32 - 25 \log \theta$, with the exception of a sector near $\theta = 90°$ (which corresponds to direct spillage over the edges of the main reflector). The recommended level is exceeded by a few decibels in that sector. With the second irradiating system, other than a sector near $\theta = 90°$, the recommended level is exceeded a little in a sector next to $\theta = 30°$ (which corresponds to spillage over the subreflector). On the receiving and transmitting frequencies $\phi_{0.5} = 24'$ and $16'$, respectively.

The reflector system is installed on a 360°-rotary azimuth-elevation angle rotary base, the structure of which consists of a rigid foundation with the azimuth shaft, the part that rotates by azimuth and elevation angle, and electrical equipment and auxiliary equipment. The main bearing construction, which absorbs all stresses from the rotary part, is a stationary base, to which is fastened the vertical shaft of the azimuth steering mechanism, the gear rim and bearing members of the azimuth steering mechanism. The rotary part is a mount with housings, on which all main mechanisms of the drive and electrical power equipment are installed.

11.10 THE ANTENNA OF THE MOSKVA EARTH STATION [11.10]

The antenna was developed to meet the following basic requirements: the gain in the $\pm 1°$ sector of angles is not less than 35 dB; T_n at an elevation angle of 5° is not more than 70 K; the first sidelobe level of the radiation pattern does not exceed −20 dB; operation with circular polarization with an ellipticity factor of not less than 0.7; the beam is set manually by azimuth and elevation angle within the following limits: approximately ±90° and 0–60°, and fine ±5°; the possibility of outfitting the antenna with a tracking drive for operating in the automatic tracking mode on two axes within ±2.5°; simplicity and structural manufactureability. A serial parabolic reflector with a diameter of 2.5 m and a focal length of 0.75 m (the angular aperture is 160°) is used in the antenna.

For reducing the sidelobe level, the reflector is irradiated by a single-reflector system with little shading by the irradiator, and the field diminishes rapidly toward the edges of the aperture. This was possible by virtue of the size of the reflector, which is sufficient for achieving a given gain with the reflector underirradiated. An irradiator in the form of a double-pass logarithmic spiral is used for achieving circular polarization. The dimensions of this spiral were selected for achieving a given ellipticity factor and for achieving a diminishing aperture field distribution with the edges of the reflector irradiated at the −16-dB level. The spiral is driven by a coaxial symmetrizer. A quarter-wave transformer in the form of inserts in the central conductor and adjustment screws is used for matching.

Measurements have shown that the characteristics of the antenna meet the given requirements: the gain on 3657 MHz was 37.5 dB (the effective area is 0.61) and in the ±1° sector it is not less than 35 dB, the ellipticity factor does not exceed 0.85, and the input standing wave ratio of the irradiator in the working band does not exceed 1.2. The width of the radiation pattern at the 3-dB level is close to 2.5°, and the first sidelobe levels do not exceed the prescribed −20 dB.

The specified beam-pointing limits are ±2.5°, i.e., within just one width of the radiation pattern at the 3-dB level the beam can be rotated without moving the main reflector and by moving the irradiator in the transverse direction by ±40 mm. The loss of gain here does not exceed 0.5 dB [11.10]. Because the arc of the circle in a small rocking angle is close to a chord, the transverse motions in the vertical and horizontal planes are made by rotating the support pipe of the irradiator relative to two orthogonal hinges near the vertex of the reflector for the purpose of simplifying the structure.

The irradiator is scanned by a tracking electric drive with the minimum angular step, which is 12 arc minutes. The electric drive is controlled by a two-coordinate extremal automaton, which performs successive extremum search on two axes. A step is worked out structurally with the aid of a combination of reducers and a "maltese cross." The latter eliminates uncontrollable rotations due to motor spindown after the voltage is cut off.

For an earth station antenna in stationary execution the mount is built as a combination of a horizontal welded frame, lying on the ground and erected at approximately the prescribed azimuth, to which is fastened an inclined frame. Its angle of inclination is set for the prescribed elevation angle. A square frame, to which the reflector is fastened, is hinged to the inclined frame. If the antenna is not equipped with an irradiator scanning system, then rotations of this frame, accomplished with a rigging screw, make it possible to steer the antenna by azimuth within ±7.5°. The antenna is steered by azimuth within ±5° by turning the rigging screw of the rod that supports the inclined frame. The antenna mount for a transportable station (see Figure 11.16) has set limits of 0–90° within which the reflector can be rotated by azimuth, so that the reflector can be set in the zenith transport position. Manual fine adjustment by elevation angle is accomplished within ±10°, and preliminary setting by azimuth is accomplished by positioning the chassis on which the antenna is mounted.

11.11 RECEIVING ANTENNAS OF THE EHKRAN SYSTEM

Two types of antennas are used in the Ehkran system: PA (expansion unknown) for operation in the low-noise TV repeaters, and PP (expansion unknown) for operation with large TV repeaters or with telecenters [11.11]. The following main requirements are imposed on PA and PP: frequency range 702–726 MHz, rotary polarization with an ellipticity factor of 0.7 on the axis of the radiation pattern, gain 21 dB and 28 dB, respectively (relative to a crossed dipole), the widths of the radiation pattern of PA in the horizontal and vertical planes are identical, and the radiation pattern of PP is two times wider in the vertical plane. Arrays of panels of the "wave channel" type are used as the antennas of PA and PP for the purposes of simplification and cost cutting. Each of the panels consists of 30 orthogonally crossed dipoles, an active element in the form of a double-turn cylindrical spiral, fed by coaxial cable, and a director, all fastened to a supporting tubular boom. All current conducting elements are made of aluminum alloy strips, which simplifies manufacturing and eliminates hard to obtain pipes.

The PA antenna is assembled as a cophasal array, consisting of four wave channel panels, arranged in pairs in two stages at the corners of a square with a 125-cm side. For achieving rotary polarization, a pair of panels set up on one diagonal is rotated 90° counterclockwise and is shifted by $\lambda_{mid}/4$ forward relative to the second pair of panels. A panel of a cophasal array is connected to a coaxial feed through a section of cable of equal length and a quarter-wave matching transformer. The transformer is used for matching four parallel sections of cable with the wave impedance of the feed. A parallel correcting short-circuited quarter-wave loop is added to the transformer to improve matching. The PA antenna is designed to be set up at elevation angles of 0–70° and by azimuth within ±180°.

The PP antenna is a uniform cophasal array of 32 wave channel panels, assembled as four stages, each with eight panels. This arrangement makes it possible to satisfy the requirement of double the width of the radiation pattern in the vertical plane. The PP antenna is designed to be set up in the same ranges of elevation angles and azimuths as the PA, and semioperational orientations within ±7° relative to any initial position are possible.

Field measurements that were conducted showed that the antennas have the prescribed characteristics. The traveling wave ratio in most of the working frequency band is not less than 0.75, and the gains of the PA and PP range from 20.8 dB to 21.5 dB and 27.5 dB to 28 dB, respectively. The fact that the gain of the PP antenna increased relative to the gain of the PA antenna by less than 9 dB, as it should have when the signals of eight PA antennas were combined, is explained by the influence of the bearing structure, placed in the middle of the active area to guarantee rigidity. The measured ellipticity ratios in the working frequency band are 0.77–0.85 for the PA antenna and 0.8–0.84 for the PP antenna.

Weather tests that were conducted disclosed that the accumulation on the antenna panels of a layer of snow or an ice crust results in a 2–4 dB decrease of gain. Therefore, the PA and PP antennas should be used basically in frostfree regions. For frosty regions a short cylindrical spiral with 2.5 turns and a passive reflector is used in the antenna arrays as a radiating element. The diameter of the spiral is 140 mm and the pitch is 92 mm. The passive reflector is built as four linear elements, fastened to the ends of cross-shaped bases in such a way that each of the elements forms the letter T with the base [11.11].

11.12 DEVELOPMENT PROSPECTS OF EARTH STATION ANTENNAS

There are two aspects of the development of earth station antennas. One is connected with the improvement of an earth station, operating in the 4/6-GHz bands, and the other is connected with the adoption of the 11/14- and 20/30-GHz bands. Three main trends can be seen at the present time in the mentioned aspects. One of them is attributed to the high precision with which satellites are held in orbit, reaching 0.1° per day. This makes it possible to significantly simplify the rotary base of an antenna and to steer the beam of a satellite precisely by swinging the subreflector in the 4/6- and 11/14-GHz bands. Investigations have shown [11.12] that when a particular rocking fulcrum is selected, the beam can be rotated to 2–2.5 of the width of the radiation pattern at the 3-dB level with less than 1 dB loss of gain. The widths of the radiation pattern at the 3-dB level of an antenna with a 7-m reflector on the upper frequencies of the 6- and 14-GHz working bands are 26′ and 12′, and for a 12-m antenna they are 16′ and 7′. By deflecting the beam by 2.5 of the width of the radiation pattern by means of rocking the subreflector, it is possible with acceptable losses to track a satellite in the mentioned bands within ±1° and 0.5° with a 7-m antenna and ±40′ and ±18′ with a 12-m antenna, respectively. Because the subreflector is small and lightweight it can be rotated by low-power tracking drives. In this case, the reflector can be set roughly manually or by a crude nontracking drive. This eliminates expensive powerful electric tracking drives, necessary for steering the beam by rotating the large main reflector. A general view of an antenna with a limited rotation rotary base is shown in Figure 11.16 as an example.

FIGURE 11.16. A 12-m antenna with a limited rotation rotary base.

The second trend is based on the use in satellite communications systems of operation on two counterrotating polarizations, which makes it possible to use each of the working frequencies of an SHF transponder twice. In order to eliminate crosstalk interference, created by the signals that are transmitted on a common frequency, it is necessary to have close to the ideal circular polarizations, i.e., an ellipticity factor of not less than

1.03–1.09. This places rigid requirements on the antenna-waveguide and on the axial field symmetry in an antenna. Operational counterrotating polarization has been adopted at stations of the 4/6-GHz bands [11.2].

The third trend is based on the outfitting of an antenna with a light guide—a system of periscopic mirrors, which transmit the signal from the irradiator to the subreflector and back [11.2, 11.8]. We will explain its operation by way of the example (Figure 11.17) of a four-mirror light guide [11.8]. For simplicity we will examine the transmission mode. Irradiator 7, a conical horn with a special finned structure on the inner walls, creates an axially symmetric radiation pattern with a shape close to the optimum gaussian, and a wave with a spherical leading edge. After being reflected from the first plane mirror 1, inclined at a 45° angle, the spherical wave, irradiated by the irradiator, changes its propagation direction by 90° and strikes the second parabolic mirror 6. The point of origin of the spherical wave is aligned with the focal point of the paraboloid, and therefore mirror 6 transforms the spherical wave to a plane wave, propagating toward the third, also parabolic mirror 5. The latter transforms the plane wave to a converging spherical wave and sends it to the fourth plane mirror 2. This mirror, like the first, changes the direction of propagation of the wave by 90°, sending it to the subreflector. The wave converges at focal point 3 and then spreads. The focal point of hyperbolic subreflector 8 is aligned with point 3. As a result the subreflector irradiates a spherical wave, emanating from its focus, the same as in a conventional Cassegrain antenna with main reflector 4. It follows from what is explained above that by using a light guide it is possible to modify the shapes of the surfaces of a subreflector and main reflector (Sections 11.2 and 11.4).

As the antenna rotates by azimuth all four mirrors of the light guide move together relative to the azimuth rotation axis, and therefore the operating conditions of the light guide remain unchanged. When main reflector 4 rotates by elevation angle, mirror 2 of the light guide turns along with it. The center of mirror 2 is aligned with the elevation axis, and therefore when the antenna rotates by elevation angle, the operating conditions of the light guide do not change. It follows from the above that when the antenna rotates by azimuth and elevation angle, the irradiation of the subreflector remains unchanged, and this means that the light guide also functions as swivels. This makes it easier to combine several frequency bands in one antenna with a light guide, because narrow-band swivels are eliminated.

It follows from the above that the feasibility of setting up the receiving and transmitting equipment right near the irradiator in a stationary room at ground level is an important advantage of a light guide antenna. This does away with the antenna-waveguide that passes through the antenna and significantly simplifies the antenna-waveguide in the section between the irradiator and receiving-transmitting equipment. All this gives a significant reduction of insertion losses.

The Mark-IV antenna, manufactured by the Japanese NEC Company, with a 32-m main reflector, is of the design shown in Figure 11.17 [1.2, 11.8]. The path length from the irradiator input to the subreflector is about 30 m. Losses in the light guide are a few tenths of a decibel. By modifying the surfaces of the main reflector and subreflector, it is possible to achieve effective areas of greater than 0.7 and 0.6 on 4 GHz and 6 GHz, respectively. A cross-correlation level low enough for counterrotating polarization operation is achieved by selecting the parabolic mirrors of the light guide with the right focal lengths and by using a finned structure of the proper design inside the horn irradiator.

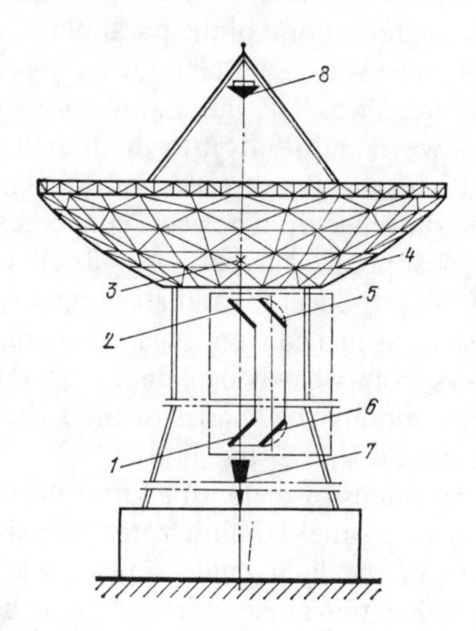

FIGURE 11.17. Diagram of an antenna with a light guide.

Along with the above-described trends in the improvement of antennas of the 4/6-GHz bands, a need arose to seek ways of reducing the levels of spurious emissions. For instance, instead of the sidelobe envelope level of 32–25 log θ, previously recommended by the CCIR (see Section 11.3), a level of 29–25 log θ is supposed to be recommended, i.e., the envelope will be reduced by 3 dB. Possibilities of achieving this much change may be based on a reduction of the level of irradiation of the aperture edges, on an increase of the surface precision of the reflector (see Section 11.2), and on a search for means of reducing the shading of the aperture. The latter can be achieved by making the subreflector and the cross section of the supports smaller, and also by giving the supports an impedance surface, so that the wave can pass around the supports [11.13]. However, shading of the aperture can be eliminated radically by moving the irradiating system out using a main reflector in the form of a nonaxially symmetric cutout [11.2, 11.7]. Antennas of this design have low levels of spurious emissions; the envelope is reduced to 22–25 log θ, but certain problems arise in the attainment of low cross-polarization levels when linear polarization is used, and also in construction and manufacturing.

All the improvements described above, made in 4/6-GHz antennas, are also being incorporated in antennas of the new 11/14 and 20/30 GHz bands, with the exception of counterrotating polarization operation. The latter in these bands involves considerable difficulties due to depolarization of a rotary-polarized wave as it passes along the path in rain. Vertically and horizontally polarized components of the field experience various amounts of delay of attenuation. As a result they end up with different amplitudes, and the phase shift between them ceases to be 90°. Therefore, it is necessary to use special expensive devices in the antenna-waveguide, which perform polarization correction by a pilot signal, transmitted from the satellite*.

Given below as an example are basic data on axially symmetric and nonaxially symmetric antennas with an 11.5-m aperture for the 20/30 GHz bands and on a 19-m axially symmetric antenna of the 12/14 GHz band for a class I station (Tables 11.4 and 11.5) [11.14, 11.15]. Twelve-meter limited

TABLE 11.4

Parameter	**Description of antenna**	
	Nonaxially symmetric	*Axially symmetric*
Aperture diameter, m	11.5	11.5
Irradiating system	Three-mirror light guide, subreflector, finned horn	−1.5
Steering sectors	Azimuth ±22° (±6° fine) Elevation angle ±5° (fine)	On X-axis: 47° ± 8° On Y-axis: ±12°

*Similar devices also have been developed for high-class stations of the 4/6-GHz band.

TABLE 11.4 (Continued)

Parameter	Description of antenna	
	Nonaxially symmetric	*Axially symmetric*
Surface precision*, mm	0.18 (mean square deviation)	0.16 (mean square deviation)
Weight* frequency band, GHz: receiving transmitting	19.4 17.7-21.2 27.5-31.0	23.9 17.7-21.2 27.5-31.0
Gain of antenna* (effective area) on frequency, GHz: 19.5 29.5	 66.3 (76%) 69.5 dB (69%)	 66.1 (72%) 69.3 dB (68%)
Noise temperature, K, at elevation angle, deg: 45 75	 13 18	13 K at elevation angle of 45° on 18.75 GHz

TABLE 11.5

Parameter	*Description*
Diameter of aperture, m	19
Frequency band, GHz: receiving receiving transmitting	 10.95-11.20 11.45-11.80 14.00-14.50
Polarization (communications channels)	Two linear orthogonal
Cross polarization	On axis: <−42 dB; within 1 dB of the level of the radiation pattern <−38 dB

*Measured values.

TABLE 11.5 (Continued)

Parameter	*Description*
Gain in receiving band, dB	65 + 20 log(f/11,375); effective area on 11,375 MHz 0.62, where f is frequency in MHz
Gain in transmitting band, dB	66.8 ± 20 log(f/14,250); effective area on 14,250 MHz 0.6
Noise temperature (in clear weather), K	81 at elevation angle of 30° on 11,375 MHz

rotation antennas are equipped with three-mirror light guide. That the working surface of the mirrors be finished and the beam of the antenna be steered with high precision are specific requirements. For example, for a 12-m antenna on 30 GHz, the radiation pattern at the 3-dB level is only about 3 arc minutes. Therefore, it is specifically the attainable beam-pointing precision that limits the maximum antenna reflector size to 12 m for 30 GHz.

11.13 SPACE ANTENNAS OF SATELLITES

The purpose of the space antennas of satellites is to receive and transmit in earth-space and space-earth links the signals of communications and telemetry systems and the signals of radio beacons and special systems. Space antennas have a significant specific feature, related to the following requirements [11.16]:

- the size of the nosecone of the booster places limits on the size of the antenna;
- when the booster is launched the antenna is acted upon by great accelerations and vibration stresses;
- antennas must have the minimum mass and must be capable of functioning properly in a deep vacuum at a pressure of 10^{-12} mmHg and less, in the thermal and radiation of the sun, solar pressure, ionizing radiation, *et cetera*, throughout the lifetime of the satellite.

In consideration of the above requirements, such materials as magnesium, titanium, aluminum, Invar, and beryllium are used for manufacturing space antennas. Composite materials, such as carbon plastics (graphite-epoxy composition) are being used more and more. With an elasticity modulus to specific density ratio of 709 (the material GY-70X-30VN1), which is 17 times greater than that of titanium, steel, aluminum, and magnesium, carbon plastics have considerably better mechanical and thermal properties: the linear expansion coefficient is close to zero, they have a low density and great rigidity. Therefore, they will become the main material for the production of the structures of mirrors, lenses, and antenna arrays [11.16].

Collapsible parabolic reflector antennas were used as first-generation global beam space antennas (the Molniya satellites), in addition to single-rod and spiral antennas, arrays of spiral elements (an array of 96 spiral elements was used for the Ehkran satellite) and horn antennas (the Molniya-1 satellite). Annular arrays of radiators (the Telstar satellite), parabolic horn antennas, connected to repeater equipment through a rotary fitting and rotating coaxially with the satellite, but in the opposite direction, so that the antenna beam is stationary relative to the earth (the Intelsat satellite), are used on rotation-stabilized satellites.

We will limit the discussion to a general description of space antennas and examine a few characteristic examples. The development of satellite communications systems required a quest for new means of separating channels. In antenna technology, this led to the appearance of systems in which working frequencies are utilized twice on the basis of polarization selection, and of systems in which space systems have multipath radiation patterns and channels are separated by space selection. Here space selection is combined with polarization selection when such is necessary. The latter should guarantee not less than 27-dB crosstalk attenuation between the channels. The following requirements stem from the above-mentioned systems problems. The solid angles, covered by individual beams, must correspond to the coverage areas on the earth, assigned for these beams, with sharp boundaries. During circular polarization operation, the ellipticity factor in each beam must not exceed 0.5 dB in order to assure adequate polarization isolation.

We will first explain how multibeam antennas are supposed to be used by way of example of the planned design of the Intelsat system [11.17, 11.8]. In the 4/6-GHz band, the antennas should have as many as six beams with high gain and good space selection. In addition to them, there should be two beams that cover two hemispheres. Polarization selection is used in addition to space selection. The requirement that the coverage zone be changed by commands from the earth is imposed on the antennas. One version of the solution of this problem is based on the formation of coverage area by an array of 85 radiator elements, arranged in a hexagonal loop. The radiators are grouped in accordance with the number of beams in six segments and can be connected to the receivers and transmitters of their own segment, or to the equipment of each of the two segments adjacent to them for switching from one area to another. Individual radiator arrays, each of which has its own reflector, are used for receiving and transmitting.

Another example is a multibeam antenna, based on a lens with a diameter of 27λ, with a focal length to diameter ratio of 1. The lens is driven by an array of 127 elements [11.19]. This design was discovered as a result of an effort to minimize spurious emissions in a beam-scanning angle of ±8.6° on 4 GHz and 6 GHz. One beam of a space-earth transmitting link in the 4-GHz band is formed with the aid of a group of 13 radiators. A group of 7 radiators is sufficient for forming a beam in a 6-GHz link for receiving signals from each area on the earth. A lens consists of two arrays, each containing 400 receiving and transmitting radiators, filling a circle with a diameter of 27λ. Each radiator consists of two independent elements, operating on orthogonal polarizations, which enables a lens to operate on two polarizations. Each receiving and transmitting radiator is connected by delay lines. To eliminate the dispersion that would cause the beam to wobble and spread, the delay lines are not of the waveguide, but of the TEM type, in which the wave propagation velocity does not depend on frequency.

A beam-generating system is used for distributing signals from six beams to the appropriate elements of the radiators. By virtue of the discrete structure of the irradiator, which is an array of 127 radiators, the positions of the beams can be changed if necessary by switching them in the right way. The beams cover approximately one-third of the visible surface of the earth in any position. Coverage can be increased to 97% of the entire globe.

In addition to the one described above, we will examine a planned system for local coverage of the major cities of the US and of all the territory [11.20, 11.21] of the country separately with narrow beams. The antenna version for this kind of communications system, by virtue of the good isolation between the beams in each of the beams intended for covering the major cities, will make it possible to use the entire 500-MHz frequency band, granted by the WARC, and divided into eight transponders. Territorial coverage is achieved with the aid of fixed contiguous beams, in which the polarization is orthogonal to the polarization in the beams intended for urban coverage. Because the spurious emission levels in each of the beams are low, frequencies can be used repeatedly. The space antenna is designed as a Cassegrain antenna with a nonaxially symmetric reflector and an extended irradiating system. The irradiator is a multistage array. An antenna of this design is compact and can be used in the 4/6, 12/14, and 20/30-GHz bands.

The elements of the array are laid out on a curved surface with a special shape, such that distortions of the shape of the partial radiation pattern and the loss of gain are minimized by moving the elements of the array out away from the focus. A 4λ-long segment corresponds to the prescribed antenna beamwidth on the focal surface. A group irradiator consisting of seven elements fits in one of the versions of the mentioned size ($4\lambda \times 4\lambda$). The excitation of the elements drops off by 8 dB from the middle to the edges. As a result, the irradiator produces a pattern that has a close to gaussian field distribution in the aperture of the main reflector, falling off toward the edges by 18 dB. The sidelobe level of this distribution does not exceed the tolerance of −35 dB. The loss of gain of the antenna as a result of this amount of decay of the distribution, in comparison with a uniform distribution, is 1.33 dB. And because the array of seven elements has a lattice structure, the irradiator contributes 1.45 dB of additional losses.

For serving different cities, the antenna has group irradiators with seven, ten, and twelve elements, which are used for making beams of unit, double, and triple width. In the latter two cases, the gain decreases by 2.64 dB and 4.06 dB. The smallest protection area between the beam, which gives 27-dB isolation, is equal to the beamwidth. Some of the elements of the irradiator array are used for forming territorial coverage beams.

Unlike the beams that are intended for urban service, which are isolated from each other, the territorial service beams overlap each other by a level of not less than −6 dB. A crosstalk attenuation of 23 dB, which is adequate for digital transmission using four-level phase modulation, is assured between beams with identical frequencies. The antenna generates a total of 46 beams. Not rotary, but linear orthogonal polarizations are used for reducing cross-polarization interference in precipitation. These polarizations are aligned in such a way that the polarization will be vertical or horizontal at each point on the earth's surface.

In conclusion, we will examine a new space antenna of a satellite, which generates area beams of a special shape, and which is exceptionally simple in design and construction [11.14]. The antenna is intended for installation on a satellite, operating in a polar satellite communications system in the 4/6 and 20/30-GHz bands. The satellite is rotation-stabilized, which made it advantageous to use a horn-reflector (similar to a parabolic horn antenna) antenna (Figure 11.18), connected to the equipment through a rotary adapter, and rotating in the opposite direction of the satellite. The aperture diameter of the antenna is set equal to 1 m to solve the problem of covering the entire territory of Japan, including the distant islands, with the beam in the 4/6-GHz band.

Here the shape of the surface of the reflector is not parabolic, but is selected such as to produce a radiation pattern with a special shape in the 20/30-GHz band, which creates a beam which in cross section follows the

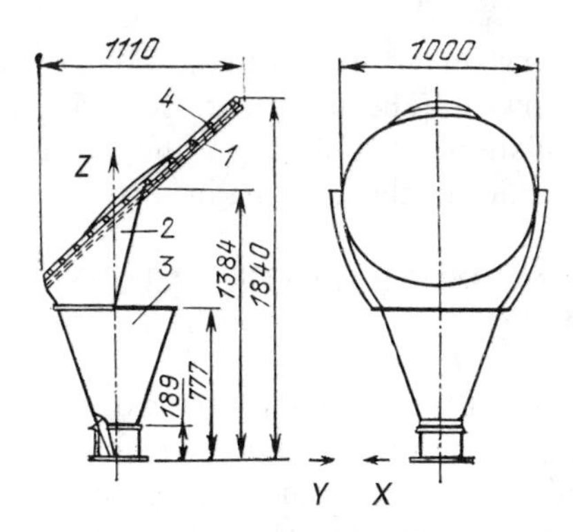

FIGURE 11.18. Space antenna of satellite.

outlines of the main islands of Japan. The calculated radiation pattern of this antenna on 19.4 GHz, shown in Figure 11.19, confirms this. Azimuth angles are plotted on the abscissa axis, and elevation angles on the ordinate axis. In both cases the direction along the *Y*-axis in Figure 11.18 is taken beyond 0°. The dashed line shows the outline of the main islands, which is completely inscribed in the radiation pattern at levels from 5 dB to 9 dB at the worst point. The gain of the antenna on 19.4 GHz on the peak line of the radiation pattern is 42.39 dB. Accordingly, the gain at the worst point is higher than 33 dB, which is high for a space antenna.

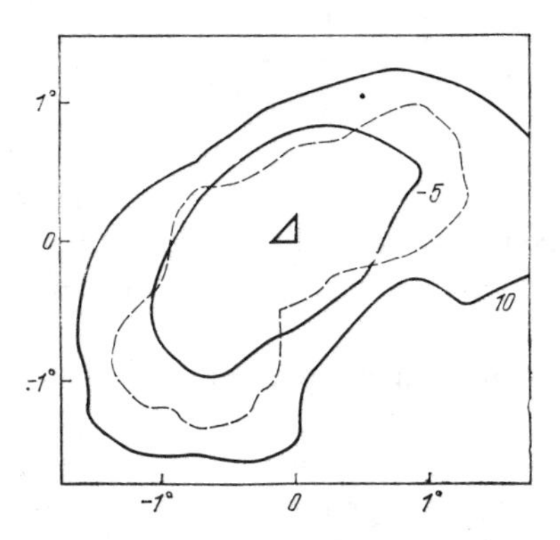

FIGURE 11.19. Calculated radiation pattern of space antenna of satellite on 19.4 GHz.

Reflector 1 of the antenna is made as an aluminum honeycomb structure with outer layers made of reinforced carbon plastic. Support 2 of the reflector and horn 3 are made of laminated carbon plastic. The support holds the reflector in a given position with the aid of rigidizing ring 4, which is bolted to the sandwich. The inner surfaces are painted white and a laminated heat insulation is applied on the outside to maintain the necessary heat properties of the antenna in space.

11.14 THE ELECTRIC DRIVE OF THE ANTENNA SYSTEMS OF EARTH STATIONS

The electric drive of the antenna systems of earth stations is built at the present time on the basis of reversible thyristor converters, which have replaced the previously used electromechanical amplifiers. This is because thyristor converters have a considerably higher efficiency, have no rotating mechanical parts, and have better performance reliability in combination with low cost. A typical schematic diagram of an electric drive, based on a thyristor converter and a dc final control motor is shown in Figure 11.20. Shown in the figure are: the input, the thyristor converter control unit, the thyristor converter, the final control motor, the velocity generator, and the stabilization circuit.

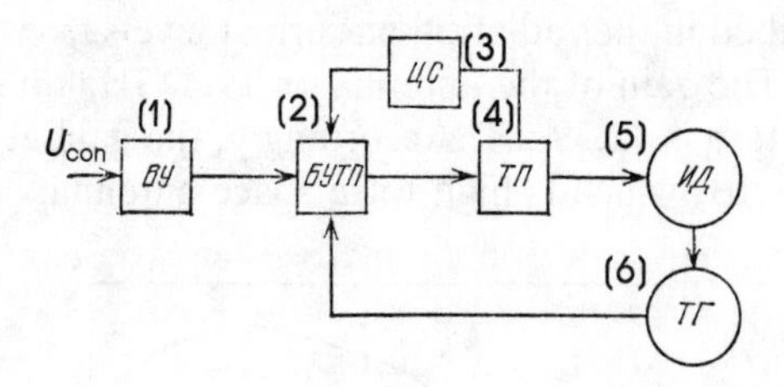

Key: **(1) Input**
(2) Thyristor converter control unit
(3) Stabilization circuit
(4) Thyristor converter
(5) Final control motor
(6) Velocity generator

FIGURE 11.20. Schematic diagram of electric drive.

As a rule, the input is a DC amplifier, the input of which receives a control signal (for example a signal from the output of a phase detector, to the inputs of which are connected master and slave selsyns). The passband (frequency correction) and the gain of the input usually can be regulated. Both these parameters have a significant effect on the stability and dynamic error of the drive.

The thyristor converter control unit controls the closing times of the thyristor valves and thereby controls the rectified voltage. The thyristor converter control unit consists of a sawtooth voltage generator, a comparator, a slave rectangular pulse generator, and a matching unit. The input of the thyristor converter control unit receives three signals: control voltage from the output of the input unit, feedback voltage from the output of the stabilization circuit, and voltage from the velocity generator output. Algebraic summation of these three voltages and of the sawtooth voltage takes place in the comparator. The zero control of the comparator controls the operation of the slave rectangular pulse generator, starting it when zero voltage appears at the output of the adder.

The thyristor converter consists of a three-phase transformer (Figure 11.21) and controlled thyristor valves. In Figure 11.21 are shown the final control motor and reactor. Thyristor converters intended for controlling the motor operate in both the valve and inverter modes for improving the dynamic parameters of the electric drive as a whole. In the valve mode, energy comes from the main to the motor, and in the inverter mode (deceleration of the motor), it goes from the motor into the main. Matched control satisfies the condition

$$\alpha_{\text{v}} + \alpha_{\text{i}} = 180^\circ \tag{11.22}$$

where α_{v} is the closing angle of the thyristors in the valve mode, and α_{i} is the closing angle of the thyristors in the inverter mode. When condition (11.22) is satisfied the mean rectified and inverter voltages are equal, but their instantaneous values can be considerably different. In this case, short-circuiting current can appear in the thyristor converter, which is limited by installing in the power circuit of the final control motor reactors, so designed as to satisfy the condition $I_{\text{s.c}} \approx 0.1 I_{\text{nom}}$, where $I_{\text{s.c}}$ is the short-circuiting current and I_{nom} is the nominal current.

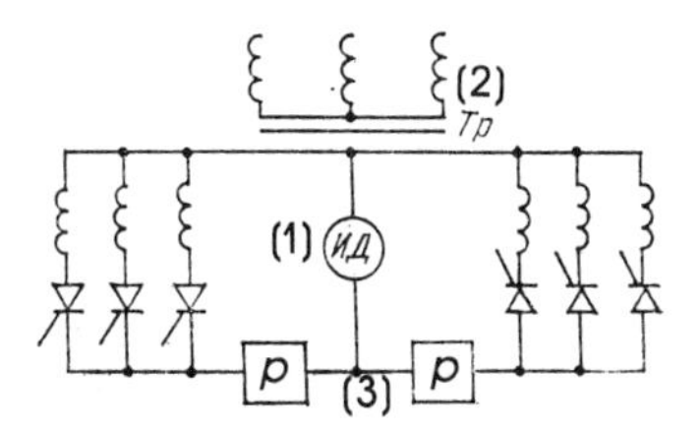

Key: (1) Final control motor
(2) Transformer
(3) Reactor

FIGURE 11.21. Diagram of thyristor converter.

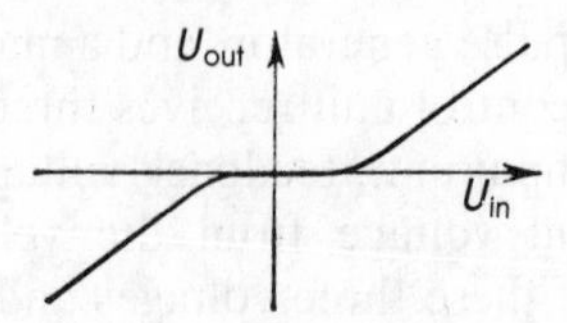

FIGURE 11.22. Output response of thyristor converter.

In the case of separately controlled thyristor converters, only one group of valves operates at any given time, and therefore the conditions for the appearance of $I_{s.c}$ do not exist. In view of this, reactors are not used for separately controlled thyristor converters. In this case, however, the output response of a thyristor converter is of the form shown in Figure 11.22, and there is virtually no area of insensitivity in a circuit with matched control.

The stabilization circuit of the electric drive is a negative feedback circuit, as a rule, relative to the first and second derivatives of the motor current. A resistor usually is connected in the power circuit for achieving feedback, and the voltage from this resistor is supplied to differentiating circuits. A thyristor drive of the EPS-U type was developed for earth stations with an antenna of the TNA-57 type (12 m in diameter). More detailed information on a thyristor electric drive, on its operating principles, and on methods of calculations is contained in the literature [11.22].

Chapter 12
Reliability Calculation of Satellite Communications and Broadcasting Systems

12.1 THE RELIABILITY CHARACTERISTICS OF SYSTEMS

In connection with the ever-increasing importance of satellite communications systems as a component part of the EASS of the Soviet Union, the problem of guaranteeing the reliability of these systems is exceedingly urgent. High reliability is especially important for satellite systems, which are virtually the only means of delivering information. This applies primarily to satellite distribution systems (for example, to a TV program distribution system), in which information from a transmitting station is transmitted through a space repeater of a satellite directly to many earth stations. A failure in this kind of system of transmitting station or of space repeater would result in the simultaneous interruption of the transmission of information to dozens or even hundreds of stations.

Satellite telephone and telegraph communications systems, as a rule, operate in parallel with earth facilities. Here partial mutual substitution is possible in the event of a failure. However, the fact that satellite systems are used for long-distance communications with remote and inaccessible regions of our country necessitates that rigid reliability requirements also be imposed on them.

From the standpoint of reliability analysis, a satellite communications system is a repairable redundant long-life system. The reliability of a satellite communications system, just as of any other communications system, is nearly completely characterized by the reliability characteristics of its channels for each kind of information that is transmitted. For example, the reliability of a telephone system will be characterized by the reliability characteristics of an arbitrary unit telephone channel between a pair of subscribers served by the system; the reliability of an audio broadcasting program distribution system will be characterized by the reliability characteristics of a one-way radio broadcasting channel; but if a satellite communications system is intended for transmitting TV programs and audio broadcasting programs simultaneously [17], then the reliability of such a system will be characterized by the reliability characteristics of a TV channel, by the reliability characteristics of a radio broadcasting channel, *et cetera*. The basic reliability characteristics of the channels of satellite communications systems are the following.

Readiness coefficient K_r. The probability that a channel is serviceable at an arbitrary moment of time during the steady state (stationary) process of operation of a system. The readiness coefficient tells the fraction of the total time of observation during which a channel is serviceable during the steady state operation of a system.

Mean accrued operating time to failure T. The ratio of operating time t_{ac} to the mean number of failures n_{ac} during that time.

Mean downtime τ. The mean time of compulsory unscheduled existence of a channel in an unserviceable state.

When two characteristics are known the third is determined uniquely by the relation:

$$K_r = T/(T + \tau) \quad \textbf{(12.1)}$$

Sometimes the mean number of failures (n_{ac}) of a channel during a certain period of time (of trouble-free operation) figures as a characteristic for a graphical comparison of the operation of channels (with equal accrued operating times).

12.1.1 Attainable Channel-Reliability Level

Satellite systems are built in such a way (Figure 12.1) that the reliability of any channel is determined by the reliability of the equipment of two earth stations 1 and 2 with connecting links 1 and 2 and that belong to them (if there are any), and by the reliability of the space segment. (*Comment.* Here the concept "space segment" refers to a transponder with all the systems of a satellite that enables it to function in orbit, and in consideration of redundancy.) The attainable reliability level of satellite systems and of their main elements is shown in Table 12.1 by way of example of the Intelsat international system [12.4]. The reliability characteristics of the channels of Soviet systems, both at earth stations, and of a channel as a whole are of the same order of magnitude. The generalized reliability characteristics of the main subsystems of Soviet earth stations are shown for clarity in Figure 12.2 in the form of a diagram.

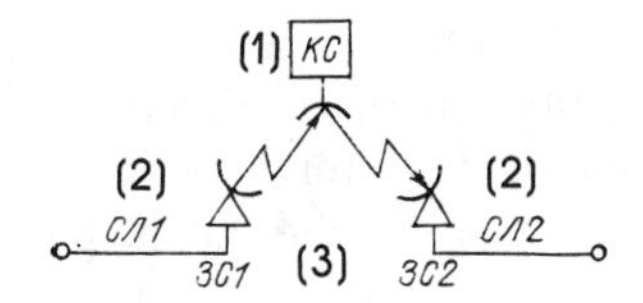

Key: (1) Space segment (3) Earth station
(2) Connecting link

FIGURE 12.1. Diagram of a satellite channel.

TABLE 12.1

Component of a satellite communications system	K_r
Space segment	0.99999
Earth station	0.99949
including: antenna	0.99983
transmitter	0.99994
low-noise parametric amplifier	0.99996
power (power supply)	0.99988
Entire channel	0.99894

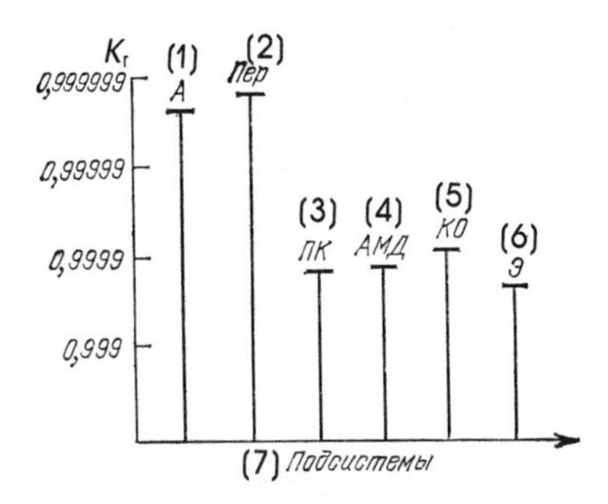

Key: (1) Antenna
(2) Transmitter
(3) Receiving complex
(4) Multiple access equipment
(5) Channeling equipment
(6) Power supply
(7) Subsystem

FIGURE 12.2. Reliability characteristics of main subsystems of a station.

At the present time, it is recommended in international documents [8] that the readiness coefficient of a telephone channel be 0.998. The readiness coefficient should be determined between the ends of a hypothetical reference circuit, consisting of one earth-space-earth link. At earth stations (not used in a space diversity system), this circuit should include one pair of modulation and demodulation equipment. Connecting links to earth stations are not connected to the reference circuit.

A fairly high channel-reliability level in a space segment is achieved by virtue of the long guaranteed lifetime of the satellites that are used ($T_{g.1}$ = 5–7 years) and of the use in satellite repeaters of backup transponders and of the strategy that is used for deploying the satellites in a system. A deployment strategy is the procedure used for replacing defective or worn-out satellites with new ones in orbit. The objective of a deployment strategy is to minimize interruptions of communications, connected with the need to replace satellites and to move earth stations for operation through a new (the next) satellite repeater.

Different satellite deployment strategies are used, depending on the purpose of the system and on how its operation is organized. The following strategies are most often used:

Strategies "with a backup satellite". When a standby satellite is in orbit along with the main satellite, ready to replace the worn-out satellite when needed. The standby satellite can be loaded in the same way as the main satellite, but by information of less importance. The information transmitted through the main satellite in this case has priority; when the main satellite breaks down this information is transmitted through the standby satellite by interrupting the transmission of the less important information. The standby satellite can also be on the launch pad, ready for launch.

The "prestart" strategy. When the next satellite is inserted in orbit as soon as the previous one uses up a certain amount of time (exhausts a certain fraction of its service life).

A strategy. Whereby the next satellite is launched only after a failure is discovered in the previous one (for example a failure of just one transponder of a repeater).

Other satellite deployment strategies also are used.

At earth stations the necessary reliability level is achieved by making extensive utilization of redundification of the main subsystems of a station. An example of redundified TV and telephone transponder at a two-way earth station is shown in Figure 12.3. In the transmitters, the power amplifier (the ASM rack) and drive (ASV rack) are duplicated simultaneously, and in the TV transponder the modulator half-set (the ASMD rack) as well. This most often is loaded standby, i.e., the standby set is in the same operating mode as the main set. Automatic switching to standby takes place in the switching rack (the ASP rack) when the SHF power drops or decreases by 3 dB relative to the nominal power.

The same method of duplication is used in a receiving complex [4]. The wideband low-noise amplifier (rack M), the SHF signal converter, and 70-MHz IF signal amplifier (rack V) are on loaded standby simultaneously, and in the TV transponder, the FM IF signal amplifier half-set (rack P) as well. Switching to the standby set takes place automatically in the event of simultaneous fading of the sync pulses at the output of the working sets of rack P and of rack RS, in which the video signal is time division multiplexed by sound signals. The corrector unit in the video channel and the sound amplifier unit in the sound channel (see Figure 12.3(a)) also have a loaded standby, but switching to the standby sets is accomplished manually.

In the multiple access equipment of a telephone transponder (see Figure 12.3(b)), the transponder equipment has unit-wise loaded duplication with automatic switching to standby. In the individual equipment racks of channels of several channel units, operating in parallel, there is one standby, which automatically replaces any of the working units in the event of a failure, i.e., the automatic sliding standby principle is used. The sliding duplication principle is also used for duplicating channeling equipment. As a rule, the equipment of a telephone transponder has several sets of channeling equipment, operating in parallel. If any of them fails, a standby set is connected in its place. Switching is done manually.

Years of operating experience demonstrate the ways of enhancing the reliability of satellite communications systems. Chief among them are the following:

- extend the active service life of the satellite repeaters that are used and perfect their deployment strategies;
- use satellites in a geostationary orbit in satellite communications systems, which decreases downtime due to imprecise steering of the antennas of earth stations to a satellite, and also downtime due to failures in the antenna steering and electric drive systems of earth stations;
- perfect the equipment of earth stations, and particularly power supply systems by increasing the number of independent energy sources and by using automatic duplication;
- shorten downtime due to failure on the part of the technical personnel of a station to observe the rules of technical operation of equipment by automating all switching processes as much as possible [12.3], develop modern highly reliable test and duplication systems (with clear indication on light panels), and convert to unattended automatic earth stations;

- improve the reliability of connecting links and put earth stations as close as possible to the sources (consumers) of the information, transmitted by satellite.

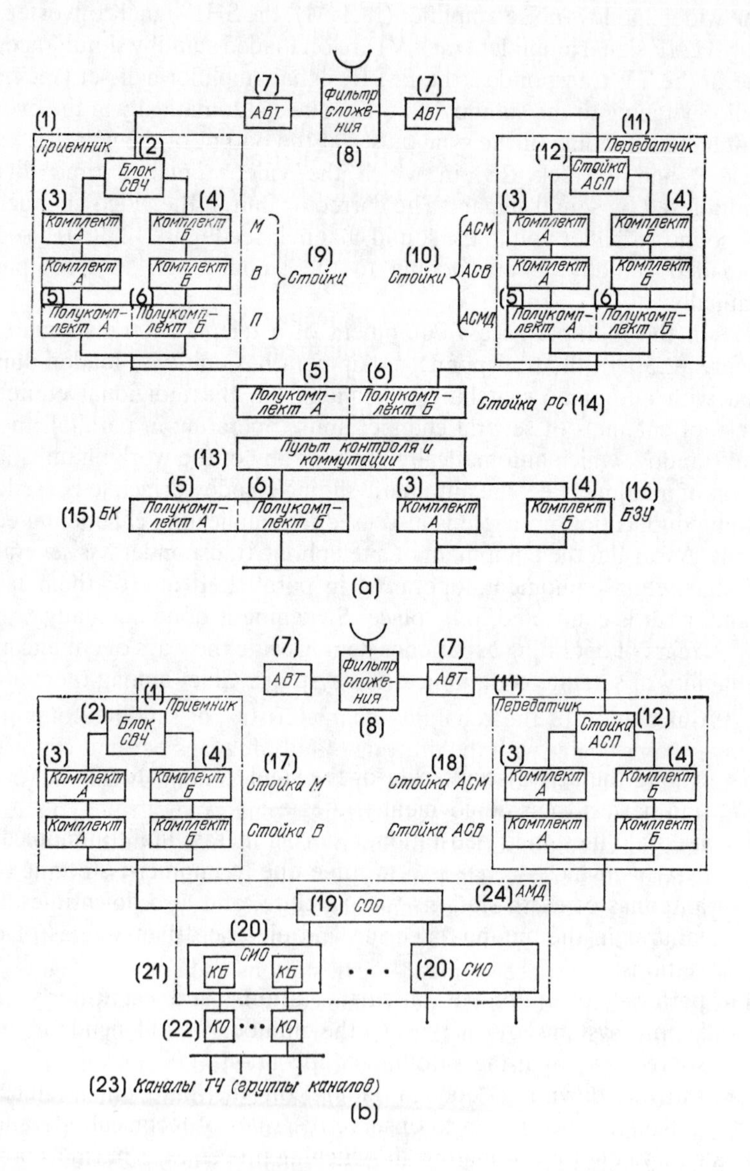

FIGURE 12.3. Schematic diagram of duplication of main sets of two-way station: (a) of video and sound channels of TV transponder of the Orbita system; (b) of telephone channels in TDMA system.

FIGURE 12.3 (Continued)

Key:
(1) Receiver
(2) SHF unit
(3) Set A
(4) Set B
(5) Half-set A
(6) Half-set B
(7) Antenna-waveguide
(8) Combining filter
(9) M rack
V rack
P rack
(10) ASM rack
ASV rack
ASMD rack
(11) Transmitter
(12) ASP rack
(13) Test and switch panel
(14) RS rack
(15) Corrector unit
(16) Sound amplifier unit
(17) Rack M, Rack V
(18) Rack ASM, Rack ASV
(19) Transponders equipment
(20) Individual equipment rack
(21) Channel unit
(22) Channeling equipment
(23) Audio frequency channels (groups of channels)
(24) Multiple access equipment

12.2 SYSTEMS RELIABILITY CALCULATION

12.2.1 Failsafe Plans of Channels

Calculation of the reliability characteristics of satellite communications systems begins with the drafting of failsafe plans per unit channel for each kind of information to be transmitted by a system. A channel failsafe plan is drafted in accordance with the following general rule: connect the elements in a circuit in series if failure of each of them results in a failure of the channel, and connect in parallel only the elements, the simultaneous failure of which results in a channel failure.

The elements of the failsafe plan of a channel are both entire subsystems of the space segment, earth stations and connecting links, and individual units of them, the performance of which has a direct effect on the performance of a given channel. For instance, an element of a space segment should include just the repeater transponder of a satellite that is used for organizing a given channel; an element of individual equipment should include just the individual equipment that belongs to a given channel. Typical failsafe plans are shown in Figure 12.4 for the most common kinds of channels: a one-way channel (the type of channel that distributes TV programs), and a two-way channel (of the telephone channel type).

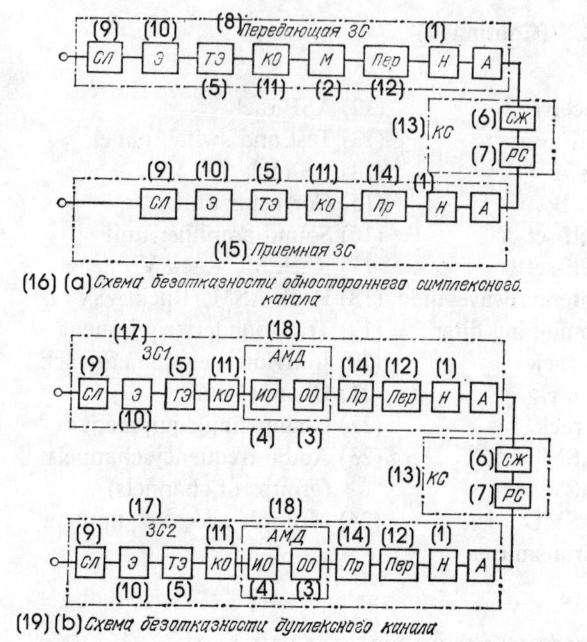

Key:
(1) Steering system
(2) Modulator
(3) Transponder equipment
(4) Individual equipment of a channel (of a group of channels)
(5) A conditional element that takes into account downtime due to improper technical operation of a channel
(6) Satellite life support system (orientation, heat control, power supply, and other systems—failure of each of them results in the complete failure of a satellite)
(7) Repeater transponder
(8) Transmitting earth station
(9) Connecting link
(10) Power supply
(11) Channeling equipment
(12) Transmitter
(13) Space segment
(14) Receiver
(15) Receiving earth station
(16) (a) Failsafe plan of one-way channel
(17) Earth station
(18) Multiple access equipment
(19) (b) Failsafe plan of two-way channel

FIGURE 12.4. Typical failsafe plans of channels of satellite communications systems: (a) one-way; (b) two-way.*

*See Figure 12.2 for other symbols.

12.2.2 Estimation of Channel-Reliability Characteristics

After the failsafe plans of channels have been drafted, the reliability of their individual elements is calculated. Basic formulas for calculating the reliability characteristics of the elements of a channel at earth stations in consideration of duplication and restoration (repair) conditions are given in Table 12.2. The elements of the duplication plans represented in the table are assumed to have exponentially distributed operating time with the parameter λ and repair time with the parameter μ. During an examination of the sliding standby plan (the fifth plan in Table 12.2), it was borne in mind that units $1,2, \ldots, \nu$ belong to different channels. Interrelationship among them exists only through a common standby and repair. We are interested in the reliability characteristics of one of ν, an arbitrary fixed working unit, i.e., the formulas derived for the reliability characteristics take into consideration the effect of sliding standby and repair specifically for the unit which we selected from among ν units. In the estimation of the reliability characteristics of a channel in a space segment, we will simply point out an approach to the determination of these characteristics in view of the unwieldiness of the calculation formulas that were derived, and of differences in the strategies used for deploying satellites.

The readiness coefficient of a channel in a space segment, in view of the fact that repeater transponders on communications satellites have about the same reliability, is

$$K_{ri} = \sum_{j=1}^{N} \left(\frac{j}{N} \right) P_{(j)} \tag{12.2}$$

where N is the nominal number of repeater transponders in the space segment of a system, and $P_{(j)}$ is the probability of the existence of j $(j = 1, \ldots, N)$ serviceable repeaters in a space segment at an arbitrary time of operation of a system. The probability is $P_{(j)} = f(T_{a.s}, \theta)$, where $T_{a.s}$ is the guaranteed active service life of the satellites used in a system; and θ is the algorithm of the satellite deployment strategy utilized in a system. We note that the choice of θ poses a rather difficult independent problem.

For estimating K_r, in practice, it is usually sufficient to find the probabilities $P_{(N)}$, $P_{(N-1)}$, $P_{(N-2)}$, or even the first two of them. The probabilities $P_{(j)}$ for $j < N - 3$ are negligible as a rule. This is explained by the fact that before two and more transponders break down in a satellite repeater, the satellite experiences a complete failure, or it exhausts its resources and further operation becomes impossible, or it is replaced with a new satellite (this situation occurs most often).

TABLE 12.2

Duplication plan	*Repair conditions*	K_{ri}	T_i	τ_i	*Comment*
A series system of n different repairable elements (diagram: λ_1, μ_1 … λ_n, μ_n)	In the event of failure the system is turned off	$\left(1+\sum_{j=1}^{n}\frac{\lambda_j}{\mu_j}\right)^{-1}$	$\left(\sum_{j=1}^{n}\lambda_j\right)^{-1}$	$T_i\sum_{j=1}^{n}\frac{\lambda_j}{\mu_j}$	—
Duplicated system Loaded standby (diagram: λ,μ; λ,μ)	Unlimited repair	$\frac{1}{1+Y_s}$	$\frac{1}{\lambda}\frac{1^*}{2Y}$	$\frac{1}{2\mu}$	$Y=\frac{\lambda}{\mu};\ Y_s=\frac{Y^2}{1+2Y}$
	Limited repair	$\frac{1}{1+Y_s}$	$\frac{1}{\lambda}\frac{1^*}{2Y}$	$\frac{1}{\mu}$	$Y=\frac{\lambda}{\mu};\ Y_s=\frac{2Y^2}{1+2Y}$
Duplicated system Unloaded	Unlimited repair	$\frac{1}{[illegible]}$	$\frac{1}{[illegible]}\frac{1^*}{[illegible]}$	$\frac{1}{[illegible]}$	$Y=\frac{\lambda}{[illegible]};\ Y_s=\frac{Y^2}{2(1+Y)}$

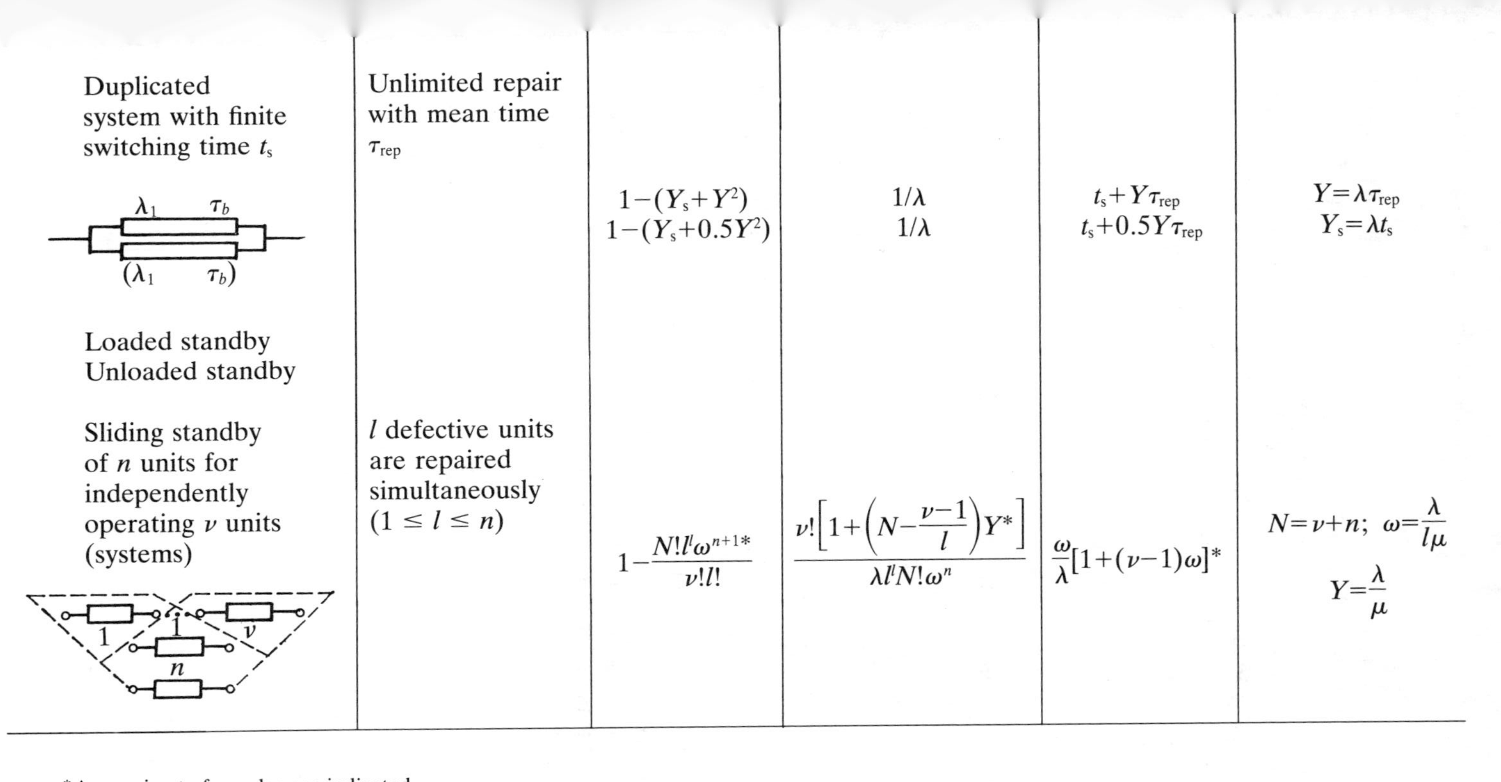

Duplicated system with finite switching time t_s	Unlimited repair with mean time τ_{rep}				
Loaded standby		$1-(Y_s+Y^2)$	$1/\lambda$	$t_s+Y\tau_{rep}$	$Y=\lambda\tau_{rep}$
Unloaded standby		$1-(Y_s+0.5Y^2)$	$1/\lambda$	$t_s+0.5Y\tau_{rep}$	$Y_s=\lambda t_s$
Sliding standby of n units for independently operating ν units (systems)	l defective units are repaired simultaneously $(1 \le l \le n)$	$1-\frac{N!l^l\omega^{n+1}}{\nu!l!}$*	$\frac{\nu!\left[1+\left(N-\frac{\nu-1}{l}\right)Y^*\right]}{\lambda l^l N!\omega^n}$	$\frac{\omega}{\lambda}[1+(\nu-1)\omega]^*$	$N=\nu+n;\ \omega=\frac{\lambda}{l\mu}$; $Y=\frac{\lambda}{\mu}$

*Approximate formulas are indicated.

The mean downtime τ_i of a channel due to a failure of the space segment is determined by the time necessary for preparing, launching, and inserting a satellite at a given point of the orbit, and for connecting it to the communications system. The mean accrued operating time to failure of a channel in a space segment can be determined using relation (12.1). The reliability characteristics of an entire channel are calculated on the basis of data obtained on the reliability characteristics of the elements of the failsafe plan of a channel as follows:

- the readiness coefficient is

$$K_r = \prod_{(i)} K_{ri} \quad (i=1, \ldots, k) \tag{12.3}$$

- the mean accrued operating time to failure is

$$T = \left(\sum_{(i)} \frac{1}{T_i} \right)^{-1} \quad (i = 1, \ldots, k) \tag{12.4}$$

- the mean downtime is

$$\tau = T \sum_{(i)} \frac{\tau_i}{T_i} \quad (i = 1, \ldots, k) \tag{12.5}$$

where i is the current index of an element; k is the number of elements in the failsafe plan of a channel.

In the reliability calculation practice of satellite communications systems it is also possible to encounter elements of a channel, the plans of the construction (duplication) of which were not examined in this section. In this case, it is necessary to refer to sources [12.1] and [12.2] to find the calculation formulas.

Chapter 13
Basic Technical Data on Typical Satellite Communications Systems

13.0 INTRODUCTION

The high technical-economic characteristics of satellite communications systems that have been achieved to date explain the extensive adoption of many different satellite communications systems in commercial operation. The rates of increase of the number of these systems and of their carrying capacities are great. More than 75 communications satellites, approximately 20 more than in 1977, were operating in a geostationary orbit by the end of 1980 [13.1].

Several regional (operating in the interest of several countries, located in a given geographic region) and national systems are being developed, or have already been developed, in addition to the Intersputnik and Intelsat international satellite communications systems. Examples of regional satellite communications systems are the Arabsat systems, operating in the interest of several Arabic countries; of Southeast Asia, and the Palapa satellite communications system; national systems have been developed in the USSR, US, Japan, Nigeria, and other countries. Shown schematically in Figure 13.1 is the location and position in a geostationary orbit of communications satellites of the main satellite communications systems, and given in Table 13.1 are basic technical characteristics of three types of satellite communications systems.

13.1 INTERNATIONAL SATELLITE COMMUNICATIONS SYSTEMS

13.1.1 The Intersputnik System [13.1]

The Intersputnik system was developed in 1971. Earth stations of this system were built in Bulgaria, Hungary, East Germany, Poland, the

USSR, Czechoslovakia, Afghanistan, Vietnam, Laos, Algeria, Cuba, and Mongolia by the end of 1981. In accordance with the Intersputnik Organization Charter, earth stations are the property of the countries that build them, and the space segment is leased from one of the members of the Organization. Today, from the USSR the Organization leases channels on satellites of the Gorizont type (the registration names are Statsionar-4 and Statsionar-5). One leased transponder is used for exchanging TV programs among countries that are in the satellite communications system.

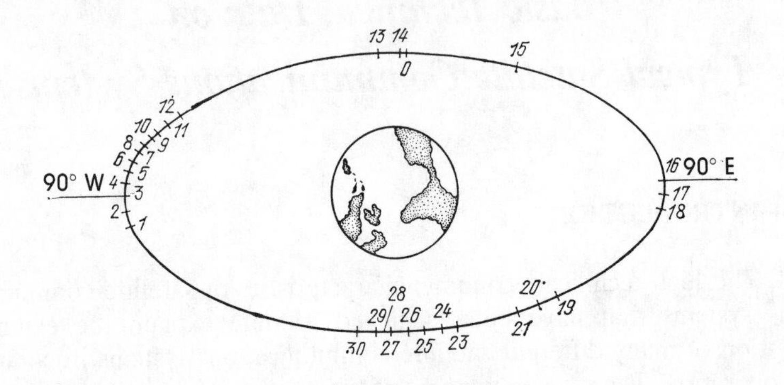

FIGURE 13.1. Some geostationary satellites of satellite communications systems: 1=Satcom, 85.7° west longitude; 2=Comstar, 87° west longitude; 3=Westar, 91° west longitude; 4=Comstar, 95° west longitude; 5=Westar, 99° west longitude; 6=Anik B, 109.3° west longitude; 7=Anik C, 112.5° west longitude; 8=Anik A, 114° west longitude; 9=Satcom, 119° west longitude; 10=Westar, 123.5° west longitude; 11=Comstar, 128.3° west longitude; 12=Satcom, 130° west longitude; 13=Intelsat IV, 179° east longitude; 14=Intelsat IV, 174° e.l; 15=Statsionar 7, 140° e.l; 16=Statsionar 6, 90° e.l; 17=Palapa, 83° e.l; 18=Palapa, 77° e.l; 19=Intelsat IV, 63° e.l; 20=Intelsat IV, 57° e.l; 21=Statsionar 5, 53° e.l; 22=Statsionar 9, 45° e.l; 23=OTS, 10° e.l; 24=Statsionar 4, 14° w.l; 25=Intelsat V, 18.5° w.l; 26=Intelsat IV, 21.5° w.l; 27=Intelsat V, 24.5° w.l; 28=Statsionar 8, 25° w.l; 29=Intelsat IV, 31.3° w.l; 30=Intelsat V, 34.5° w.l.

Frequency modulation with a peak frequency deviation of ±13 MHz is used for transmitting TV programs. The sound accompaniment of TV program transmissions is transmitted both on the 7.5-MHz-FM subcarrier with a frequency deviation of ±150 kHz, and by time division multiplexing of a TV transponder by inserting width-modulated pulses in the intervals of time that correspond to line blanking pulses.

Telephone messages are transmitted by means of multiple access by FDM and FM of each carrier by the signal of the audio frequency channel. A more detailed description of the Gradient-N multiple access equipment with a frequency division multiplex transponder, used in the Intersputnik system, is given in Chapter 18.

Technical Characteristics of Typical Earth Stations of the Intersputnik System

Quality of earth station (G/T), dB/K, more than:	
existing earth stations	$29 + 20(\log f)/3.8$
newly constructed and reconstructed earth stations	$31 + 20(\log f)/3.8$
Antenna gain, dB, transmission in 6025–6225 MHz band, more than	54
Polarization:	
transmission	Left-hand rotation
reception	Right-hand rotation
Working band, MHz:	
transmission	6025–6225
reception	3700–3900
EIRP, dBW:	
audio frequency channel	51.8
television	84.6
service telegraph	41.8
Transmitter frequency instability, kHz:	
of TV transponder	±50
of telephone transponder	±2
Signal-to-noise ratio, dB, at input of demodultor of audio frequency channel in 90-kHz band	10

13.1.2 The Intelsat System

There were 103 permanent member nations of the Intelsat Organization by the end of 1981. Capital investments of the member nations, proportional to the demand of a country for all kinds of communications (telephone, telegraph, facsimile, data transmission, audio, and TV broadcasting), organized within the framework of the system, constitute the financial foundation of the Organization. The annual income from the operation of the system is divided among the member nations

TABLE 13.1

	Parameter for system			
Parameters	*Intersputnik (satellite of the Gorizont type)*	*National USSR (satellite of the Gorizont type)*		*Intelsat IV*
Kinds of information transmitted	Telephone, television	Telephone, television		Telephone, television
Working frequency band, GHz	6/4	6/4	14/11	6/4
Stabilization method	Triaxial	Triaxial		Rotary
Weight of satellite, kg	>1000	>1000		730
Power of power sources, W	>1000	>1000		570
Service areas	Global; hemisphere	Hemisphere; area; narrow beam	Narrow beam	Global; narrow beam
Number of transponders	2	6	1	12
Transponder bandwidth, MHz	36	36	36	36
Number of antenna beams	2	4	4	6
Polarization	Rotary	Rotary		Rotary
G/T, dB/K	−17	−17 −11	−6	−18.6
EIRP, dBW	27 31	31 34 36	40	22.5 34.5
Carrying capacity (of one-way audio frequency channels)	200/ transponder	3500		9000

TABLE 13.1

Parameter for system					
Intelsat IVA	*Intelsat V*		*Anik A*	*Anik B*	*Anik C*
Telephone, television	Telephone, television		Telephone	Telephone	Telephone, television
6/4	6/4	14/11	6/4	6/4 14/11	14/11
Rotary	Triaxial		Rotary	Triaxial	Rotary
790	1010		270	440	520
710	1220		300	840	925
Global; hemisphere; narrow beam	Global; hemisphere; narrow beam	Narrow beam	Canada	Canada	Canada
20	21	6	12	12 6	16
36	36 72	72 240	36	36 72	54
7	5	2 (western and eastern)	1		4
Rotary	Rotary	Linear	Linear	Linear	Linear
−11.8	−18.6 −11.6 −8.6	0.0−Eastern beam 3.3−Western beam	−7	−6 −1 —	+1
2 6 9	−23.5/26.5 26/29 29	41.4−Eastern beam 44.4−Western beam	33	36 47.5	48
1000	22000	—	—		

TABLE 13.1 (cont'd)

Parameters	*Satcom*	*Westar*	*Westar advanced (future)*	
Kinds of information transmitted	Telephone, television	Telephone, television	Telephone,	television
Working frequency band, GHz	6/4	6/4	6/4	14/11
Stabilization method	Triaxial	Rotary	Triaxial	
Weight of satellite, kg	460	300	2130	
Power of power sources, W	770	300	1200	
Service areas	US, Alaska, Hawaii	US	US, Alaska, Hawaii	US
Number of transponders	24	12	12	4
Transponder bandwidth, MHz	34	36	36	225
Number of antenna beams	2	2	2	7
Polarization	Linear	Linear	Linear and rotary	
G/T, dB/K	−5 −10	−6	−7 −12.5 −12.5	
EIRP, dBW	32 26	33	33 28 26	42 50.3
Carrying capacity (of one-way audio frequency channels)	14400	6000	15000	

TABLE 13.1 (cont'd)

Comstar	*Palapa*	*SBS*	*OTS*
Telephone, 6/4	Telephone, television 6/4	Telephone, television 14/11	Telephone, television 14/11
Rotary	Rotary	Rotary	Triaxial
810	300	460	—
760	300	1000	520
US, Puerto Rico, Alaska, Hawaii	Indonesia	US (without Alaska and Hawaiian Islands)	Europe and North Africa 7.5°×4.5° (2 Each) 2.5°×2.5° (1 Each) 5°×3.5° (1 Each)
24	12	10	
36	36	43	40/120/5
4	1	1	5
Linear −8.8	Linear −7	Linear 2 . . . −2	Rotary —
33	32	40 . . . 43.7	—
14400	7000	12000	20000

in proportion to their capital investments. The Intelsat Organization is responsible simply for the condition and operation of the space segment of the system, including earth stations, from which the satellite is controlled. The earth stations are the properties of the member countries of the system, and their characteristics must correspond to the paremeters, adopted in the system as standards. The satellite channels are leased.

Three standard earth stations are used within the framework of the system: A and B in the 6/4-GHz band and C for the 14/11-GHz band. The basic technical parameters of the standard earth stations of the Intelsat system are given in Table 13.2.

Satellites of the Intelsat IV and Intelsat IVA types are operating in the system at the present time, and of the Intelsat V types since the end of 1980. A satellite of the next generation, of the Intelsat VI type, is under development, and it will go into operation around 1993. The carrying capacity of one Intelsat IVA satellite is about 6000 two-way telephone channels and two TV programs, and the carrying capacity of an Intelsat V satellite is 11,000 telephone channels and two TV programs. The carrying capacity of the Intelsat VI satellite is being planned on the level of 60,000–80,000 telephone channels.

Only the 6/4-GHz band is used on Intelsat IV and Intelsat IVA satellites, and on Intelsat IVA satellites, fequencies are reutilized by means of space division multiplexing of the eastern and western beams of the satellite antenna system. The 14/11-GHz band is used instead of the 6/4-GHz band on the Intelsat V satellite for the purpose of increasing its carrying capacity, and frequencies also are reutilized by means of orthogonal polarization.

Now only two versions of frequency multiple access are being used in systems of the Intelsat type for transmitting telephone signals: transmission on one frequency-modulated carrier of a group of standard frequency division telephone channels, and transmission on one carrier frequency of one telephone channel (one channel per carrier). The basic parameters of the signals that correspond to the first version are given in Table 13.3.

Television messages can be transmitted in the "one program per transponder" mode or in the "two programs per transponder" mode. In the latter case, a transponder is frequency-multiplexed by decreasing the deviation of each of the two frequency-modulated carriers. This kind of multiplexing of transponder is accomplished by means of reducing the signal-to-noise ratio on the video channel to 49 dB. The following message processing and modulation methods are used for transmitting telephone messages by one channel on a carrier:

TABLE 13.2

Parameter	Parameter for standard earth station		
	A	*B*	*C*
Quality (ratio of antenna gain-to-noise temperature at irradiator input G/T), dB/K	$>40.7+20\log(f/4)$ for elevation angles above 5°; f=receiving frequency, GHz	$>31.7+20(\log f)/4$ for elevation angles above 5°; f=receiving frequency, GHz	$>41.0+20(\log f)/11.2$ for elevation angles above 10° and in clear skies; f= receiving frequency, GHz
Gain of antenna, dB	≥ 57 on 4 GHz ($\phi \approx 26$–30 m); ≥ 60 on 6 GHz	≥ 50 on 4 GHz ($\phi \approx 11$–13 m); ≥ 53.2 on 6 GHz	≥ 64.2 on 11.2 GHz ($\phi \approx 18$ m); ≥ 66.3 on 14 GHz
Sidelobe level, dB	$29 - 25\log\theta$ for angles $1 \leq \theta \leq 48°$; ≤ 10 dB for angles $\theta > 48°$ relative to axis of main lobe		—

TABLE 13.2 (Continued)

Polarization	*Rotary*	*Rotary*	*Rotary and linear*
Ellipticity factor in antenna system and antenna-waveguide	<1.06	<1.06	<1.03
Frequency band, GHz: transmission reception	 5925–6425 3.7–4.2	 5925–6425 3.7–4.2	 14–14.5 10.95–11.2 11.45–11.7
EIRP, dBW	88 in TV mode (for telephone see Table 13.7)	82 in TV mode; 69 per channel in telephone in one channel on a carrier mode	95.7 (maximum)

TABLE 13.3

Number of channels per group	Global beam		Zonal beam	
	Signal frequency band, MHz	*EIRP, dBW*	*Signal frequency band, MHz*	*EIRP, dBW*
24	2.5	74.7	—	—
60	5.0	77.8	2.5	81.4
96	7.5	79.5	—	—
132	10.0	80.6	5.0	83.9
192	—	—	7.5	84.7
252	15.0	82.8	10.0	85.4

Number of channels per group	Global beam		Zonal beam	
	Signal frequency band, MHz	*EIRP, dBW*	*Signal frequency band, MHz*	*EIRP, dBW*
432	25.0	85.1	15.0	88.4
612	—	—	20.0	90.1
792	—	—	25.0	91.5
972	36.0	90.1	—	—
1872	—	—	36.0	98.6

- analog-digital conversion of the audio frequency signal of a channel by pulse-code modulation and phase keying of the carrier (PCM/PSK-2 or PCM/PSK-4).
- digital conversion of the audio frequency signal of a channel by delta-modulation (adaptive, as a rule) and phase keying of the carrier (DM/PSK-2 or DM/PSK-4);
- messages transmitted by frequency modulation of the carrier by the analog audio frequency signal of the channel (companding may be used, especially for transmitting speech messages, for the purpose of increasing the carrying capacity of a system).

The first of these methods, implemented in SPADE equipment, is utilized in the Intelsat system. In this equipment, channels are also made available by request, which significantly increases the effective carrying capacity of a satellite transponder and provides significant economic benefits for the countries in which there is light traffic on some communications routes and it is not profitable to lease group channels. Time division multiple access at a rate of 120 Mb/s (TDMA-120) using the Intelsat V satellite is supposed to begin to be adopted in the Intelsat system after 1983. Digital speech interpolators are supposed to be used in the TDMA-120 system, which makes it possible to increase the carrying capacity of a transponder to 1500 two-way telephone channels.

Basic Technical Parameters of the SPADE System

Passband of audio frequency channel, Hz	300–3400
Transmission rate on digital channel, kb/s	64
Type of analog-digital conversion	seven-digit PCM with A-87.6 companding, sample frequency 8 kHz
Modulation	four-phase
Carrier level control	Suppression in pauses of speech signal
Carrier separation, kHz	45
Noise band of channel, kHz	38
Channel signal-to-noise ratio in channel band on intermediate frequency, dB	15.5
Error probability on digital channel:	
nominal	10^{-6}
threshold	10^{-4}

Basic Technical Characteristics of the TDMA-120 Equipment

Nominal line rate, Mb/s	120, 832
Modulation	four-phase, absolute
Demodulation	Coherent
Combination for frequency regeneration:	
carrier	176 symbols
clock	$(0, \pi, 0, \pi, \ldots)$
Duration of start-address combination (Unique Word)	24 symbols
Prevention of reverse operation	By start-address combination
Output phase precision of modulator, deg	±1
Output signal level precision of modulator, dB	±0.1
Maximum carrier frequency deviation (from nominal), kHz	±50
Difference of carrier frequencies between individual packets, kHz	6
Maximum difference of levels of received packets, dB	5
Energy characteristic of modem (E/N_0), dB, for error probability:	
$2 \cdot 10^{-2}$	6.3
$1 \cdot 10^{-4}$	10.0
$1 \cdot 10^{-6}$	12.6
$1 \cdot 10^{-7}$	14.0
Error probability in reception of start-address combination for $E/N_0 = 7$ dB	10^{-8}
Frame duration, ms	2

13.1.3 The Inmarsat System [13.3]

The Inmarsat Organization was founded in June 1979 for the purpose of communicating with seagoing ships, located anywhere on the world ocean. The Organization uses special Marisat satellites, operating simultaneously in the US Navy communications system. The capabilities of two existing Marisat satellites, serving the regions of the Indian and Atlantic Oceans, are supposed to be used. A similar satellite, serving the Pacific Ocean area, is already being used in the Marisat system. Today about 700 ships are outfitted with the equipment of the Inmarsat system. In the future it is planned to convert to new specially developed Marecs satellites.

Special repeaters, installed on Intelsat satellites (Intelsat-MCS) are also supposed to be used.

The earth stations of the system are equipped with 13-meter antennas and operate in the 6/4-GHz band. The 1.5/1.6-GHz band is used in a satellite-ship link. 6 → 1.5 GHz (shore → ship) and 1.6 → 4 GHz (ship → shore) frequency conversion is utilized in a repeater transponder. Frequency modulation in combination with companding is used for making a telephone channel. The band of one channel is 28 kHz. The subjective quality of a channel corresponds to a noise power of 25,000 pW. Time division multiplexing is used for organizing telex communications, and in this case a 1.2-Kb/s digital stream is transmitted on one carrier, which corresponds to 22 telex channels in a shore → ship link. Time division multiple access is used for telex communications in a ship → shore link. The line rate of the channel is 4.8 Kb/s. Basic data on the three different satellites that are used (or which are planned for use) by the Inmarsat organization are given in Table 13.4. Four standard ship stations have been established. Their basic parameters are given in Table 13.5.

TABLE 13.4

Type of satellite	*Marisat*	*Marecs*	*MCS*
Utilization	Jointly with the US Navy	Special satellite	Jointly with fixed satellite service
Stabilization	Rotary	Triaxial	Triaxial
G/T, dB/K:			
ship→shore	−21	−17	−18.6
shore→ship	−17.5	−12.1	−15
EIRP, dBW:			
ship→shore	27	34.2	32.6
shore→ship	18.8	19.5	20
Passband of transponder, MHz	4	5	7.5
Capacity of speech channels:			
shore→ship	8	40	30
ship→shore	14	50	100

TABLE 13.5

Standard	*A*	*B*	*C*	*D*
EIRP, dBW	37	29	19	37 – 47
G/*T*, dB/K	−4	−12	−19	+5
G of antenna, dB	23	15	8	32
Antenna diameter, m	1.2	0.5	Nonparabolic	3
Kinds of information transmitted and received	Speech, telex	Low-quality speech, telex	Telex only	Multichannel telephony, telex, high-speed data

13.2 THE WESTERN EUROPE REGIONAL SATELLITE COMMUNICATIONS SYSTEMS

The European Space Agency (ESA) is engaged in the development of satellite communications systems of the West European governments. This organization had eleven member countries at the end of 1980. The Western Europe satellite communications system, based on the European Communication Satellite (ECS) is intended for supplementing the Western European cable and radio relay ground communications network, in which respect it is fundamentally different from the national satellite communications systems of the United States, which form closed networks. It is presumed that in the very near future, half of the telephone and telex traffic, and virtually all of the TV broadcasting traffic of the Eurovision system in Western Europe will pass through the ECS satellite (an experimental satellite and prototype of the ECS is called the Orbital Test Satellite (OTS) and is in orbit at this time [13.2]). The characteristic features of the ECS are the utilization of the 14/11-GHz band, polarization multiplexing and the use of a parametric amplifier on the satellite.

One large earth station with a 17-meter antenna (with parameters close to standard C of the Intelsat system) and several small stations with a 3–5-antenna are supposed to be built in each of the member countries of the system. The service area of the ECS covers the European continent and the North African coast. The system is intended for the transmission of TV programs and telephone communications, and will be used as a technical means of national communications channels. It should connect switching centers of the member nations of the European Conference of

Postal and Telecommunications Administrations (ECPT), as well as radio and TV centers of the member nations of the European broadcasting media through national channels. One global beam is supposed to be used for transmitting TV programs and three zonal beams for telephony and data transmission.

13.3 NATIONAL SATELLITE COMMUNICATIONS SYSTEMS

13.3.1 The Soviet National Satellite Communications System

The Soviet national satellite communications system, the world's first, began to be developed in 1964, when TV programs were exchanged for the first time between Moscow and Vladivostok through the Molniya satellite. About 100 receiving earth stations, which are the technical foundation of the national satellite communications system, were operating in the USSR by the end of 1981. The main purpose of these earth stations is to operate in the Orbita distribution television network, with the aid of which, along with the Ehkran and Moskva systems, two-program five-area television broadcasting is organized. At the same time, these very same stations, after being outfitted with transmitters, multiple access, and terminal equipment, are being used for telephone communications.

Three types of satellites (Molniya-3, Raduga, and Gorizont) are used in the Soviet national satellite communications system. The Molniya-3 permits operation in two transponders: television and telephony. Because the satellite is in a highly elliptical orbit its service area covers the far northeast of the country and at the same time the central regions, which offers obvious advantages for the development of direct telephone communications links between these regions. At the same time, the limited duration of a communications session using one satellite and unavoidable interruptions during the day, connected with the need to re-point the antennas of the earth stations from one satellite to another, also create indisputable inconveniences in the use of the Molniya satellite.

The Raduga permits operation in three transponders, one of which is used for transmitting TV programs in the Orbita distribution network, and the other two are used for organizing telephone communications links. Frequency division multiple access is used in one of these transponders, and several communications links with a comparatively low capacity of 12–36 audio frequency channels are made available. Time division multiple access is used in the other telephone transponder. Here, as a rule, large-capacity (more than 60 audio frequency channels) communications are established.

The Gorizont permits operation in seven transponders (six in the 6/4-GHz band and one in the 14/11-GHz band). One of these transponders is used for transmitting TV programs in the Orbita network, another is used for the Moskva distribution TV network, and the rest are intended for telephone communications links and channels for delivering radio broadcast programs. The performance characteristics of the channels that are organized in the Soviet national satellite communictions system are given in Chapter 18, and the basic technical decisions on the equipment complexes are described in Chapter 10.

13.3.2 The Canadian Satellite Communications System

A national satellite communications system, intended for servicing many remote and isolated population centers, located chiefly in the northern regions of the country, was placed in operation (and continues to be developed and improved) in Canada in 1972. The system belongs to the Telesat Canada Corporation, which is owned on a commercial basis by the government of the country and by companies engaged in the development of Canadian communications links. The system has four Anik satellites (three Anik A and one Anik B). The Anik A satellites operate in the 6/4-GHz band, and the Anik B operates in the 6/4- and 14/11-GHz bands. Two new modifications of the satellites, presently in different stages of development (the Anik C and Anik D) are being implemented at this time. Multiple utilization of frequencies by means of space and polarization multiplexing will be used in the Anik C and Anik D satellites. Different earth stations, depending on purpose, are used in the Canadian national communications system. The basic parameters of the earth stations of the Canadian national system are given in Table 13.6

Basic Technical Characteristics of the Anik D* Satellite	
Frequency band, GHz	6/4
Weight, kg	1100
Number of transponders	24
Bandwidth, MHz	36
Service area	Canada
EIRP, dBW	36

National high-capacity communications links, as can be seen in Table 13.6, are established with the aid of earth stations, equipped with a

---

*The paremeters of the Anik A, B, and C satellites were given earlier in Table 13.1.

TABLE 13.6

Purpose	*Antenna diameter, m*	*G/T, dB/K*	*Comment*
6/4-GH band			
National high-capacity communications links	30	37.5	Delta-modulation at 40 Kb/s
Medium-capacity communications	30	37.5	
links, FDMA,	10	28	
FDM	8	27.5	
Medium-capacity communications	30	37.5	
links, TDMA, PCM	10	31	
Low-capacity communications	30	37.5	
links	8	26	
	8	22	
	4.6	21.7	

TV distribution network	30	37.5	Signal-to-noise ratio on video channel ≥54 dB
	10	28	
Remote TV stations (receive only)	8	26	Signal-to-noise ratio on video channel ≥48 dB
	8	22	
	4.6	21.7	
Border TV stations (receive only)	4.5	18.5	Signal-to-noise ratio on video channel ≥45 dB
	3.6	18.5	
14/11-GHz band			
Digital communications links	8	35	Standard eight-digit PCM
TV broadcasting	8	29.5	Signal-to-noise ratio on video channel 45–48 dB
	4.5	26.5	Signal-to-noise ratio from 48–54 dB
	8	35	
Direct TV reception	1.8	16.5	Signal-to-noise ratio on video channel 40–42 dB
	1.2	13	Signal-to-noise ratio on video channel 40–42 dB

30-meter antenna and an uncooled parametric amplifier. A 960-channel group is transmitted on one carrier by frequency modulation. There are two of these stations in the network.

Frequency division multiple access in combination with frequency modulation is used for organizing medium-capacity communications links, and standard 12- and 60-channel groups are transmitted. As a rule, stations with a 10-meter antenna are used for these communications links. It is important to mention that one of the links between a station with a 30-m antenna and a station with a 10-m antenna is organized by TDMA, and the total capacity of this link is 300 two-way telephone channels.

Low-capacity communications links are organized by one channel on a carrier, and the DM/PSK modification is used. The total carrying capacity of one transponder is 360 two-way telephone channels. Transportable stations with a 3.5-m antenna, permitting operation on two telephone channels, can also operate in the network.

Stations intended for operation with the Anik C satellite transmit digital messages at a line rate per transponder of 91 Mb/s, and they also receive and transmit TV programs. These stations will have a cooled parametric amplifier (G/T = 35 dB/K), and stations intended for operating just in the TV mode will have a transistor amplifier (G/T = 29.5 dB/K). The feasibility is being studied of using the Anik C satellite for direct TV broadcasting at a station with a 1.2–2.5-m antenna with a signal-to-noise ratio greater than 40 dB on the video channel (the frequency band of the transponder is 18 MHz).

13.3.3 The National Satellite Communications Systems of the United States

The RCA communications system was the first national satellite communications system of the United States. For this system RCA leased from Telesat Canada two of 12 transponders of the Canadian Anik B satellite. The system made it possible to establish communications between the United States and Alaska. After this system was developed, it was no longer necessary to lease channels from the Intelsat system for communications and TV transmissions to Alaska.

The RCA network expanded significantly after the Satcom I satellite was launched. By 1981 this satellite communications system had three Satcom I satellites, ground stations in Alaska, in the United States, and in the Hawaiian Islands. Twenty-one earth stations with 10-m antennas are used for exchanging TV programs and telephone communications, and there are also more than 100 small earth stations which operate in just the TV program receiving mode.

The Western Union Telegraph satellite communications system was developed on the basis of the Westar I and Westar II satellites. It services with communications the main states of the United States, Alaska, the Hawaiian Islands, and Puerto Rico with earth stations located in the vicinities of New York, Atlanta, Chicago, Dallas, and Los Angeles. The company plans to expand the network on the basis of improved Westar satellites with more than 100 earth stations with 5.7- and 13-meter antennas.

The Satellite Business System (SBS) Corporation is developing a satellite communications system, the main purpose of which is wideband high-speed data transmission, principally directly between computers. The earth stations of this network will have 375 small earth stations with a 5.7-meter antenna.

13.3.4 The National Satellite Communications Systems of Japan

The first communication system (CS) and business system (BS) experimental satellites were launched in 1977–1978 by US booster rockets. A national satellite communications system began to be placed in operation in 1982 on the basis of these satellites. Plans are being made to develop a satellite communications system primarily for data transmission, which will make it possible to connect computers in different regions of the country, and also for telephone communications with the remote islands. The characteristic feature of the national satellite communications system of Japan is the use for telephone communications, along with TDMA, of one channel on a carrier with companded FM. In general, it should be pointed out that the ever-increasing number of national and even of regional satellite communications systems permit a deviation of some of the parameters of telephone channels from standards recommended by the CCITT and the CCIR, for the sake of increasing the carrying capacity of a system and for improving its economic indices. Delta-modulation with companded frequency modulation is being used increasingly for this purpose.

13.3.5 The National Satellite Communications System of France

This system is planned on the basis of the Telecom satellite, operating in the 6/4- and 14/11-GHz bands. In the 6/4-GHz band the satellite will contain two transponders, each with a 120-MHz band, and in the 14/11-GHz band there will be six transponders, each with a 36-MHz band.

Standard A Intelsat stations in Paris and a standard B in each overseas department will be used for communicating with overseas territories. Two types of stations will be used for the TV distribution network: a few transportable earth stations with a 3.5-meter antenna for urgent transmissions (EIRP=72 dBW), and a large number of stations with a small-diameter antenna (about 2 m).

A business communications network will be developed within the framework of the same program. These stations will have a 3.5-meter antenna, G/T = 25.5–27.6 dB/K (depending on the type of input amplifier), and a transmitter with a traveling wave tube output amplifier (EIRP ≈ 72 dBW). Time division multiple access equipment will be used.

Basic Parameters of Time Division Multiple Access Equipment

Transmission rate per trunk, Mb/s	24.576
Frame duration, ms	20
Maximum number of packets transmitted	64
Maximum number of packets per frame	256
Superframe duration, s	5.12
Bandwidth, MHz	27

13.4 BASIC TECHNICAL DATA ON SOVIET SATELLITE TV BROADCASTING SYSTEMS

13.4.1 The Orbita-2 System

This system uses Molniya-3, Raduga, and Gorizont geostationary satellites (the registration name of Soviet geostationary satellites is Statsionar). The signal carrier frequency in the satellite-earth link is 3875 MHz. The flux density created by a satellite at the earth's surface is not less than $2.5 \cdot 10^{-14}$ W/m^2 (-136 dBW/m^2).

The transponder of the Raduga and Gorizont satellites with the above-indicated frequency operates on a directional antenna (the aperture of the radiation pattern is 9 × 18°), and a transponder of the Molniya-3 satellite operates on a global antenna (17 × 17°), but because the Molniya-3 space transmitter has greater power (40 W), the power flux density of the signal at the earth's surface is sufficient for operating Orbita stations. Frequency modulation is used; the peak carrier frequency deviation is 15 MHz.

A TV sound signal is transmitted on a subcarrier frequency of 7 MHz with a frequency deviation of ±150 kHz, and the deviation of the carrier by the subcarrier signal ranges up to approximately ±1.5 MHz. Audio broadcast and facsimile signals (see Chapter 19) are transmitted on 7.5- and 8.2-MHz subcarriers. The quality with which sound signals are transmitted on the subcarrier corresponds to quality class I (the 10-kHz band, see Chapter 2). It is also possible to transmit TV sound signals or audio broadcast signals by time division multiplexing of the video signal. Two sound carrier pulses, duration-modulated by sound signals, are located in the interval of the blanking line pulse (in front of and behind a sync pulse, see [4]).

Either one audio program with an up to 10-kHz band, or two class-II programs with a 6-kHz band can be transmitted by time division multiplexing. However, the equipment used in the Orbita-2 system for time division multiplexing by sound signals damages (breaks up) a color subcarrier pulse that is transmitted at the same time, and this destroys color transmission at the beginning of each line, i.e., causes distortions along the left side of a frame.

The performance characteristics of a TV channel, created by the Orbita-2 system (for the entire transmitting earth station-satellite-receiving earth station link) correspond basically to CCIR recommendations for a national TV program transmission channel. The passband of the video signal is 6 MHz. The ratio of the video signal amplitude to the weighted noise is 54–55 dB, and to background interference it is 36 dB; "differential amplitude" distortions on the 4.4-MHz color subcarrier do not exceed 6 percent, "differential phase" distortions do not exceed 6°, and the time discrepancy between the brightness and color signals does not exceed 25 ns.

The receiving earth stations of the Orbita-2 system (Figure 13.2) are rather large installations, consisting of a round building containing the equipment room and auxiliary rooms, atop which the antenna is erected. The antenna station of the TNA-57 type, with a 12-meter parabolic reflector of the two-reflector design, is installed on a 360° turntable. The receiver complex contains a low-noise cooled parametric input amplifier. The standard G/T of the station (the ratio of the antenna gain G = 52 dB to the summary noise temperature $T°$ of the station) is

$$G/T = 29 \text{ dB/K}$$

The antenna is equipped with programmable and manual satellite tracking systems, and it can also be outfitted with an automatic satellite tracking complex that tracks the peak received signal. A more detailed description of the Orbita-2 station is given in Chapter 16. It should be pointed out that many Orbita-2 stations are being converted to multitransponder, multipurpose two-way stations for television communications. Also, the G/T of the station may be increased to 31 dB/K for the purpose of increasing the carrying capacity, which will be accomplished by installing new models of parametric amplifiers, including uncooled. Central transmitting stations, equipped with Gradient transmitting equipment (Chapter 15), are used for transmitting signals of the Orbita-2 system to satellites.

FIGURE 13.2. An Orbita station.

13.4.2 The Ehkran TV Broadcasting System

The further intensive development of the Orbita system as a means of delivering TV programs became uneconomical in the late 1970s, because it is too expensive to build earth stations of this type in population centers with several thousand citizens. A new technical means had to be found to increase the efficiency of the system. The radiated power of the satellite has to be increased for this purpose, which makes it possible to simplify and reduce the cost of receiver installations and to make them accessible for use in small population centers. This new means was the Ehkran system, developed in the USSR in 1976. The first satellite of this system was launched on October 26, 1976 into a geostationary orbit at a point with the coordinates 0° north latitude and 99° east longitude. The service area of the Ehkran system covers a region of more than nine million square kilometers (about 40% of the USSR), and the area covers regions of Siberia, the Far North, and a part of the Far East. About 20 million citizens live in these regions, more than 7.5 million of whom never before could receive television programs. The Ehkran system operates on 714 MHz; the USSR applied to the international registration for the second channel of the system with a 754-MHz carrier frequency.

Satellite broadcasting coexists in the 0.7-GHz band with terrestrial

TV broadcasting, in which connection the power flux density in other countries is restricted by the CCIR Radio Regulations (article 332A and Recommendation No. Space 2-10) to the -129-dBW/m^2 level, although studies, including studies which the USSR has sent to the CCIR [4, 13.3], show that this standard is too strict and should depend on the signal energy dissipation of the satellite system through the frequency spectrum. In some areas of the USSR it was possible to release the necessary number of decimeter wave channels for satellite broadcasting, and by virtue of adequate space selectivity to observe the standards on interference in other countries while achieving a higher power flux density (approximately -116.5 dBW/m^2) in the service area.

The advantage of the 0.7-GHz band is the fact that simple and inexpensive receivers can be used. For instance, a receiver noise temperature of the order of 600 K can be achieved with the aid of inexpensive transistor input amplifiers; multicomponent "wave channel" antennas have low wind resistance and a high gain. For all these reasons, a carrier frequency in the 0.7-GHz band was chosen for the Ehkran system. The modulation method in this band is prescribed by the CCIR Radio Regulations; only frequency modulation may be used in the satellite service in the interest of compatibility with terrestrial broadcasting.

The frequency deviation in satellite distribution systems should be such that the necessary output signal-to-noise ratio can be achieved in the vicinity of the threshold of the FM signal detector [16]; that is how the frequency deviation was chosen in the Orbita-2 system. However, because of the limited frequency band in the Ehkran system, a somewhat lower value of ± 9 MHz was chosen (the peak value, including sync pulses). Linear predistortions, which are standard for radio delay links and satellite links, are used for the purpose of reducing signal distortions.

In the Ehkran system, audio accompaniment is transmitted on a subcarrier frequency. This entails a certain amount of power loss (the frequency deviation of the 2-MHz audio subcarrier is less than 2 dB), but on the other hand it significantly simplifies the receiver. For simplifying the generation of a standard TV signal, the subcarrier frequency is set at 6.5-MHz, i.e., equal to the frequency separation of the video and audio carriers in terrestrial television broadcasting. For the same reasons the subcarrier frequency deviation is 50 kHz, which is standard for TV audio accompaniment. All of this entails great power losses, but the receiver can be simplified to the greatest extent.

In addition to TV video and audio signals, it is also possible with the Ehkran system to receive (with just class I sets) an audio broadcast program, transmitted on the 7.0-MHz subcarrier. A large frequency deviation of ± 150 kHz, controllable companding, and a tracking demodulator with frequency feedback are used in the radio broadcasting channel for the purpose of reducing the associated power losses (they are 0.5–1 dB) on the television channel.

The main power parameters of the Ehkran system are listed in Table 13.7. The parameters of the earth-satellite link (in the 6-GHz band) are not crucial and are not given here. The transmission of signals with the above-indicated parameters requires careful execution of all the elements of the line in order to observe the established limits of distortions. In the ±12-MHz band, it is necessary to specify that the nonuniformity of the amplitude-frequency response does not exceed 1 dB and the nonuniformity of the group delay time is about 10 ns.

TABLE 13.7

Parameter	*Symbol*	*Value*
Transmitter power, delivered to satellite antenna, W	P_{sat}	200
Transmitting antenna gain, dB:		
in main direction	$G_{tr.sat0}$	33.5
toward the edge of the area	$G_{tr.sat}$	26
Signal losses, dB:		
in open space	L_0	182
additional losses	L_{add}	2.5
Field intensity on boundary of service area, μV/m	E_0	29
Gain of receiving antenna of earth station, dB	$G_{rec.e}$	30/23
Width of radiation pattern, deg	$\phi_{rec.e}$	4.5×2.5/9×9
Losses in antenna-waveguide, dB	$\eta_{rec.e}$	1/1
Equivalent noise temperature of receiving station, K	$T_{n.e}$	800/800
Receiver input signal power, dBW	$P_{s.e}$	−106/−113
Receiver input noise power, dBW	$P_{n.e}$	−126.8/126.8
Receiver input signal-to-noise ratio, dB	$(P_s/P_n)_{in.e}$	20.8/13.8
FM gain on video channel, dB	B_{FM}	11.3/11.3
Receiver output signal-to-noise ratio on video channel, dB	$(U_{av}/U_n)_w$	54–53/48
Receiver output signal-to-noise ratio on audio accompaniment channel, dB	$(P_s/P_n)_{au.r.s}$	53–56/49

(*Comment*. Values for class I receivers are given in the numerator, and for class II receivers in the denominator.)

The transmitting earth station of the Ehkran system is equipped with a 5-kW Gradient transmitter and a 12-meter antenna. The FM signal generated at the station is beamed to the Ehkran satellite on 6200 ± 12 MHz. These signals are picked up by the satellite receiving antenna, pointed at the transmitting earth station, amplified to 200 W, and transmitted by the space antenna, which is a phased array antenna. The satellite antenna has a narrow radiation pattern. Power is supplied to the repeater and other space system from a solar panel, which delivers up to 2 kW. The need to keep the satellite antenna permanently and precisely pointed at the center of the service area and the solar batteries at the sun require the use of a triaxial orientation and stabilization system. The satellite is held at a given point in orbit by a correction system, built on the basis of liquid fuel rocket micromotors. The satellite is controlled and its orbit is corrected by radio commands from the earth.

The Ehkran satellite is inserted by a multistage rocket booster first into a near-stationary orbit, and then it "drifts" toward its parking place (99° east longitude), where it is stopped by the correcting system, and its orbital period is adjusted to diurnal. After that the satellite is activated and made ready to retransmit signals. The signals that are retransmitted by the satellite, with rotary (right-handed) polarization on frequencies of 714 ± 12 MHz (the 52nd–54th TV channels) are picked up by a network of class I and II earth receiver installations.

Class I receiver installations are intended for delivering a high-quality TV signal to local TV centers and large TV retransmitters. As a rule, they are outfitted with a "wave channel" antenna with 32 panels, each of which has a dipole, reflector, and about 300 crossed directors (for receiving circularly polarized signals), and a receiver, built as a rack, which receives, demodulates and separates the video and audio signals (Chapter 17).

Class II receiver installations are intended for delivering a TV signal to small TV retransmitters or to a cable distribution network. They consist of a simplified "wave channel" antenna with four panels, similar to the antenna panels of the class I installation, and a small fixed frequency receiver, which transfers the spectrum of the received signal from the 52nd–54th channels of the decimeter wave band to one of the first channels of the meter band and performs FM to AM conversion (Chapter 17).

A feature of the Ehkran system is the fact that the transmitting station, located in the Moscow suburbs, is outside of the service area of the Ehkran satellite and cannot pick up its signals with the necessary quality. This creates two problems: precision correction of the steering of the transmitting antenna and quality control. The first problem was solved with the aid of a high-sensitivity monitor receiver, which guarantees reliable reception at the transmitting station, *albeit* with reduced quality. The quality control problem was solved by building reference check points in the service area, which transmit quality estimate data to the control center of the system. The service area can be expanded or the received video quality improved by outfitting the receiving stations with more sophisticated antennas and a parametric amplifier.

Given the parameters that are used in the Ehkran system, the 6.5-MHz audio subcarrier in particular, direct FM-AM converter can be used for generating the AM single-sideband video signal that is received at a terrestrial broadcasting station. However, a simple analysis shows that this kind of conversion does not guarantee the standard terrestrial broadcasting AM video modulation index (m_{AM} = 87.5%) for the standard ratio of 10/1 between the video and audio carrier powers. Therefore, it was necessary to use a more cumbersome conversion system in receivers of the Ehkran system, containing a frequency detector, amplitude modulator for the video signal, a subcarrier selection filter, and a selector.

13.4.3 The Moskva TV Broadcasting System

The Basic Principles of the System. The Ehkran system, described in the preceding section, is simple and economical, and it has a deficiency—a limited service area. Because the tolerable limit of irradiation of neighboring countries, recommended by the CCIR Radio Regulations, must be observed, the service area of the Ehkran system is limited and cannot be expanded either to the entire Soviet Far East, Kamchatka, and Chukotka, or to the European part of the USSR.

Because the 4-GHz band has been adopted extensively and the Orbita-2 systems and the Molniya-2, Molniya-3, Raduga, and Gorizont (the registration name is Statsionar) satellites have been operating in this band in the USSR for many years already, it is desirable to develop a TV distribution system with comparatively simple and inexpensive receiver stations in this band. However, as was already mentioned, the 4-GHz frequency is intended both for satellite systems, and for other services, including terrestrial radio relay links. Therefore, the CCIR Radio Regulations limit the power flux density, created by a satellite on the earth's surface, to 152 dBW/m^2 at 4 kHZ (for angles of incidence smaller than 5°), which until now had been an obstacle to the use of smaller antennas and less expensive receiving stations.

The development of methods of effectively dissipating energy [4], and also of servicing uncooled low-noise parametric amplifiers has made it possible to solve this problem and to develop the Moskva system. These stations are equipped with comparatively small antennas (the reflector diameter is 2.5 m) and do not require automatic satellite tracking systems.

The Moskva system is suitable for servicing any region of the USSR, including the European part and the Far East, without the risk of interfering with terrestrial services.

The main difference between the Moskva system and all the preceding systems is the fact that the service area was selected in consideration of the principles that are used for organizing TV and audio broadcasting in the USSR, i.e., such that there are two or three time zones in the area that is serviced by one satellite. This enables TV viewers to watch the most interesting programs at a convenient viewing time.

Unlike the Orbita system, which operates through a standard transponder of a multitransponder satellite, and the Ehkran system, which operates through a special TV satellite, the Moskva system operates through a special transponder of a multitransponder multifunctional Gorizont satellite—a transponder which, in combination with the needs of the specialized distribution network, has higher power and operates on a spot-beam antenna, corresponding to the serviced broadcasting area.

Parameters of the Specialized Transponder of the Gorizont Satellite

Total peak power of space transmitter, delivered to satellite antenna, W	40
Gain of satellite transmitting antenna, dB	30
Frequency band, MHz	40
Tolerable deviation of position of satellite from nominal, deg:	
by longitude	±0.5
by latitude	±1.2

For a space repeater with the above-mentioned parameters the integral power flux density at the earth's surface is approximately −120 dBW/m^2. We note that several space segments, located at the points 14° west longitude, 53°, 90°, 140° east longitude, applied for the Statsionar satellite, may be used for covering the entire territory of the USSR.

Performance Characteristics of Video, Audio Accompaniment and Radio Broadcasting Channels

Upper frequency of video channel, MHz	6
Output signal-to-weighted-noise ratio of video channel $[(U_{av}/U_n)_w]$, dB, not less than (usually 55 dB)	53
Output signal-to-background-ratio of video channel	35

Nonlinear brightness signal distortions, %, not more than	15
"Differential gain" distortions, %, not more than	10
"Differential phase" distortions, deg, not more than	8
Transient response distortions in medium time range, %, not more than	5
Transient response distortions in long time range, %, not more than	10
Transient response distortions in short time range, nonuniformity of the amplitude-frequency response of video channel	GOST 19463–74 stencil
Upper frequency in audio signal sectrum, kHz	10
Signal-to-weighted-noise ratio on TV audio signal (broadcasting) transmitting channel $[(P_s/P_{au.r.s}]$, dB	57
Noise immunity of TV audio signal (broadcasting) transmitting channel relative to audible crosstalk interference, dB, not less than	70
Harmonic coefficient on 800 Hz	2%
Nonuniformity, dB, of amplitude-frequency response of TV audio signal (broadcasting) transmitting channel on frequencies, Hz:	
from 50 to 100 and from 8000 to 10000	+1.8 − 4.5
from 100 to 200 and from 6000 to 8000	+1.8 − 3.6
from 200 to 6000	+1.8

The performance characteristics given above are the top-to-bottom characteristics and apply to the Moskva system as a whole. The size of the Moskva earth station antenna was selected on the basis of several conflicting factors. On the one hand, the antenna must be as large as possible in order to have the highest gain and to guarantee a given receiver input signal level. On the other hand, to simplify the receiving stations it is desirable to eliminate the sophisticated steering system, and in this case the antenna radiation pattern must be rather wide in consideration of the possible instability of the satellite in orbit.

In consideration of the above factors, and also of the effort to simplify the station and to make it as inexpensive as possible, a 2.5-meter diameter was selected. The width of the radiation pattern of this antenna is ±1°. At unattended stations, the operational instability of orientation of an earth receiving antenna can exert a certain amount of influence when the radiation pattern is of such a width. Therefore, a simple automatic satellite tracking system (the satellite is tracked by moving the irradiator around in the stationary main reflector) is used in certain cases.

An uncooled parametric amplifier with a noise temperature of 80 K is used as the input at a Moskva station. By virtue of the power characteristics of a communications link in the Moskva system, it is possible to transmit by FM one video signal, two audio signals, and a facsimile signal. The frequency deviation of the video signal should be 12–13 MHz in order to achieve the necessary TV signal quality. A frequency deviation of ±1 MHz is necessary for the TV sound and audio broadcasting signals. TV sound signals and audio broadcasting signals are transmitted by frequency modulation on subcarriers of 7 MHz and 7.5 MHz. The frequency deviation of the subcarriers is ±150 kHz. Variable companding is used for improving the quality of audio signal transmission channels [13.7].

Energy Ratios in Moskva System

Equivalent isotropically radiated space transmitter power, dBW	43
Cumulative signal propagation losses in satellite-earth link L_0, dB	198
Receiving antenna gain of Moskva station, dB	37.5
Equivalent noise temperature of Moskva receiving station T_n, K	200
Noise band of receiving station R_n, MHz	37
Receiver input signal-to-noise ratio $[(P_s/P_n)_{in}]$, dB	12.5
Upper video signal frequency F_v, MHz	6
Frequency deviation assigned for video signal without sync pulses, $f_{d.r}$, MHz	9.0
Frequency modulation gain on video channel B_{FM}, dB	13.0
Outut signal-to-weighted-noise ratio of video channel $[(U_{av}/U_n)_w]$, dB, not less than	53–57
Frequency deviation assigned for audio subcarrier $f_{d.sub}$, MHz	1
Subcarrier frequency f_{sub}, MHz	7
Subcarrier frequency deviation $f_{d.s}$, kHz	150
Upper frequency of audio signal band $F_{u.sou}$, kHz	10
Frequency modulation gain on audio channel B_{FMsou}, dB	38
Output signal-to-noise ratio of audio broadcasting channel $[(P_s/P_n)_{au.r.s}]$, dB	57

The transmitter of an earth transmitting station of the Moskva system is similar to the transmitter of the Orbita and Ehkran systems, but a special TV signal processing and dispersion signal insertion unit, and the transmitting equipment of the audio accompaniment and audio broadcasting channels are installed in front of the frequency modulator. The signal retransmitted by the satellite is picked up by Moskva earth receiver installations and from there goes to a small (10 W, or more often 100 W) TV transmitter, which delivers the received program to subscribers. The receiving station can also operate on a cable distribution network. Audio broadcasting signals can be delivered to a local radio relay network, or to a UHF FM radio broadcasting transmitter.

Facsimile transmission is also possible (on the 8.2-MHz subcarrier frequency). The capability of receiving newspapers is an advantage of the Moskva system, because connecting links are not necessary when a Moskva station is built next to the typography building. In the Moskva system the transmitted signal is subjected to special processing. This processing consists in the limitation of spikes that occur on the leading edges of video signal pulses, subjected to standard linear predistortions. This makes it possible, by increasing the TV signal frequency deviation, to improve the picture quality by 2–3 dB (see [4]).

To make the Moskva system electromagnetically compatible with terrestrial and other satellite communications systems operating in the 4-GHz band, it is necessary to meet the standards of the CCIR on the spectral power flux density at the earth's surface, created by the space transmitting station; the flux density may not exceed 152 dBW/m^2 in any 4-kHz band. When this requirement is satisfied there is no need to worry that the Moskva system will interfere with terrestrial services and, in particular, with radio relay links, even when the video subjects are exceedingly unfavorable in relation to the scattering of energy. The minimum power flux density, measured in any 4-kHz band, obviously will be achieved when the signal power is uniformly distributed in the entire frequency band that is reserved for it.

As is known, the FM TV signal spectrum is nonuniform, and for this reason the flux density in the worst 4-kHz band in certain cases may be only 2–3 dB lower than for an unmodulated carrier. Under these conditions, considering that the power flux density of an unmodulated carrier at the earth's surface is 120 $dBW/m^2 \cdot 4$ kHz, it was necessary to use artificial dissipation of power with the aid of a dispersion signal.

A triangular dispersion signal is used in the Moskva system, becuase the FM spectrum of this kind of signal is the most uniform. The frequency of the dispersion signal was set equal to approximately 2 Hz. This choice, on the one hand, is explained by the need to effectively select the dispersion signal in the receiver using narrow-band frequency feedback. On the other hand, this choice of frequency does not detract from the effectiveness of dispersion. In fact, the noise power on a telephone channel of a radio relay link is standardized per minute on the average, and the mean minute interference power may be assumed not to depend on the dispersion signal frequency. The deviation assigned for the dispersion signal is selected such as to provide the necessary amount of dissipation of the carrier, and in the Moskva system is ±4 MHz with a little to spare.

In the receiver, the dispersion signal is selected with the aid of an instrument with frequency feedback. A schematic diagram of the instrument is shown in Figure 13.3. By selecting the dispersion signal frequency below the video signal spectrum, it is possible to filter it out and to close the frequency feedback circuit on just the dispersion signal. Here the receiver band is calculated to pass the FM signal, modulated just by the useful message. This solution makes it possible to avoid losses of noise immunity, which would appear when the receiver passband is expanded. The residual dispersion signal is taken out by "clamping" the video signal level.

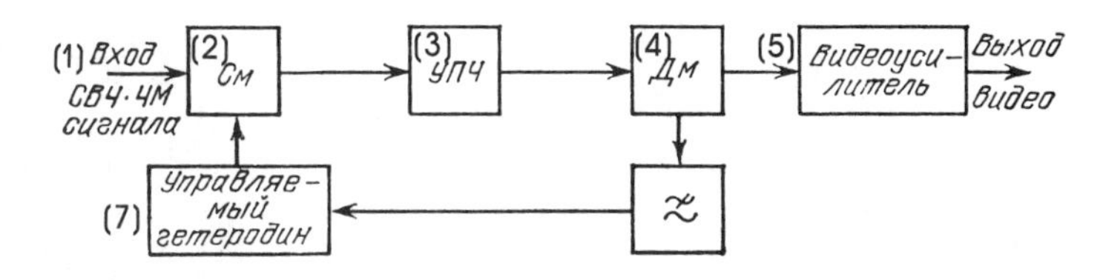

Key: (1) SHF FM signal input
(2) Mixer
(3) IF amplifier
(4) Demodulator
(5) Video amplifier
(6) Video output
(7) Tunable heterodyne

FIGURE 13.3. Schematic diagram of dispersion signal selector in receiver of Moskva station.

Because the frequency bands coincide, it is necessary to consider the possible influence of a radio relay link on a Moskva station. Electromagnetic compatibility in this respect is achieved by putting the central frequency of a transponder of the Moskva system in the "window" of the frequency plan of many Soviet radio relay links. The closest carrier frequency of a radio relay link is 9 MHz away from the central frequency of a Moskva transponder. In addition, by selecting a good site for installing a Moskva station, it is possible in most cases to avoid interference, even from nearby radio relay stations. Experience shows that by choosing the right site for a Moskva station, it is possible, by virtue of its small antenna, to utilize the shielding properties of natural barriers, various buildings and facilities.

Chapter 14

Ways and Means of Transmitting Audio Broadcasting Programs, Television Audio Signals, and Facsimile Through Satellites

14.1 TRANSMISSION METHODS

The requirements on channels intended for the transmission of audio broadcasting programs were formulated in Chapter 2. High-speed newspaper facsimile transmission requires a wider frequency band (most often the band of a 60-channel group, i.e., approximately 240 kHz), but on the other hand the signal-to-noise ratio is considerably lower (22–24 dB). Therefore, the transmission of facsimile and audio broadcasting signals requires a channel with approximately the same carrying capacity, so that the methods described below for transmitting audio broadcasting signals are also applicable to the problem of facsimile transmission. We will examine possible methods of transmitting audio broadcasting signals in satellite systems.

Audio Broadcasting Program Transmission on Subcarrier Frequency. This transmission technique is used not only in satellite communications, but also in radio relay links, where up to four audio broadcasting programs are transmitted in this manner. The subcarriers are located above the frequency band of a television message, i.e., 6 MHz higher. In that range, as is known (see Chapter 4), the spectral density of fluctuation noises G_n is the greatest (Figure 14.1), and the transmission of audio programs with the necessary high signal-to-noise ratio by means of conventional frequency division multiplex channels, i.e., by transferring the audio program spectrum upward on the frequency scale and keeping one sideband, has a negative effect on the power characteristics of a TV transponder.

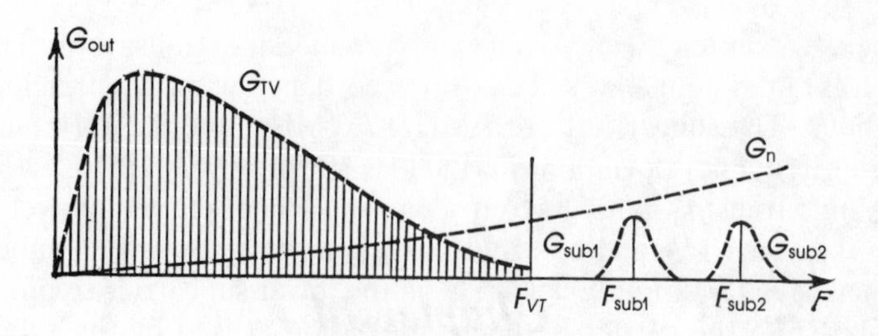

FIGURE 14.1. Output signal spectrum of demodulator of television transponders with audio signals transmitted on subcarriers.

Therefore, it is customary to use double-frequency modulation, i.e., the audio program signal frequency-modulates the subcarrier (in the 6.5–8.5[11] MHz range), and then this modulated signal is added to the TV signal and is fed to the input of the main frequency modulator 2 (Figure 14.2), the spectra of the FM subcarriers (G_{sub1} and G_{sub2} in Figure 14.1); the reverse conversions take place on the receiving end. The demodulation of the frequency-modulated subcarrier yields a power gain, which makes it possible to reduce the level of the subcarriers (in other words to reduce the carrier frequency deviation by subcarrier frequency signals) and, consequently, to reduce losses by way of the transmission of a TV message. This gain increases as the subcarrier frequency deviation increases. The increase of the subcarrier frequency deviation is limited by the onset of the threshold FM gain, when the power gain is lost for that reason. Noise suppressors, usually the familiar compander systems [14.1], are used in a TV audio signal transmitting channel for achieving power gain.

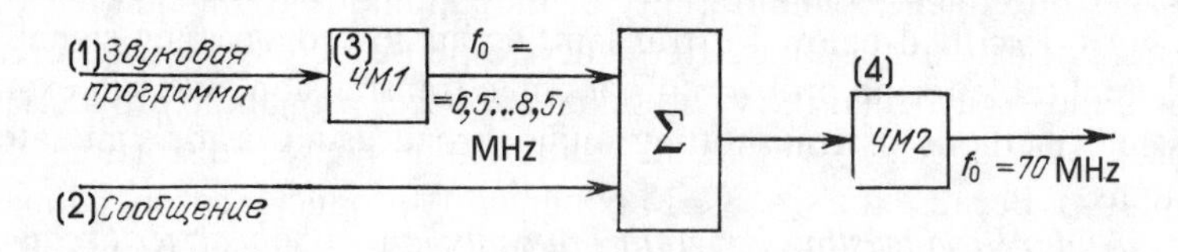

Key: **(1) Audio program (3) Frequency modulator 1**
(2) Message (4) Frequency modulator 2

FIGURE 14.2. Schematic diagram of signal generator for transmitting audio broadcasting signals on subcarriers in a television transponder.

The parameters of audio broadcasting, TV audio accompaniment, and facsimile channels in the Orbita and Moskva distribution systems described in Chapter 13, and of an audio broadcasting transmitting channel in the Ehkran system, are selected in accordance with the principles set forth above. The subcarriers are 7 MHz, 7.5 MHz, and 8.2 MHz, and the subcarrier frequency deviation is ±150 kHz (from ±0.6 MHz to ±1.5 MHz for facsimile transmission). In contrast to the one described above, in the Ehkran system, TV audio signals are transmitted with a subcarrier frequency deviation of only ±50 kHz on the lower subcarrier frequency of 6.5 MHz. Here the carrier frequency deviation caused by the subcarrier had to be increased to ±2.8 MHz, which for a carrier frequency deviation from a television message of ±9 MHz means that the power losses for transmitting TV messages are significant, more than 2 dB. These parameters were selected because the subcarrier frequency (6.5 MHz) and the subcarrier frequency deviation (±50 kHz) are the same as the separation of the TV video and audio signal carriers and the signal carrier frequency deviation, adopted in the USSR for terrestrial telecasting. This simplifies the generation of the standard TV signal for terrestrial systems, thereby making it possible to eliminate the demodulation of the audio signal (see Chapter 17). A similar method of transmitting a TV audio signal was adopted during the drafting of the telecasting plan in the 12-GHz band at the WARC in 1977.

When TV audio signals are transmitted by the above-described method, the video and audio signals are separated by filters, and the low-pass filter must pass the video signal with the minimum amplitude-frequency and phase-frequency distortions, and suppress the TV audio signals. The development of this kind of filter poses problems for an audio subcarrier frequency of 6.5 MHz or lower. Furthermore, because the transmission responses of a channel are nonlinear for modulating a message (due to the nonuniformity of the amplitude-frequency response and nonlinearity of the phase-frequency response), crosstalk interference occurs, and in particular the video signal interferes with audio broadcasting signals. This effect is similar to "differential phase" distortions (see Chapter 2). Because the size of FM nonlinearity products is proportional to the frequency of these products [16], the low-frequency components of the video signal have a greater effect on the audio signal subcarrier than on the color subcarrier; requirements on the tolerable interference on audio signal channels are also very strict (see Chapter 2). Therefore, in cases when audio programs are transmitted on subcarrier, the requirements on the linearity of the channel for the modulating message are strengthened considerably.

It is necessary to consider not only bias cross-modulation, which causes interference on the audio signal transmitting channel, but also amplitude modulation of the subcarrier due to tailing off toward the edges of the amplitude-frequency response of a circuit's spectral components, attributed to the subcarrier, for the maximum carrier frequency deviations, attributed to the video brightness signal. This effect can lead to the onset of the threshold in the subcarrier channel [14.2], and the need to eliminate it places rigid requirements on the bandwidth of the HF circuit and on the uniformity of the amplitude-frequency response in its band.

When many audio broadcasting programs are to be transmitted they are put on several FM subcarrier frequencies in the entire band of a transponder of a satellite or radio relay communications system (i.e., not to be added to TV signals, but in place of them, Figure 14.3). Because of the above-described nonlinear effects, and in particular of audible cross-talk interference between channels, on which, as was mentioned in Chapter 2, especially strict standards are established (−70 dB), the number of audio broadcasting channels is small when this method is used (8–10). This problem is exacerbated by the need to take into consideration a situation, when an interfering program is transmitted simultaneously and simultaneously in phase on many channels, and the other program is transmitted on the channel that is subjected to the interference. For the purpose of increasing the carrying capacity of a transponder when this transmission method is used, the subcarrier level and the frequency deviation usually are increased as the subcarrier frequency increases (see Figure 14.3) for the purpose of offsetting the increase of the output noises of the frequency detector.

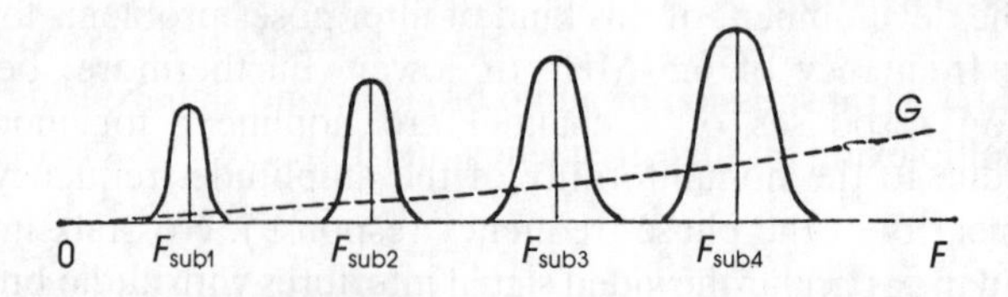

FIGURE 14.3. Transmission of audio broadcasting signals on FM and frequency multiplex subcarriers.

Transmission of Audio Broadcasting Programs Using HF FM and Frequency Multiplexing. In this version, a satellite transponder is multiplexed by several audio broadcasting signals, each of which is transmitted on a separate carrier with the aid of frequency modulation, and is separated on high frequency (Figure 14.4). This very same method is used extensively in multiple access satellite telephone communications systems (Chapter 7), because it is the simplest way of organizing several communications links between different earth stations through one transponder of a satellite repeater. The final stage of this kind of transponder (the power amplifier-transmitter) usually operates on a traveling wave tube or a klystron, and when one FM TV program is being transmitted (or one FM signal, modulated by a multichannel telephone message), it develops the nominal power with high efficiency in the nonlinear mode, essentially in the clipping mode. In the case examined here, when several FM signals with carrier frequencies f_1, f_2, . . . are amplified in a common nonlinear link nonlinearity products occur: combination oscillations, the strongest of which have frequencies $2f_1 \pm f_2$, $2f_2 \pm f_1$. An analysis of this phenomenon is given in Chapter 7. For the output stage to operate in a more or less linear mode, it is necessary to load the space repeater so that the cumulative capacity of the transponder for all signals will be 3–5 dB less than nominal for the saturation mode. It should be pointed out that nonlinear noises are caused not only by nonlinearity of the amplitude response of an output vacuum instrument, but also by the AM-FM conversion that takes place in it. This can cause audible crosstalk interference, the tolerance on which is extremely strict on audio broadcasting channels.

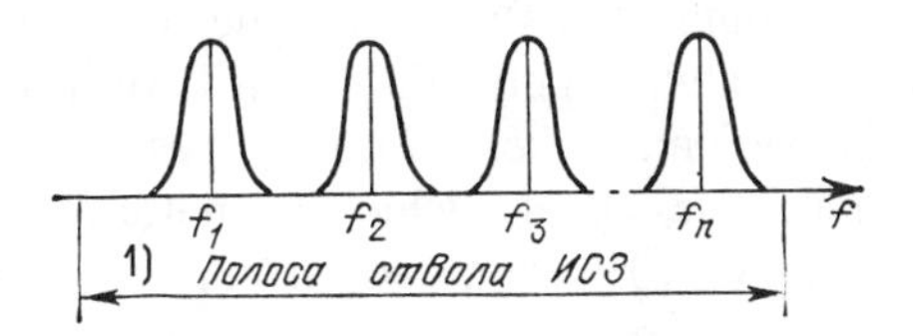

Key: **(1) Satellite trunk band**

FIGURE 14.4. Transmission of audio broadcasting signals using frequency division multiplexing on high frequency and FM.

Calculations and experiments, carried out in consideration of all the circumstances described above, show that in a conventional transponder of a satellite communications system with a band of about 35 MHz, intended for the transmission of FM TV signals, it is possible to transmit not more than 8–10 top-quality audio programs instead of TV messages, and this method is not used very extensively for this reason. The fact that radio broadcasting programs can be exchanged between different earth stations in different directions, which is important, for example, in international satellite communications systems, is an advantage of the examined method. Another advantage of the method is that it is compatible with telephone communications systems, i.e., some of the carriers can be modulated by audio broadcasting programs, and the rest by multichannel telephone messages. For these reasons, this method is used for exchanging radio broadcasting programs in the Intersputnik system, *albeit* with some modification; the audio broadcasting signal first is converted to discrete form, and then it is transmitted by means of phase modulation of the carrier. The transmission rate is approximately 480 Kb/s, and the quality is high-class (the Gradient-V equipment). In this equipment, an audio broadcasting channel is the power equivalent of ten telephone messages, transmitted by frequency modulation on separate carriers, so that one decade of telephone channels is taken away during the transmission of broadcasting programs.

The transmission of audio broadcasting programs on a group of channels of any telephone communications systems (the AV-2/3 equipment) is a familiar method. In this case, an audio broadcasting channel is created by occupying the band of two or three audio frequency channels; however, to achieve the necessary noise immunity, audio broadcasting is transmitted by increasing the level (for example for certain multiplexing systems—a power of 923 μW instead of 32 μW for an ordinary audio frequency channel). We will not examine this method in detail, not only because it is generally ineffective, but also because it obviously is unsuitable for satellite communications systems, due to the fact that it is not adaptable for one-way satellite networks. As is known, the latter are best developed in the USSR and make possible successful reception of audio broadcasting and facsimile signals, which by nature are one-way, have inexpensive public Moskva, Ehkran, and Orbita TV receiving stations.

Transmission of Radio Broadcasting Programs on Time Division Multiplex Channels. Time division multiplexing (TDM) telephone signal transmission is a familiar method and is described in many textbooks. For transmitting audio broadcasting signals, the sampling frequency (digitization frequency) must be higher than in the case of audio frequency channels, in accordance with Kotel'nikov's theorem and the upper frequency of a message. Each sample must be transmitted with high precision (with a large number of distinct levels), so that the signal-to-quantization-noise ratio will be high. In the case of transmission with the aid of PCM, i.e., using a certain number of binary sequences, this means that it is necessary to have a large number of digits, usually 13 or 14. In the case of nonlinear sampling, the equivalent quality is achieved with only 9 or 10 digits [14.3]. Another version is possible, in addition to the usual version for time division alternation of samples of each program (Figure 14.5(a)), whereby several samples of one channel (of one program) are transmitted successively, and then of the second, *et cetera* (Figure 14.5(b))*. In this

*Shown in Figure 14.5 is the transmission of signals using amplitude-pulse modulation. Presently, every pulse is most often replaced with a group of binary sequences (a version of PCM).

(the second) case, several samples of each program are memorized and transmitted successively in the time t of a subframe. Whereas in the first version, frame duration T_1 is determined by the quantization frequency, and in the second version, frame duration T_2 can be many times longer. This provides the necessary spaces π (see Figure 14.5(b)) between channel groups of different stations and between frames without heavy losses of carrying capacity, and makes it possible to transmit programs from different earth stations or to consolidate with telephone communications systems on a common transponder, using so-called time division multiple access (Chapter 7).

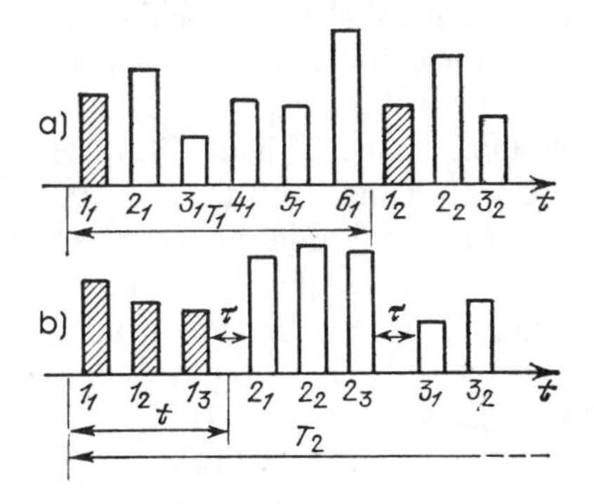

FIGURE 14.5. Versions of sampling during transmission of audio broadcasting signals using time division multiplexing.

Transmission on TDM channels requires additional equipment (analog-digital and digital-analog conversion equipment), and when the method illustrated in Figure 14.5(b) is used, also time companders which transform signals in time (storage and reading of pulses). This transmission technique is used in the Orbita-RV equipment, described below in Section 14.3.

14.2 FACSIMILE TRANSMISSION THROUGH SATELLITES

Newspapers are delivered to subscribers in other cities by transporting the entire necessary circulation, which is an unwieldy and expensive operation, often accompanied by delays. The local printing of newspapers from matrices delivered by airplanes has been practiced extensively for a long time. However, this method also involves delays. Therefore, newspapers are being transmitted on communications channels photoelectrically instead of being shipped. If the transmission rate is high enough, the newspaper image transmission operation takes only a few minutes, and the printing of the necessary number of copies of a newspaper on the receiving end of the communications link takes place in virtually the same time as on the transmitting end. An electrical copy of a newspaper is made in the following way:

The original of a newspaper (an especially sharp copy of it) is fed into the cylindrical camera of the scanner of a facsimile transmitter. An image is expanded into lines and frames much as in television, but mechanical scanning is used. In the camera, there is an optical device which projects the image of the paper onto a photoelectric cell, which converts light signals to electrical signals. The cylindrical camera moves along its axis, and the optical device rotates; that is how scanning is accomplished. The output signal of the photoelectric cell has just two levels, one of which corresponds to the white field of the paper, and the other to the black parts of letters and photographic images (Figure 14.6). Half-tone photographs are made on a newspaper original with the aid of black dots, uniformly distributed across the field, and of different sizes, which gives the impression of different optical densities. The electrical facsimile copy then is transmitted on city and long-distance communications links to a receiving facsimile machine, located in a typography building. In the receiving machine, the electrical signals are reconverted to light signals, and then a photocopy of the newspaper is made with the aid of large-format photographic film and a scanning mechanism. A matric for printing is made on the basis of this photocopy.

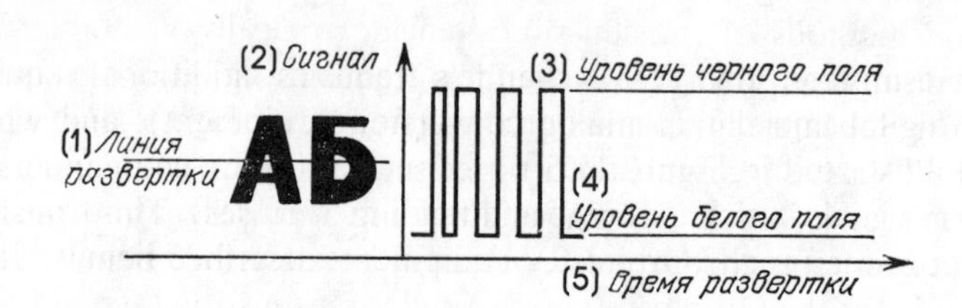

Key: **(1) Delay link**
(2) Signal
(3) Level of black field
(4) Level of white field
(5) Scanning time

FIGURE 14.6 Generation of electrical facsimile signal.

Different transmission rates are achieved, depending on the scanning speed; the necessary passband of a communications link depends on this (and also on the sharpness of the picture). The transmission rate usually is such that the output signal of the newspaper transmitter fits in the band of one standard (secondary) multiple access group: (12-channel, 60-channel, *et cetera*). The Gazeta-2 machine with a drum speed of up to 3000 rpm is used widely in the USSR; for this machine one strip (page) of a newspaper is transmitted in approximately two minutes, and the output signal (with one vestigial sideband and suppressed carrier) fits in the spectrum of a 60-channel group (312–552 kHz), i.e., the necessary passband of the communications channel is 240 kHz. In this band the necessary signal-to-noise ratio is usually 22–24 dB. It is obvious that facsimile transmission through circular satellite systems releases many terrestrial communications channels on different routes.

One of the distinguishing features of satellite communications links is a Doppler frequency shift (see Chapter 1). In application to the problem of newspaper transmission, it is more appropriate to talk not about a frequency shift, but about a change of the signal propagation time between the transmitting and receiving stations, which in photocopies that are made of newspapers when the conventional constant receiver scanning frequency is used (scanning is synchronized at the beginning of a session) causes a page to slant due to a shift of the time of the beginning of each line. A quantitative analysis of this effect is a simple matter [14.4]; it is sufficient to establish here that in facsimile transmission through a satellite in an elliptical orbit, the continuous compulsory synchronization of the receiver is necessary, much as is done in television; in the case of geostationary satellites, intolerable distortions do not occur in facsimile machines, operating in the conventional mode [14.4, 14.5].

The methods of transmission coincide basically with the methods described in the previous section for transmitting audio broadcasting programs. Subcarrier transmission on a frequency higher than the spectrum of a TV message is most often used in the USSR (see Figure 14.1). Of course, in view of the great bandwidth of a facsimile signal, its parameters are a little different from those of audio broadcasting signals. In a TV transponder of the Orbita system, facsimile transmission is conducted on a higher subcarrier frequency of 8.2 MHz with a subcarrier deviation of ±250 kHz; the carrier frequency deviation of a TV transponder, attributed to the subcarrier, is ±1.5 MHz. The same kind of facsimile transmission is possible and is beginning to be used in the Moskva system, i.e., with reception at a comparatively small station. In this case, the receiving station can be located right next to the typography building, which eliminates connecting links.

Facsimile transmission in discrete form should be considered a promising method. By nature, a facsimile signal is already digitized and has just two levels (white and black); all that needs to be done is to digitize the signal in time. Discrete transmission meets the most rigid quality requirements on newspaper transmission (for example, for the high-quality offset printing that is used at the present time). The necessary transmission rate is approximately 2048 Kb/s, but when redundancy elimination techniques are used it can be reduced to 513 Kb/s and less.

A facsimile signal can be transmitted "within" a telephone communications system in the band of a 12- or 60-channel group, just as is done in terrestrial links. In this transmission method, the circular nature of facsimile signals, which is extremely favorable for satellite systems, is not utilized, and a separate communications channel is necessary for every transmission route. Therefore, this transmission technique is not used in satellite communications systems.

14.3 THE ORBITA-RV EQUIPMENT

There is a need in the USSR for transmitting many audio broadcasting programs of different quality classes (top, I, II) from Moscow to different points of the country, to local air and wire broadcasting centers, and facsimile signals to local typography offices. The Orbita-RV system was developed and adopted for this purpose. In this system messages are transmitted through a transponder of a Gorizont or Raduga satellite, located in a geostationary orbit (the international registration name is Statsionar), and the Orbita-RV system makes only one-half of the capacity of a transponder (the other half is used by a TDMA system; see Chapter 18).

In the Orbita-RV system, 25 audio broadcasting and facsimile signals are transmitted simultaneously; programs can be received at any Orbita station with the necessary equipment (Figure 14.7). Signals are transmitted by TDM; the separation between audio broadcasting and facsimile signals, on the one hand, and a telephone communications signal, on the other hand, also is time division—a 125-μs frame is divided in halves between Orbita-RV and TDMA signals (Figure 14.8). Audio broadcasting signals are converted at the input of the Orbita-RV transmitting complex to discrete signals; nonlinear (13-segment) instantaneous companding is used for this purpose, as a result of which the 9-bit coding, used on class I and II channels, is equivalent to 13-bit coding.

The sampling frequency is 21.33 kHz for class I, and 14.22 kHz for class II channels (for higher-class channels the digitization frequency is 32 kHz with 10-bit nonlinear coding). The transmission rate at the analog-digital converter output of a class I channel is 192 Kb/s, and for a class II channel it is 128 Kb/s. Ten class I channels or fifteen class II channels (or six higher-class channels) are transmitted in a standard 2048 Kb/s stream. Class I channels have a bandwidth of 50–10,000 Hz, and class II channels

50–6000 Hz. The signal-to-accompanying-noise ratio is about 44 dB, and the signal-to-noise ratio of a vacant channel (in a pause) is not less than 70 dB. This may be assumed to be as good as the 57 dB signal-to-noise ratio, established for an uncompanded analog signal. All the other characteristics of signal distortions on channels do not exceed one-third of the specifications for channels of the corresponding class (see Chapter 2).

In all, two 2048-Kb/s streams are set aside for audio broadcasting; before these streams are transmitted through a line, they are subjected to noise-immune coding with a redundancy ratio of 3:4. This guarantees the reception of sequences with an extremely low-error probability (better than 10^{-9} for undetected errors), so that cracks occur on a channel approximately once an hour, which is considered acceptable*.

A 2048-Kb/s stream for facsimile transmission is added to the coded audio broadcasting signal stream, which in principle makes it possible to transmit simultaneously with high quality (the necessary reliability is approximately 10^{-4}) up to four facsimile signals from Gazeta-2 equipment, converted to discrete signals. The combined output stream of the transmitting complex of the Orbita-RV equipment thus has a rate of approximately 10 Mb/s. However, because TDM with a telephone communications system is used, the transmission rate is doubled to 20 Mb/s in the simplest time compander on a delay link. Two 10-Mb streams (direct and delayed) are fed to two inputs of a four-phase phase modulator, and the modulator output is blocked for one-half of the clock time. Transmission is conducted with the aid of a four-phase phase modulator, so that most of the power of the signal fits in approximately a 10-MHz band, i.e., in a small fraction of the transponder band. This affords an additional margin of noise immunity and reduces signal distortions in the communications link, which is very important for a large one-way network, where it is exceedingly undesirable to practice retransmission of a newspaper when it is received with poor quality at one station. We note that in a TDMA, telephone communications system transmission is conducted at 40 Mb/s in the same satellite transponder and with reception in the same, but specially equipped, Orbita stations (see Chapter 7).

The phase-modulated signal modulator of the Orbita-RV system at receiving stations is coherent; the carrier frequency is regenerated by a demodulating system. The power losses of the demodulator and the error probabilities $P_{er} = 10^{-5}$ are approximately 2 dB relative to the potential noise immunity.

---

*For detected errors, 10^{-6} is acceptable; these errors are excluded and interpolated.

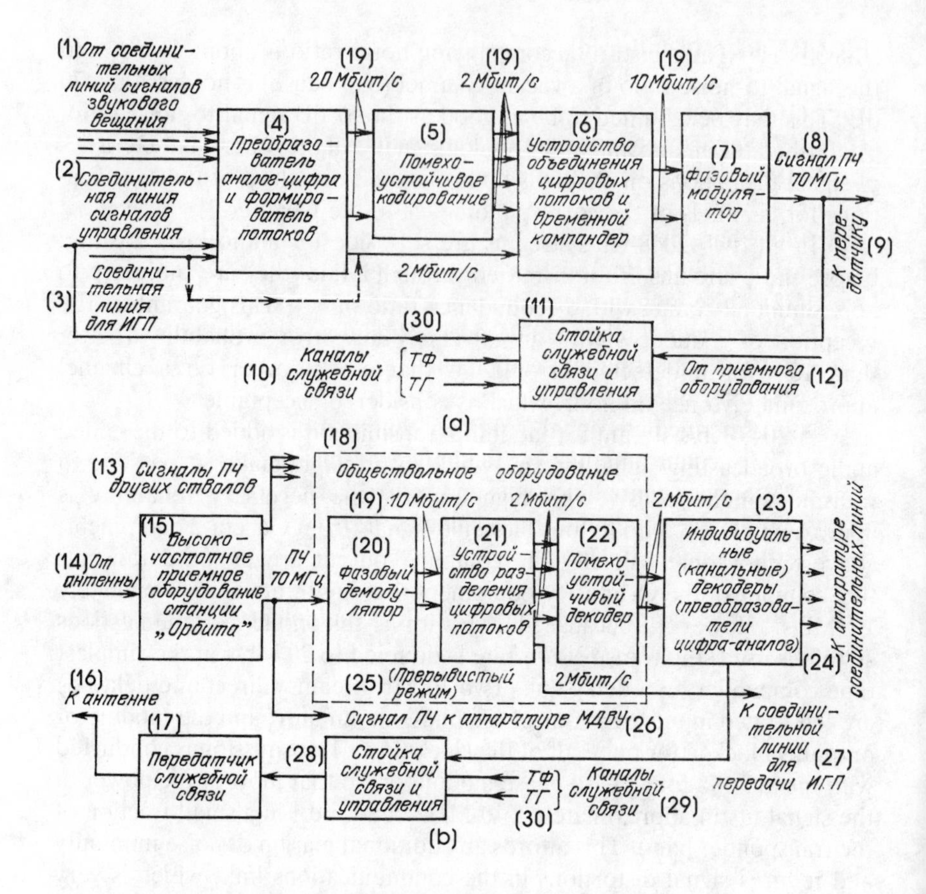

Key: (1) From connecting links of audio broadcasting signals
(2) Connecting link of control signals
(3) Connecting link for facsimile
(4) Analog-digital converter and stream generator
(5) Noise-immune coding
(6) Digital stream combiner and time compander
(7) Phase modulator
(8) 70-MHz IF signal
(9) To transmitter
(10) Service communications channels
(11) Service communications and control rack
(12) From receiving equipment
(13) IF signals of other transponders
(14) From antenna
(15) Orbita channel receiving equipment

FIGURE 14.7. (Continued)

(16) To antenna
(17) Service communications transmitter
(18) Transponder equipment
(19) Mb/s
(20) Phase demodulator
(21) Digital stream divider
(22) Noise-immune decoder
(23) Individual (channel) decoders (digital-analog converters)
(24) To connecting link equipment
(25) Intermittent mode
(26) Intermediate signal to TDMA equipment
(27) To connecting link for facsimile transmission
(28) Service communications and control rack
(29) Service communications channel
(30) Telephone Telegraph

FIGURE 14.7. Diagram of the Orbita-RV equipment: (a) transmitting complex; (b) receiving complex.

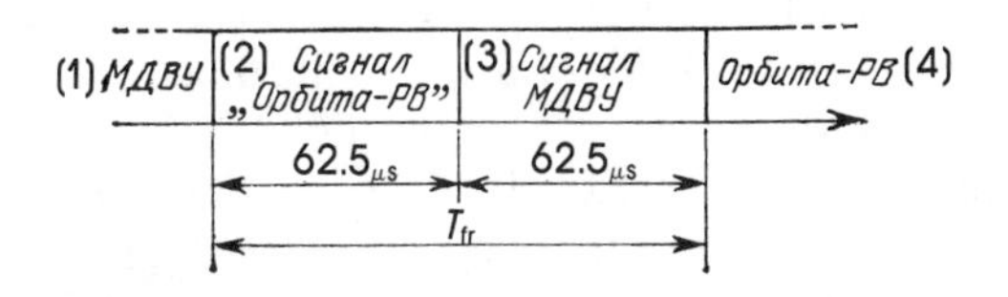

Key: (1) TDMA equipment
(2) Orbita-RV signal
(3) TDMA signal
(4) Orbita-RV

FIGURE 14.8. Frame of combined Orbita-RV TDMA system.

The intermittent operation of a link significantly complicates the operation of the demodulator, of which fast entry into synchronism on the carrier and clock frequencies is required; therefore, a preamble is inserted in every signal transmission clock interval of the Orbita-RV system, during which time the demodulator must be synchronized, much as is done in TDMA equipment (see Chapter 18).

A distinguishing feature of the Orbita-RV equipment is the fact that it uses unsecured* channels, which is extremely effective, not only for satellite communications systems (see Chapter 1), but also for one-way systems. The number of channels processed through a satellite is determined in this case by the number of programs that are being transmitted simultaneously, and not by the number of receiving stations, and it does not depend on the configuration of the earth network. The number of individual (channel) receiver sets at receiving stations is determined only by the number of programs that are received at the same time at that station, and not by the number of radio programs or the number of terrestrial channels necessary for solving the same problems. Any individual set (decoder) of a receiving station can be switched to the necessary time position (i.e., for receiving any program) by command from the program distribution center (these commands are transmitted in the common stream with information sequences), or by command from local radio broadcasting switching equipment.

The stations of the system that receive a limited number of central audio broadcasting programs can operate in the secure** (uncontrolled) mode. The receiving stations of the system that operate on unsecured audio broadcasting channels, or which receive a facsimile signal, are equipped with independent two-way telephone and telegraph service communications equipment, which also assures transmission of acknowledgments of the control system. The telephone service communications system operates on the TDMA principle, affording both radial communications, and "each with each" communications.

The transmission of audio broadcasting signals in digital form, in addition to high multiplexing efficiency, affords virtually perfect identity of channels, which permits any pair of them to be used for transmitting stereophonic programs. In this case, it is desirable to transmit on channels outgoing signals A and B of the left and right channels of a stereo pair; conversion to the sum ($M = A + B$) and difference ($S = A - B$) signals for compatibility with monophonic transmission systems is accomplished right in the UHF FM broadcasting transmitters. The connecting links between program sources (consumers and satellite communications earth stations), outfitted with Orbita-RV equipment, can be made both analog and digital; in the latter case, the analog-digital and digital-analog converters are installed not at an earth station, but at the audio broadcasting program and facsimile source and consumer.

---

*Alternate translation is "unassigned".
**Alternate translation is "assigned".

Chapter 15
The Gradient, Gelikon, and Grunt Transmitting Equipment

The Gradient transmitting complex is intended for operation as a part of a satellite communications and broadcasting earth station for the purpose of transmitting television, multichannel telephony, and other kinds of information in satellite links. It is manufactured to operate in the 5975–6225-MHz band in two models: 10 kW and 3 kW. The 10-kW transmitters are used at earth stations of the Ehkran TV broadcasting system, and the 3 kW transmitters are used in transportable stations of the Mars-2 type.

A schematic diagram of the Gradient transmitting complex, which explains its operating principle and configuration, is shown in Figure 15.1 [15.1]. The complex consists of two driver racks, one modulator rack, two transmitter test racks, two power amplifier or power stage racks, a switching and power meter rack, two high-voltage power units, two control, interlock and alarm racks, and an interrack connector and bunched connector set. In this configuration, the Gradient transmitting complex performs the following functions: generates the FM signals of the transmitting complex in the modulator rack, converts IF FM signals to SHF signals, and preamplifies them to 3 W in the driver rack, amplifies the SHF FM signal to the necessary output power level (3 kW or 10 kW) in the power stage rack, and the 3 kW and 10 kW models differ structurally only in the high-voltage equipment unit.

The Gradient transmitting complex is installed in every earth station. Special combiners are used for combining large groups of transmitters at an earth station, a distinguishing feature of which is that they have no resonant elements. This kind of combiner has small losses (0.1–0.3 dB) and good crosstalk attenuation between transponders (not less than 35 dB). The main specifications of the Gradient transmitting complex are listed in Table 15.1

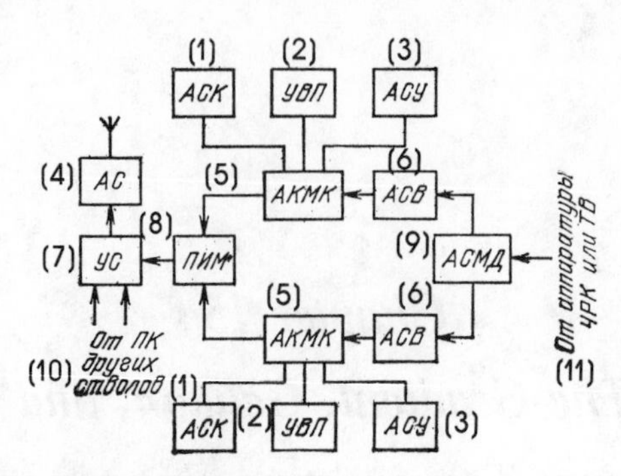

Key: **(1) Transmitter test rack**
(2) High-voltage power unit
(3) Control, interlock and alarm rack
(4) CIA rack (Power amplifier)
(5) Power stage rack
(6) Driver rack
(7) Combiner
(8) Switching and power meter rack
(9) Modulator rack
(10) From transmitting complexes of other transponders
(11) From FDM or TV equipment

FIGURE 15.1. Schematic diagram of Gradient transmitting complex.

The Gelikon transmitting complex [15.2] is installed in satellite communications earth stations and is intended for the transmission of the same kinds of information as the Gradient transmitting complex. It is smaller, uses less power, can be switched alternately to different frequency groups in a selected frequency band, and has an air-cooled power stage. The main specifications of the Gelikon transmitting complex are listed in Table 15.2. Shown in Figure 15.2 is a schematic diagram of the Gelikon transmitting complex, which consists of one preamplifier rack, a power amplifier rack, load cooling rack, high-voltage rectifier TPS-18, and a plug-in control panel.

IF signals on 70 MHz are supplied to the input of the preamplifier rack. The IF spectrum is transferred in this rack to the band of the transmitter output frequencies and are amplified to the level necessary for operating the power amplifier. It contains a frequency converter and a traveling wave tube preamplifier. On-line tuning of the rack to one of the frequency channels is accomplished by connecting the appropriate crystal microoscillator to the master generator. The main amplifier element of the

TABLE 15.1

Characteristic, unit of measurement	*Parameter*	*Comment*
Nominal output power, kW:		
Gradient-OPP transmitting complex	3	—
Gradient-K transmitting complex	10	
Total lettered frequency band, MHz	5975–6225	—
Passband, MHz, of each lettered transponder	±17	At 1-dB level
Efficiency of output klystron, not less than, %	25	In saturation mode
Gain of power stage, dB	40	—
Long-term heterodyne frequency instability, not more than	$\pm 5 \cdot 10^{-7}$	—
Suppression of parasitic emissions at output of transmitting complex, not less than, dB:		
of second harmonic	40	—
of third harmonic	20	—
Output standing wave ratio of transmitting complex	1.1–1.2	—
Standby switching time, not longer than, ms	200	—

TABLE 15.1 (Continued)

Characteristic, unit of measurement	*Parameter*	*Comment*
Nonuniformity of group delay time response, not more than, ns	5	In 15-MHz band relative to lettered transponder frequency
"Differential gain" distortion, not more than, %	7	For FM frequency deviation of ±11 MHz
"Differential phase" distortions, not more than, deg	4	For FM frequency deviation of ±7 MHz
Power stage cooling	Water	—
Power source voltage of anode and collector, kV	12	—
Operating mode	Around the clock	In stationary heated rooms at ambient air temperature of 5–40° C
Power voltage of transmitting complex, V	380 ± 38	Triphase, AC, 50 ± 5 Hz

power amplifier is a klystron. The plug-in control panel is intended for remote control of the transmitter and of the signal light display that shows the condition of its individual components in different operating modes.

The Grunt transmitter is intended for transmitting one of the following kinds of information: telephone signals using the Gruppa multiple access FDM equiment, and low-capacity (few channels) telephone signals using Gradient-N equipment. The Grunt transmitter operates in the 5975–6275-MHz band on the frequency of one of six wideband satellite transponders. The central frequencies of these are listed in Table 15.3.

TABLE 15.2

Characteristic, unit of measurement	*Parameter*	*Comment*
Nominal output power, kW	3	—
Frequency band, MHz	6000–6250	On-line tuning
Cooling of power stage	Air	—
Input power, kV·A	25	—

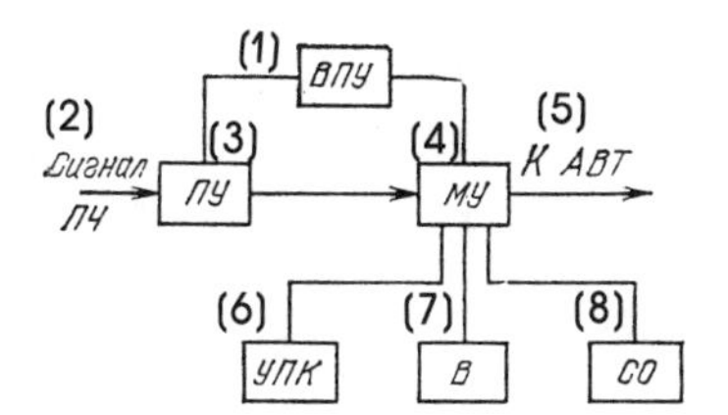

Key: (1) Plug-in control panel
(2) IF signal
(3) Preamplifier
(4) Power amplifier
(5) To antenna-waveguide
(6) Processing equipment
(7) Rectifier
(8) Load cooling rack

FIGURE 15.2. Schematic diagram of transmitting complex.

TABLE 15.3

Number of transponders	*6*	*7*	*8*	*9*	*10*	*11*
Central frequency of transponder, MHz	6000	6050	6100	6150	6200	6250

Shown in Figure 15.3 is a schematic diagram of the Grunt transmitter, which explains its operating principle and the makeup of the equipment. The transmitter consists of two half-sets and common auxiliary units. Each half-set has a preamplifier rack and a power amplifier rack. Each half-set may be both on-line and on standby. The half-set on standby gets all the necessary voltages and the input signal. If the working set fails or its output power falls by 6–8 dB, automatic switching to the standby set occurs, and the waveguide switch of the backup system switches the output signal of the working set to the antenna-waveguide of the standby set, to an equivalent antenna.

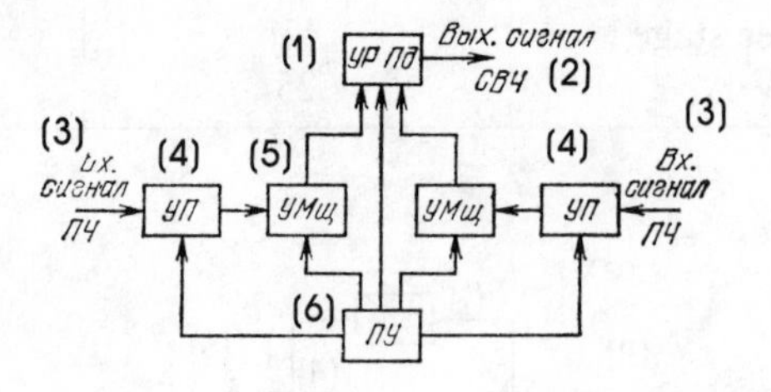

Key: **(1) Backup system**
(2) Output SHF signal
(3) Input IF signal
(4) Preamplifier
(5) Power amplifier
(6) Control console

FIGURE 15.3. Schematic diagram of Grunt transmitter.

Control of the transmitter may be both remote and local (for tuning and for preventive maintenance work). Its performance is monitored by meters, signal lights, and digital indicators. The output power can be monitored at a plug-in control console, also with digital readout. Switching from one transponder to another is accomplished by switching master frequency generators in the preamplifier rack. The main specifications of the Grunt transmitter are listed in Table 15.4.

TABLE 15.4

Characteristic, unit of measurement	*Parameter*	*Comment*
Nominal output power, W	200	In transponder
Overall frequency band, MHz	5975–6275	—
Nonuniformity of amplitude-frequency response of transponder, dB	1	In ±18.5-MHz band
Level of intermodulation products, dB	−30	Four signal method
Nonuniformity of group delay time response, ns	5	In 15-MHz band
Time of automatic switching from working set to standby, ms	300	—

Chapter 16
The Orbita-2 Receivers

The Orbita-2 receivers [16.1], intended for operation in the 3.7–4.2-GHz band, are designed either for receiving just TV signals (the single-transponder version) or for the simultaneous reception of TV signals, multichannel, and low-capacity telephony, as well as other kinds of information (the multitransponder version), depending on the purpose of the earth station. Shown in Figure 16.1 is a schematic diagram of the Orbita-2 receiving equipment for the single-transponder version.

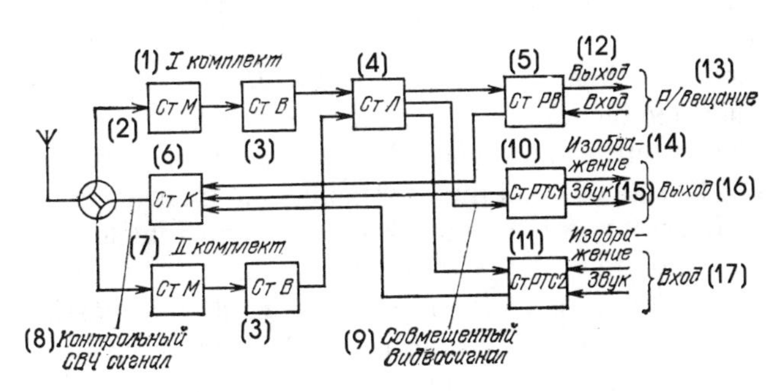

Key:
(1) Set I
(2) Rack M (low-noise amplifier)
(3) Rack V
(4) Rack P (IF)
(5) Rack RV (broadcasting)
(6) Rack K (test)
(7) Set II
(8) SHF test signal
(9) Combined video signal
(10) Rack RS1 (separation)
(11) Rack RS2 (separation)
(12) Output and Input
(13) Broadcast
(14) Video
(15) Audio
(16) Output
(17) Input

FIGURE 16.1. Schematic diagram of the receiving equipment of an Orbita-2 earth station.

The signal picked up by the antenna from the satellite passes through a coaxial switch to the input of rack M of a receiver (set I or II, depending on the position of the switch), which contains a low-noise amplifier. The SHF signal, amplified in the low-noise amplifier, passes from the output of rack M to the input of rack V, which contains a frequency converter and preamplifier, after which is installed rack P, where the IF signal receives most of the amplification and is demodulated. TV and radio broadcasting signals, and also service (telephone-telegraph) communications also are separated there. The TV signal is separated in racks RS (RS1 and RS2) into a video signal and audio accompaniment signal, which go into a connecting link. The radio broadcasting signal, amplified and demodulated in rack RV, also goes into a connecting link.

All the equipment of the Orbita-2 receiving complex is completely redundant for the purpose of increasing its performance reliability. There is a capability of automatic switching from one set to the other (the sets are equivalent) by alarm signals, generated at the output of an automatic redundancy unit, installed in the test rack (rack K). The performance of the receiver can also be checked completely and in operation, and its performance characteristics can be evaluated at this rack by connecting the output of the test rack with the aid of a coaxial switch to the input of the set to be tested.

A simplified schematic diagram of an Orbita-2 three-transponder receiver is shown in Figure 16.2. Here the signals picked up by the antenna pass from the output of the waveguide to the input of a wideband filter, which passes the signals of all the transponders in the working frequency band and supresses interference outside of it (by approximately 20 dB for a frequency deviation of ±37.5 MHz relative to the middle frequencies of the outer bands) for protecting the receiver input from possible strong out-of-band interference. Wideband filter *Y* has low losses (0.1–0.2 dB) and affords good matching (standing wave ratio ≥1.2) with the input of the low-noise amplifier in the working frequency band.

The SHF signals of the transponders go from the output of the wideband filter to the input of the waveguide switch of the backup and test system and from there to the low-noise amplifier of the selected set of rack M. The HF path of rack M has the same bandwidth as the wideband filter and is suitable for amplifying the signals of all the transponders that are received. In this respect, the configuration of the multitransponder version is no different from the single-transponder version.

The signals from these transponders are separated with the aid of a bank of racks of the type V, tuned to the appropriate frequency band (racks V1, V2, V3), in which circulators and bandpass filters are installed for this purpose. The output of any rack M can be switched to the input of

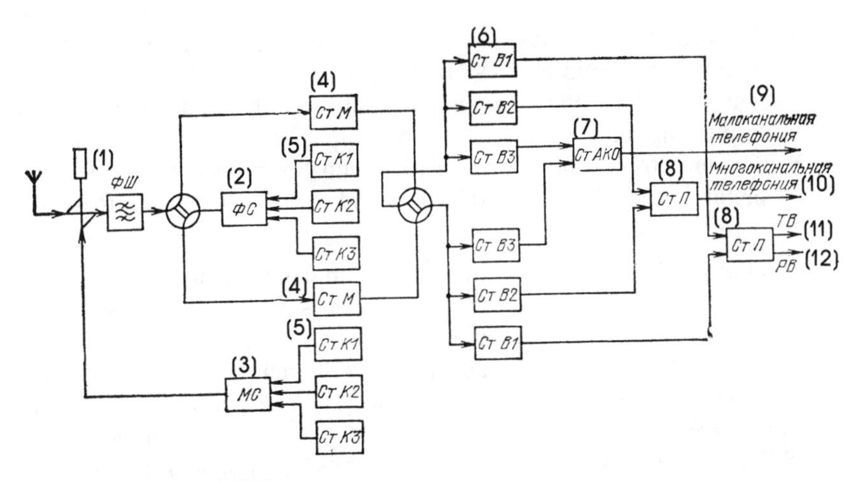

Key: **(1) Wideband filter**
(2) Waveguide switch
(3) Combining bridge
(4) Rack M (low-noise amplifier)
(5) Rack K (test)
(6) Rack V
(7) Channeling equipment rack
(8) Rack P
(9) Low-capacity telephony
(10) Multichannel telephony
(11) TV
(12) Radio broadcasting

FIGURE 16.2. Simplified schematic diagram of the Orbita-2 three-transponder receiver.

each set of racks V for the purpose of increasing the reliability of the equipment. The transponders of the receiver are, in principle, identical in terms of their parameters (only the frequency-dependent elements are different) and can be used for transmitting both TV and telephone signals. In practice, each is used as a rule for transmitting a particular kind of information. For instance, a receiver, similar to the one shown in Figure 16.2, is used for handling 3 transponders.

- the first transponder is used for receiving TV signals in the 3775 ± 17-MHz band and therefore the signal, amplified and converted in rack V1 and then processed in rack P, is sent to the video and audio signal separation equipment;

- the second transponder is used for receiving low-capacity telephony signals in the 3875 ± 17-MHz band; here the signal from rack V3 is sent into channeling equipment, for example, and from there to connecting links;
- the third transponder is used for receiving multichannel telephony signals in the 9875 ± 17-MHz band and in this case are sent from rack II to processing equipment.

Thus, in the general case the conversion from the single-transponder version to the multitransponder version consists in the quantitative expansion of the corresponding frequency-dependent equipment (rack V) and the addition of the necessary transponder equipment to each transponder, depending on the purpose of a given transponder (racks P, RS, RV, and channeling equipment rack). This also applies to test equipment racks (rack K), of which there is the same number as transponders, and a special combining bridge is used for combining SHF test signals.

Main Specifications of the Orbita-2 Receiver

Working frequency band, GHz	3.4–3.9
Effective noise temperature at receiver input, K	80–90
Passband of one transponder, MHz*	34
Frequency deviation of received FM signal:	
in TV mode, MHz	±15
in low-capacity telephony mode**, kHz	30
Passband:	
of video channel, MHz	6.5
of TV audio signal channel, Hz	50–1000
Weighted signal-to-noise ratio, dB:	
on TV channel	56–57
on TV audio signal channel	70
in low-capacity telephony mode***	59 (49)
Differential phase, deg	5
	Differential
gain, %	8
Automatic standby switching time, ms	250–30
Receiver readiness coefficient	0.998

*At 1-dB level.

**Frequency modulation of individual carriers.

***In racks in which the carrier is not suppressed in a pause.

All racks of the Orbita-2 receiver, with the exception of racks M and V, are manufactured as typical cabinets measuring 600 × 225 × 2000 mm. Because equipment of the Orbita-2 type is very popular at earth stations in one configuration or another it is worthwhile to examine in greater detail the individual racks of this equipment.

Main Electrical Parameters of Rack M

Frequency range, GHz	3.4–3.9
Low-noise amplifier input noise temperature, K	70–80
Passband at 1-dB level, MHz	250
Gain, dB	40

Rack M contains the following: a Dewar flask containing a cooled preamplifier*; an uncooled preamplifier unit; a pumping generator unit in which there are two klystrons; a stabilized power unit; a monitor panel on which there are controls and instruments for monitoring bias voltages; diode current indicators; and an instrument for checking the pumping power. The wideband low-noise amplifier is a four-stage preamplifier.

Main Electrical Parameters of Rack V

Working frequency range, GHz	3.4–3.9
Noise coefficient, dB	12
Passband at 3 dB level, MHz	50 ± 1
Nonuniformity of amplitude-frequency response in 34 MHz band, dB	0.5
Gain of IF preamplifier, dB	23

Racks V are manufactured for a given frequency and are intended for converting an SHF signal and for preamplifying an IF signal (70 MHz). The heterodyne frequency path of rack V consists of a crystal oscillator with a frequency multiplier, a filter for suppressing heterodyne noises and a power amplifier.

Main Electrical Characteristics of Rack P

Passband of IF circuit at 1-dB level, MHz	34
Selectivity with ±30-MHz frequency deviation, dB	50
Gain of IF circuit, dB	55

---

*Abbreviation PU also used for parametric amplifier.

An AGC system holds the IF output level within an input level fluctuation of ±10 dB, dB	±1
Nonuniformity of the group delay time response in the band (in MHz), ns:	
±15	8
±11	2
Differential phase, deg	5
Differential gain, %	4.5

Rack P is intended for amplifying the IF signal, for generating the working frequency band and for a given selectivity, and also for demodulating the FM signal; rack P contains two interchangeable sets of functional units as well as two panels: switching and instruments, which both sets share in common. After being demodulated, the video frequency signals go into a TV and radio broadcasting signal dividing filter. The radio broadcasting signal passes through the switching panel to the radio broadcasting rack for further processing; the TV signal passes to the video amplifier panel, and from one of its outputs it goes to rack RS (separation) through the switching panel, and from the other output to rack K (test). In rack P, for increasing the noise immunity of the FM signals in the threshold region, there is sudden frequency deviation (SFD), which operates when the frequency deviation is 15 MHz and the high upper modulating frequency is up to 6 MHz.

Differential distortions of color TV signals, which occur due to the nonlinearity of the phase-frequency responses of the HF signal path, are corrected by two phase correctors, the first of which (a three-stage corrector) corrects the phase-frequency response of the Orbita-2 receiver, and the second (a two-stage corrector) the phase-frequency response of the space repeater. By regulating the second corrector on-line with the aid of two regulators (on the front panel), it is possible to eliminate the instability of the phase-frequency response of the space repeater, for example as it degrades or when it is replaced with a new one. With a "bias" regulator, the linear component of the group delay time response can be changed within ±12 ns (Figure 16.3(a)), and with a "parabola" regulator, the parabolic component of the group delay time response can be adjusted within ±5 ns relative to an initial parabolic component of the order of 15 ns (Figure 16.3(b)).

Racks RS (RS1 and RS2) are intended for TDM of a video signal by audio signals and for regenerating synchronized signals. Rack RS1 contains both receiver and transmitter sets, and rack RS2 contains a receiver set. The transmitter complex performs time-pulse multiplexing of the video signal by the audio signal, and duration-modulated audio carrier pulses are

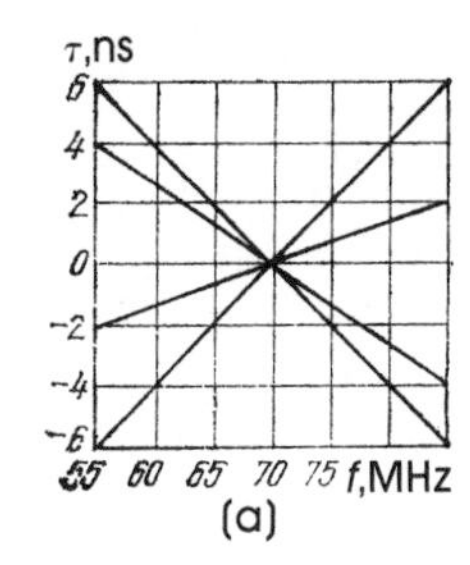

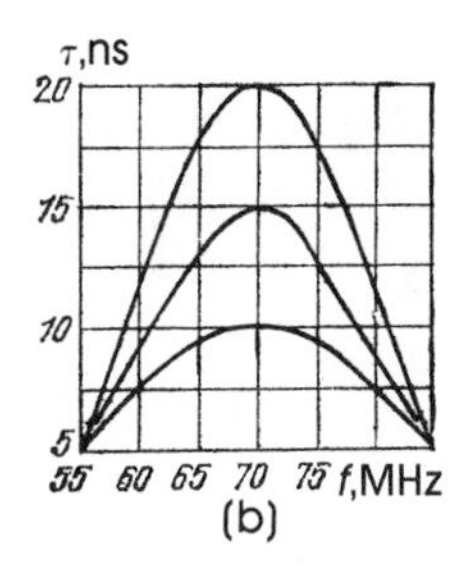

FIGURE 16.3. Components of the group delay time response: (a) linear; (b) parabolic.

put in the intervals of the blanking pulses. The receiver complex performs the opposite operation, i.e., it separates video and audio signals and also converts sound pulse signals to an analog signal. Unlike the Orbita-1 equipment, which has one TV audio channel with a 6-kHz band, the Orbita-2 racks RS can transmit and receive one TV audio channel with a 10-kHz band or two channels, each with a 6-kHz band. A compander with a compression ratio of 3 is used in racks RS for increasing the noise immunity of the audio signal.

Rack RV (radio broadcasting) is intended for transmitting radio broadcasting signals on a subcarrier by means of FDM of a TV transponder. A 7.5-MHz FM subcarrier of a radio broadcasting channel with a frequency deviation of 150 kHz passes from rack P to the input of rack RV, where it is first additionally filtered and amplified, and then it is fed to a noise-immune demodulator with frequency feedback. A compander and a frequency predistortion filter are used for improving the performance characteristics of the radio broadcasting channel in rack RV.

Racks K [test] are intended for on-line performance testing of the receiver, for measuring its performance characteristics, and also for accomplishing automatic or manual standby switching. The units that are a part of the test rack generate the following test signals: an SHF or IF FM signal, and a crystal-stabilized IF signal. The rack has extra inputs for connecting test instruments with which all necessary tests are conducted.

Chapter 17

The Ehkran Earth Receiver Installations

17.1 GENERAL INFORMATION

The earth receiver installations of the Ehkran system are designed for receiving an FM signal, transmitted by the space repeater of a satellite in the 702–726-MHz band [17.1]. As is known from Chapter 13, peak deviations of the carrier by video and audio accompaniment subcarrier (6.5 MHz) signals of 9 MHz and 2 MHz, respectively, were selected for this system, and the nominal field intensity, created by the Ehkran satellite space repeater on the edge of the service area is 29 μW/m. Several versions of receiver installations of two reception quality classes have been developed and are in series production today in accordance with these data [17.2].

Class I receivers are high-quality and reliable FM receivers, intended for delivering central TV programs to large and medium TV transmitters, containing video and audio modulators (commercial grade-receiver installations). Class II receivers are simplified FM signal receivers, intended for converting an FM signal to a standard TV signal with an AM video carrier and an FM audio carrier for further distribution among subscribers through a small repeater or through a cable distribution network (a subscriber receiver).

17.2 THE PPST1-78 COMMERCIAL SATELLITE TELEVISION RECEIVER

The PPST1-78 commercial satellite television receiver [17.3] is intended for delivering TV and radio broadcasting signals to large TV and radio transmitters. In the Ehkran system, the PPST1 receives FM signals from a satellite in the decimeter wavelength band and generates an output video signal, a TV sound accompaniment signal, and radio broadcasting signals. (Shown in Figure 17.1 is a simplified schematic diagram of a commercial receiver.)

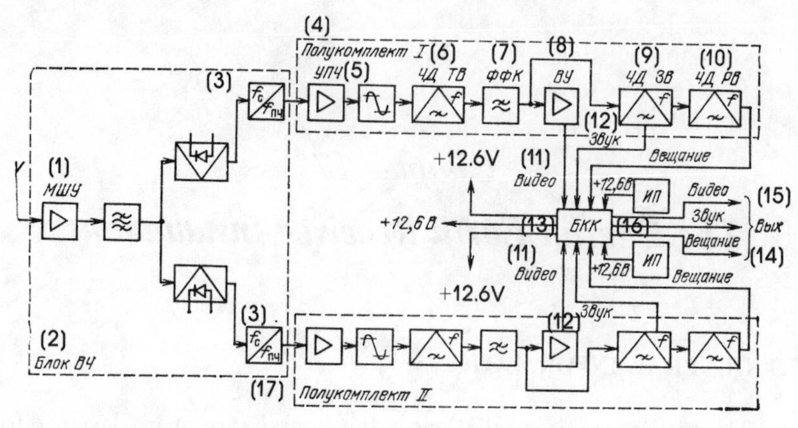

Key: (1) Low-noise amplifier
(2) HF unit
(3) f_s/f_{IF}
(4) Half-set I
(5) IF amplifier
(6) TV frequency detector
(7) Phase-correcting filter
(8) Input
(9) Audio broadcasting frequency detector
(10) Radio broadcasting frequency detector
(11) Radio broadcasting frequency detector
(12) Audio
(13) Monitor and switching unit
(14) Broadcast
(15) Out
(16) Power source
(17) Half-set II

Overall dimensions	440 × 240 × 165
Weight, kg	5

FIGURE 17.1. Simplified schematic diagram of a commercial receiver.

Main Specifications of the PPST1-78

Nominal receiving frequency, MHz	714
Passband, not less than, MHz	24
Nominal input signal level, dBW	−106
Effective noise temperature of input low-noise amplifier, K	500–800
Output signal-to-weighted-noise ratio, dB	53
Peak TV signal frequency deviation, MHz	9
Peak sound accompaniment subcarrier signal frequency deviation, MHz	2
Sound accompaniment signal subcarrier frequency, MHz	6.5
Radio broadcasting signal subcarrier frequency, MHz	7
Output signal voltage on 75 Ω load:	
of TV video signal, V	0.9–1.8
of TV audio accompaniment, radio broadcasting signals, MV	775
Overall dimensions, mm	700 × 340 × 1,345
Weight, kg	140

From the antenna a signal goes to the input of an HF unit, consisting of a low-noise amplifier with GT362B transistors, bandpass filter and a diode HF switch. The amplified signal is connected by the HF switch to the input of the frequency converter of the first or second half-set. There it is converted to a standard IF signal (70 MHz). This signal is amplified in an IF amplifier, is amplitude-limited, and with a voltage of about 300 mV is supplied to a TV video frequency detector, from which the TV video and audio subcarrier signals are separated in phase-correcting filters. The video signal is amplified in the input amplifier to 1 V and is sent into a connecting link through a monitor and switching unit, capable of switching simultaneously both HF input signals and output signals (video, audio accompaniment and radio broadcasting signals of the first and second half-sets, which are identical). The audio subcarrier is demodulated in the audio broadcasting frequency detector, and the radio broadcasting signal is demodulated in the radio broadcasting frequency detector and is also sent through the monitor and switching unit to the corresponding connecting links. Each half-set and the HF unit receive power from separate power sources with a voltage of ±12.6 V, connected to the switching units and assemblies in the monitor and switching unit.

17.3 THE SUBSCRIBER RECEIVER OF THE EHKRAN SYSTEM

The subscriber receiver (unit PA) receives signals, frequency-modulated by TV and audio accompaniment signals in the decimeter band, and converts them to AM video and FM audio signals in the UHF band.

Main Specifications of the PA Subscriber Receiver

Nominal receiving frequency, MHz	714
Passband, MHz	24
Effective noise temperature of input low-noise amplifier, K	500–800
Nominal input signal level, dBW	−109
Output signal-to-weighted-noise ratio, not less than, dB	48
Peak frequency deviation by TV signal, MHz	9
Peak frequency deviation by audio accompaniment signal, MHz	2
Audio TV signal subcarrier frequency, MHz	6.5
Output signal voltage, not less than, mV	40

A schematic diagram of the subscriber receiver is shown in Figure 17.2. Here, just as in the preceding case, the FM signal from the antenna goes to a low-noise amplifier of similar design and with similar parameters. From there the amplified signal is transferred in a frequency converter from the decimeter band to the 70-MHz intermediate frequency. Then the signal is amplified in the IF amplifier, is limited, and goes into a frequency detector, from the output of which come video and audio accompaniment subcarrier signals. These signals are separated in an amplitude modulator unit and are converted to a standard TV signal with an AM video carrier and FM audio accompaniment carrier with a frequency separation of 6.5 MHz between the carriers. All the elements of the receiver are powered by their own compact ±126-V power source rectifier, and the receiver itself is powered from the 220-V DC main.

A schematic diagram of the amplitude modulator, in which a standard TV signal is generated, is shown in Figure 17.3 [17.4]. Here the cumulative video and audio subcarrier signal with a frequency of 6.5 MHz is supplied to the input of a stage with a divided load; the FM audio accompaniment signal is separated by a bandpass filter and goes to the input of a balanced loop mixer, at the second input of which is supplied the video carrier signal from the heterodyne, and then the FM audio accompaniment signal is separated by bandpass filter, tuned to f_{car} + 6.5 MHz, and it is supplied through a buffer stage to the output of the subscriber receiver;

the buffer stage, along with the bandpass filter, suppresses the video carrier by not more than 25 dB.

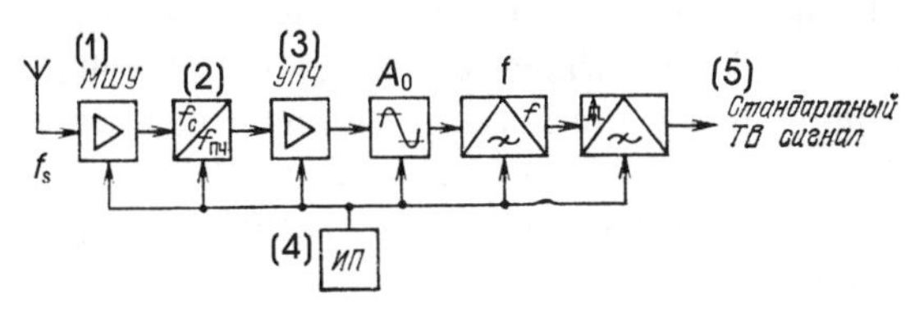

Key: (1) Low-noise amplifier
(2) f_s/f_{IF}
(3) IF amplifier
(4) Power source
(5) Standard TV signal

FIGURE 17.2. Schematic diagram of a subscriber receiver.

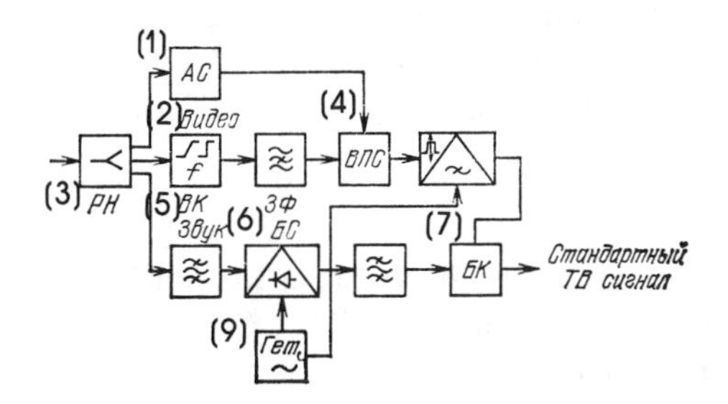

Key: (1) Amplitude selector
(2) Video
(3) Divided load
(4) DC component regeneration
(5) Frequency regeneration
(6) Trap
Balanced mixer
(7) Buffer stage
(8) Standard TV signal
(9) Heterodyne

FIGURE 17.3. Structural diagram of amplitude modulator.

In the video signal after the divided load, the signal, which predistorted on the transmitting end, is subjected to frequency regeneration. A trap suppresses the FM audio subcarrier by approximately 30 dB. From the output of the trap, the video signal passes through a dc component regenerating stage to an amplitude modulator, assembled as a bridge on D311A diodes (on one of the arms of the bridge). The second arm receives the heterodyne signal. The bridge is balanced by the voltage of the video signal; dc component regeneration is accomplished with the aid of a controllable level regulator, for which control pulses are generated by an amplitude selector (AS) from the video synchronizing pulses. The AM video signal passes through the buffer stage along with the audio accompaniment signal, which also goes there, to the receiver output.

17.4 THE EHKRAN-KR-10 COLLECTIVE TELEVISION PROGRAM RECEIVING STATION

The Ehkran-KR-10 collective TV receiving station is intended for delivering color TV broadcasts to large settlements, located in the service area of the Ehkran system, by means of the reception of FM signals from a satellite and generation of a standard 10-W TV signal on the frequency of one of 12 TV channels of bands I, II, and III (a receiving-transmitting station).

Main Specifications of the Station

Nominal receiving frequency, MHz	714
Receiver passband, MHz	24
Peak deviation by video signal, MHz	9
Peak deviation by audio subcarrier signal, MHz	2
Intermediate frequencies, MHz:	
on receiving channel	70
on transmission channel	38
Noise coefficient, converted to receiver input, not more than, KT	1.5
Receiver selectivity on mirror channel with a frequency deviation of 128 MHz from the carrier in low-noise amplifier, not worse than, dB	60
Receiver selectivity on adjacent channel with a frequency deviation from the carrier +25 MHz, not worse than, dB	26
Input mismatch attenuation in 7000–728-MHz band, not less than, dB	14
Nominal input signal level, dBW	−114

Transmitter output power on frequency of one of TV channels, not less than, VT	10
Transmitter output signal-to-noise ratio for nominal input signal, not less than, dB	52
Receiving antenna gain, dB	24
Crosstalk distortions between carriers by video and audio frequencies (for a power ratio of 10:1), not more than, dB	−49
Time separation of brightness and color signals, ns	50
Power voltage from alternate current (AC) main, V	220 ($^{+10\%}_{-2\%}$)
with frequency, Hz	50 ± 2
Effective radius of station, km	6–7
Input power, W	450

The station is serviceable in ambient temperatures of from 5° to 40° C at 95% relative humidity at 30° C. The Ehkran-KR-10 set includes the following: a remote control console; receiver-transmitter; a remote control circuit lightning arrester; an antenna feed (receiving and transmitting antennas); and a monitor TV set. The type of receiver-transmitter and antenna feed is determined by the number of the nominal standard TV channel and by the kind of polarization of the transmitting antenna. The nominal frequencies of the channels with the corresponding frequency bands are given in Table 17.1.

The design and operating principle of the "Ehkran-10" station are explained by the schematic diagram in Figure 17.4. The SHF signal on 714 MHz, picked up by the antenna from the satellite, goes to an output, consisting of a low-noise HF amplifier (low-noise amplifier), assembled on transistors and frequency converters, in which the signal is preamplified and converted to an IF signal (70 MHz). In the IF signal path, consisting of an IF amplifier and a phase corrector (FK), the IF signal is amplified to 250 mV, necessary for normal operation of the frequency demodulator. The signal, amplified to that level, goes into a video signal demodulator, consisting of an amplitude limiter and the frequency detector itself. The video signal, along with the audio subcarrier (6.5 MHz) goes from the output of the frequency detector unit to a filter unit, where the video and audio signals are separated, and the group delay time response of the video channel is corrected. The signal from the output of the filter unit goes to the input of variable video corrector 1 for precorrection of the group delay time response of the transmitter.

TABLE 17.1

Channel number	Frequency band, MHz	Channel number	Frequency band, MHz
1	48.5–56.5	7	182.0–190.0
2	58.0–66.0	8	190.0–198.0
3	76.0–84.0	9	198.0–206.0
4	84.0–92.0	10	206.0–214.0
5	92.0–100.0	11	214.0–222.0
6	174.0–182.0	12	222.0–230.0

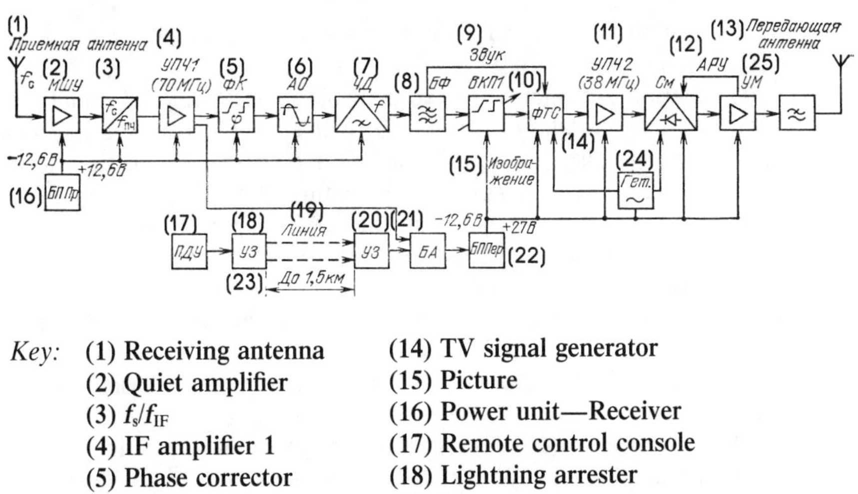

Key:

(1) Receiving antenna
(2) Quiet amplifier
(3) f_s/f_{IF}
(4) IF amplifier 1
(5) Phase corrector
(6) Amplitude limiter
(7) Frequency detector
(8) Filter unit
(9) Sound
(10) Variable video corrector 1
(11) IF amplifier 2
(12) AGC
(13) Transmitting antenna
(14) TV signal generator
(15) Picture
(16) Power unit—Receiver
(17) Remote control console
(18) Lightning arrester
(19) Line
(20) Lightning arrester
(21) Automation unit
(22) Power unit—Transmitter
(23) Up to 1.5 km
(24) Heterodyne
(25) Power amplifier

FIGURE 17.4. Schematic diagram of the "Ehkran-KR-10" collective receiver station.

The corrected video signal and the audio subcarrier signal go into the TV radio signal generating unit, where a complete TV signal is generated on another intermediate frequency of 38 MHz. These signals are amplified in IF amplifier 2 and are fed into the frequency converter of the transmitter, consisting of a mixer and heterodyne. The heterodyne, which generates a 38-MHz reference frequency for the TV signal generating unit and one of 12 reference frequencies for the mixer, is equipped with a crystal frequency stabilizer with a thermostat for the purpose of increasing the stability of the output frequency of the transmitter. A complete TV signal with a frequency of 38 MHz is transferred in a frequency converter to the frequency of one of 12 TV channels and is supplied with a voltage of 350 mV to an FM output power amplifier, where it is amplified to 10 W. The output filter of the power amplifier suppresses the second and third harmonics by 46–49 dB. The complete TV radio signal goes from the filter to the transmitting antenna with a circular horizontally or vertically polarized radiation pattern.

The station can operate in two control modes: local and remote. In either of these two modes of control, the collective TV receiving station is initially in the standby mode, i.e., the receivers, power, and thermostat heater units are on. The power amplifier is turned on either automatically or manually only when a signal from the satellite appears at the receiver input. Lightning arresters are installed between the output of the remote control console and input of the receiver-transmitter to protect the equipment from atmospheric discharges through the remote control line. Because stations of the Ehkran-KR-10 type have become very popular recently in the Ehkran system, we will examine some of its basic elements in greater detail.

The low-noise amplifier (the circuit amplifier unit) is intended for the low-noise preamplification of the SHF signal that comes from the antenna and gives the receiver mirror channel selectivity. A schematic diagram of the HF amplifier unit is shown in Figure 17.5. As can be seen, the unit consists of bandpass filters and a two-stage amplifier, assembled as a common emitter circuit on GT362A transistors. The amplifiers are mounted on a 2 mm thick foil dielectric circuit board with strip lines and mounted radio components. A three-stage bandpass filter with inductive coupling loops is installed at the amplifier input, and a two-stage filter is installed at the output. The specifications of the amplifier unit and of other units of the receiver are given in Table 17.2. Given in Table 17.3 are the nominal frequencies, generated by the heterodyne unit, for generating the 38-MHz intermediate frequency and one of the reference frequencies of the output frequency converter. The output reference frequencies are given in Table 17.3.

The heterodyne unit contains two crystal oscillators: reference and intermediate frequency. The plan of connection of the amplifiers and multipliers in the unit depends on *N* TV channels, and the signal passes either through all three stages, or just through the output stage. For instance, the signal on the frequencies of the sixth to twelvth TV channels passes through three stages, and the frequency is tripled in the second stage.

The automation unit is intended for turning on the receiver and transmitter power units in the local and remote control modes, for generating confirmation signals for the remote control console, and also for protecting the power units from voltage surges on the main. The main purposes of the unit are to connect to and disconnect from the remote control console the collective TV receiving station and to turn off the collective TV receiving station when the threshold voltage (25 ± 1 V), corresponding to an increase of the main voltage to 250 V, is exceeded.

TABLE 17.2

Characteristic of unit, unit of measurement	For unit					
	HF amplifier	*IF*	*IF amplifier 1*	*Phase corrector*	*Amplitude limiter*	*Frequency detector*
Purpose	Preamplification of received signal and selectivity assurance	Conversion of frequency spectrum of received signal to IF signal and amplification	Main amplification of IF signal	Equalization of group delay time response of FM signal path	Suppression of parasitic amplitude modulation of FM signal	Separation of video signal from FM IF signal
Gain, dB	16	30	22	6±1*	1	
Nonuniformity of amplitude-frequency response, not more than, dB	1	2	2**	0.3	0.3	0.5***
Input mismatch attenuation, dB	14	14	26	26	30	30
Output mismatch attenuation, dB	—	24	26	26	26	30
Noise coefficient, not more than, KT	2.7	16	—	—	—	—

TABLE 17.2 (Continued)

Characteristic of unit, unit of measurement	For unit					
	HF amplifier	*IF*	*IF amplifier 1*	*Phase corrector*	*Amplitude limiter*	*Frequency detector*
Purpose	Preamplification of received signal and selectivity assurance	Conversion of frequency spectrum of received signal to IF signal and amplification	Main amplification of IF signal	Equalization of group delay time response of FM signal path	Suppression of parasitic amplitude modulation of FM signal	Separation of video signal from FM IF signal
Input current, mA	9	80	—	—	—	—
Change of output level when input level changes by 6 dB, dB	—	—	1.5	—	—	—
Range of control of group delay time response in 70±17 MHz band, ns	—	—	—	4···12	5****	8*****

*Maximum coefficient for nominal signal.
**IN 70 ± 20-MHz band.
***The end-to-end video amplitude-frequency response relative to the signal on 1.5 MHz.
****The nonuniformity of the group delay time response of the amplitude limiter.
*****Nonuniformity together with the amplitude limiter.

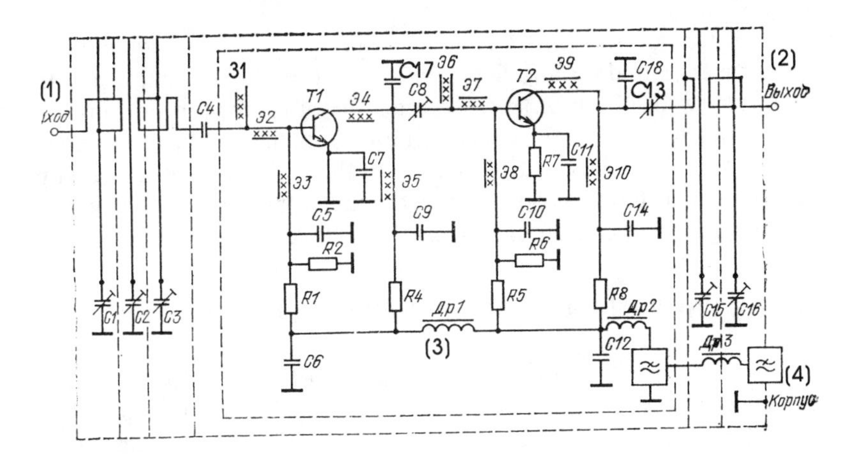

Key: (1) Input
(2) Output
(3) Choke coil
(4) Chassis

FIGURE 17.5. Schematic diagram of the HF amplifier unit.

TABLE 17.3

Channel	*1*	*2*	*3*	*4*	*5*
Frequency, MHz	87.75	97.25	115.25	123.25	131.25
Channel	*6*	*7*	*8*	*9*	*10*
Frequency, MHz	213.25	261.25	289.25	237.25	245.25
Channel	*11*	*12*			
Frequency, MHz	253.25	261.25			

The safety system, intended for reducing dangerous voltages that occur in the remote control line to a safe level for the remote control console and collective TV receiving station as a whole, contains an RB-5 lightning arrester with a 400-V ignition voltage, fuses, a dividing transformer and a pulse voltage limiter. Pulse voltages of more than 400 V at the input of the safety system ignite the RB-5 discharger and the line is grounded. If the charging unit exceeds the tolerances for the discharger the fuses burn out.

The Ehkran-KR-10 station is designed as functionally complete units:

a receiver-transmitter in the form of a three-section rack; the power amplifier is mounted on a radiator on the back wall of the rack, and the second side (wall) of the rack serves as a radiator for the power transistors of the power unit. The receiver-transmitter is connected to the receiving antenna through a connector on the top of the rack, and the transmitting antenna is connected to a connector on the radiator of the power amplifier. The remote control console is designed as a desk-top unit.

17.5 THE EHKRAN-KR-1 COLLECTIVE RECEIVER STATION

The Ehkran-KR-1 collective receiver station is intended for delivering TV broadcasts to small population centers, located in the reliable reception area of the Ehkran system. A schematic diagram of the Ehkran-KR-1 collective receiver station is similar to that examined in the preceding section. However, the output power of the transmitter is not 10 W, but 1 W. Neither station requires the constant attendance of servicing personnel; the receiver is turned on from the remote control console at a distance of up to 1.5 km. The effective radius of the Ehkran-KR-1 receiver station is 2–2.5 km.

17.6 THE EHKRAN SATELLITE TELEVISION STATION

The Ehkran satellite television station is intended for delivering TV broadcasts to weather stations, guard camps, geological expeditions and construction-assembly trains, located in Siberia, on the shores and islands of the Arctic Ocean, in the Far East, *et cetera*. The station includes the following: a receiving antenna with a 16–18 or 25 dB gain; a receiver-transmitter; a distributor; and a Shilyalis color TV monitor.

TV broadcasting signals, received from the Ehkran satellite, are amplified in the receiver of the satellite television station, are demodulated and then a standard TV radio signal is generated on the first or fourth radio channel. From there it is amplified to 1 V and is fed to the distributor, to the outputs of which one to eight TV sets can be connected through cable links. The described satellite television station can be equipped with a compact transportable antenna, so that it can be used in expeditions, mobile mechanized columns and other organizations, working in the field.

Basic Specifications of the Ehkran Satellite Television Station

Receiver noise coefficient, not more than, units	1.5
Nominal input signal level, dBW	−115

Video channel output signal-to-weighted-noise ratio, not less than, dB	49
Nonlinear TV signal distortions, not more than, %	10
Time separation of brightness and color signals, not longer than, ns	30
Input power, not more than, W	30
Power, V:	
mains	220
battery	12

Chapter 18

Multiple Access Equipment

18.1 THE GRADIENT-N EQUIPMENT

The Gradient-N equipment utilizes the FDMA principle, whereby the signal from each TV channel is transmitted on a separate carrier within the common passband of a transponder by means of frequency modulation of that carrier. Its schematic diagram is shown in Figure 18.1.

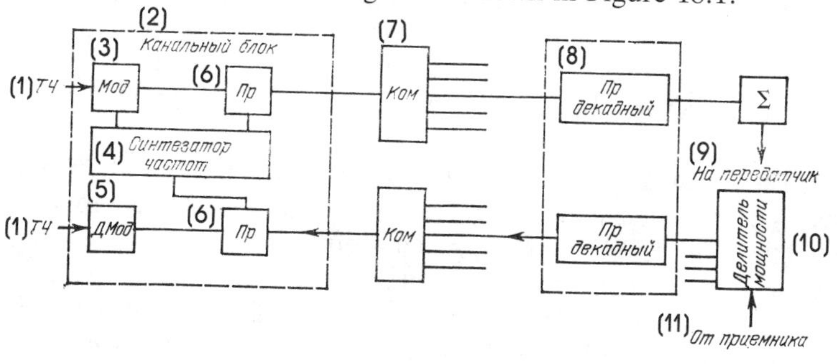

Key: (1) Audio frequency
(2) Channel unit
(3) Modulator
(4) Frequency synthesizer
(5) Demodulator
(6) Converter
(7) Switchboard
(8) Decade converter
(9) To transmitter
(10) Power divider
(11) From receiver

FIGURE 18.1. Schematic diagram of the Gradient-N equipment.

Basic Specifications of the Gradient-N Equipment

Frequency separation between adjacent carriers, kHz	160
Noise band of one channel, kHz	90

Nominal peak carrier frequency deviation, corresponding to test level, kHz	30
Nominal signal-to-noise ratio on intermediate frequency, corresponding to the necessary performance characteristics of audio frequency channel, dB	10
Performance precision of automatic frequency control system of receiver:	
relative to nominal, kHz	±1
with frequency deviation, kHz, within range	±300
AGC precision of receiver, dB	±0.5
for input levels	Within the range of −3 dB to +20 dB relative to nominal
Performance precision of automatic frequency control system:	
of transmitter, kHz	±1 kHz
for initial frequency deviation, kHz	±30
Performance precision of AGC system of transmitter, dB, relative to pilot signal level	±1.0
Number of carriers in transponder	200

The equipment consists of two types of racks: transponder equipment rack and individual equipment rack. The transponder equipment rack contains 70-MHz IF rack amplifier equipment of the receiving and transmitting channels; rack generator equipment; automatic gain and frequency control systems of the receiving and transmitting channels; group (decade) amplifier and converter receiver and transmitter equipment (for the carrier frequency group); receiving-transmitting equipment for channeling telegraph service communications; and receiving-transmitting equipment of the pilot signal and general signal channel. The individual equipment rack includes individual telephone channel units, each of which contains a modulator and demodulator, and also a common heterodyne, with the aid of which they are tuned simultaneously to a selected communications channel.

The entire transponder passband is divided into 21 frequency groups (decades), each with ten carrier frequencies. The frequency band occupied by a decade is 1.6 MHz. The decades in which the carriers of telephone channels are located are called working, and numbers from 1 through 20 are assigned to them in order of increase of the central frequency of a decade. Decades 1–10 are located in the lower half of a transponder, and decades 11–20 are in the upper half, and between the tenth and eleventh decades there is a service decade, in which all auxiliary and service signals are transmitted.

Two carrier frequencies, by means of which a two-way telephone channel is formed, are shifted relative to each other by 17.6 MHz and are placed, successively, one in the lower and the other in the upper half of the transponder. Thus, the pairs of frequencies of two-way channels are in the following decades: 1–11, 2–12, 3–13, 4–14, 5–15, 6–16, 7–17, 8–18, 9–19, 10–20. The carrier frequencies of one one-way service channel in the service decade are arranged symmetrically with respect to the middle frequency of the decade, and the frequencies of the pilot signal and of the general service communications selector are also arranged symmetrically. The telegraph channel carriers are separated by 80 kHz, and the telegraph channel frequency net is shifted by 40 kHz relative to the telephone channel carrier net. The transponder frequency plan is illustrated in Figure 18.2, the working decade plan is shown in Figure 18.3 and Table 18.1, and the service decade plan is shown in Figure 18.4.

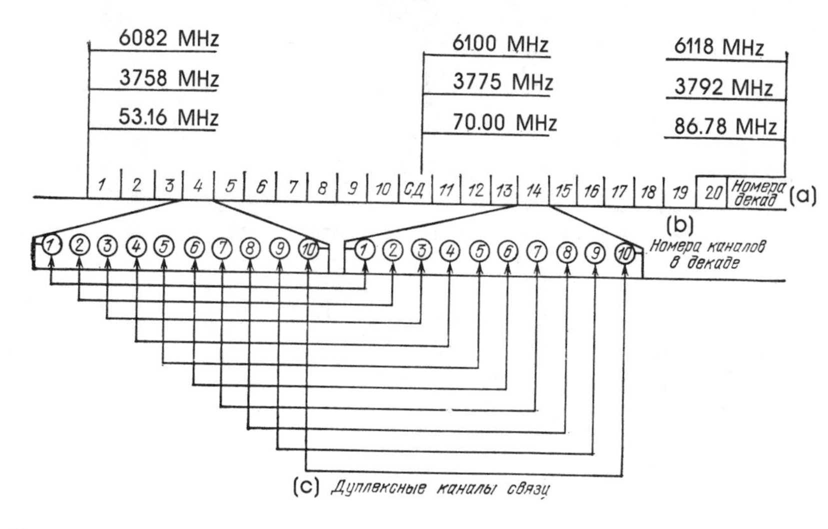

Key: (a) Number of decade
(b) Numbers of channels in decade
(c) Two-way communications channels

FIGURE 18.2. Transponder frequency plan.

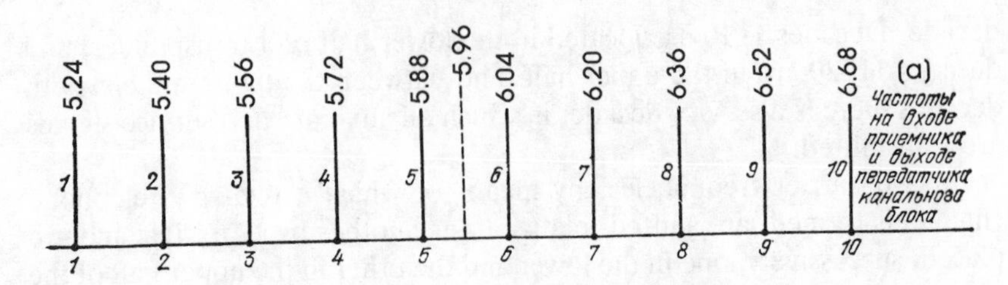

Key: (a) Frequencies at receiver input and transmitter output of channel unit

FIGURE 18.3. Frequency plan of working decade.

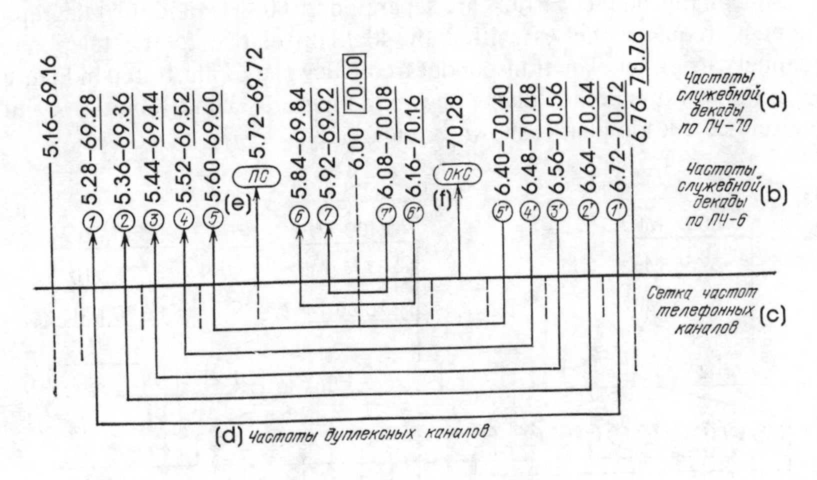

Key: (a) Frequencies of service decade on IF-70
(b) Frequencies of service decade on IF-6
(c) Telephone channel frequency net
(d) Two-way channel frequencies
(e) Power source
(f) Common signal channel

FIGURE 18.4. Frequency plan of service decade.

The receiver input of the transponder equipment rack receives the entire transponder signal, picked up by the receiver on intermediate frequency in the 70 ± 17-MHz band (from the output of the transponder filter of the "step-down" converter); the transponder equipment rack

TABLE 18.1

Number of decade	*Middle frequency of decade, MHz*	*Heterodyne frequency, MHz*	*Number of decade*	*Middle frequency of decade, MHz*	*Heterodyne frequency, MHz*
1	54.0	48	12	73.2	67.2
2	55.6	49.6	13	74.8	68.8
3	57.2	51.2	14	76.4	70.4
4	58.8	52.8	15	78.0	72.0
5	60.4	54.4	16	79.6	73.6
6	62.0	56.0	17	81.2	75.2
7	63.6	57.6	18	82.8	76.8
8	65.2	59.2	19	84.4	78.4
9	66.8	60.8	20	86.0	80.0
10	68.4	62.4	Service	70	64.0 — transmitter
11	71.6	65.6	—	—	76.0 — receiver

contains 20 working decade receiving converters, the input of each of which gets the entire transponder signal spectrum through a power divider. A receiving decade converter selects the frequency band corresponding to one of the frequency decades with the aid of a decade bandpass filter, connected to its input, and it transfers it from the 70 ± 17-MHz intermediate frequency to the 6 ± 0.8-MHz frequency. The transponder equipment rack also contains 20 transmitting decade converters, each of which transfers ten telephone channel carriers of one decade from the 6 ± 0.8-MHz band to the appropriate part of the 70 ± 17-MHz band. A bandpass filter is connected to the output of each transmitting decade converter, and it attenuates the heterodyne frequency component and limits the out-of-band spectrum of the decade converter. Each decade converter receives its own heterodyne frequency, and the heterodyne frequency of receiving and transmitting decade converters, corresonding to a given decade, is identical. The receiver and transmitter decade converters are structurally combined into a single unit.

The telephone channel unit (individual equipment rack) contains a superheterodyne receiver with double frequency conversion and a transmitter (a frequency modulator) with subsequent frequency transfer. The first frequency converter of the channel receiver and frequency converter of the channel transmitter receive the same heterodyne frequency from a channel frequency synthesizer. The receiver and transmitter of the channel unit can be tuned to receive and transmit one of the following ten frequencies in the 6 ± 0.8-MHz band: 5.24, 5.40, 5.56, 5.72, 5.88, 6.04, 6.20, 6.36, 6.52, 6.68 MHz. Tuning is accomplished by changing the frequency of the channel heterodyne. Its frequency is changed discretely in steps of 0.16/0.08 MHz. Frequency can be controlled either manually with a toggle switch, located on the front panel of the unit, or automatically by signals from the channel control equipment.

Between the decade converters and channel units there are switches, which connect the inputs of the channel receivers of the individual equipment rack to the outputs of the receiving decade converters of the tranpsonder equipment rack and the outputs of the transmitters of the channel units to the input of the transmitting decade converters. These switches can be controlled either manually or automatically from the channel control equipment. Thus, a channel unit is tuned to any of the two-way channels in two steps: by tuning the channel unit within a decade and by connecting the tuned channel unit to any decade.

The frequency of a received channel is transferred in the receiving part of a channel unit by binary frequency conversion from 6 ± 0.8 MHz to 2 MHz, on which frequency the demodulator itself, which for the purpose of improving its threshold properties is designed as a synchronous phase detector, also operates. The modulator of a channel unit operates on 31

MHz, which then is transferred with the aid of the channel synthesizer to 6 ± 0.8 MHz.

The equipment has automatic frequency control systems, the reference signal for which is the pilot signal, radiated by the central network station, in the receiver and transmitter for correcting the instability of the generators of the transmitter and receiver of an earth station, the heterodynes of the space repeater and also the Doppler effect. Here the heterodyne frequency of the receiver ("step-down" converter) of an earth station is changed by the signal of the phase-locked loop in the receiver, and the heterodyne frequency of the transmitter ("step-up" converter) is changed by the signal of the phase-locked loop in the transmitter.

The automatic gain control system in the receiver also operates on the pilot signal from the control station, maintaining a constant level at the output of the pilot signal reciever. The AGC voltage is supplied to the transponder IF amplifier, which is a part of the "step-down" converter of the receiver of an earth station, changing its gain appropriately. The received pilot signal levels and one of the channels are compared at the transponder equipment rack for achieving automatic gain control on the transmitting end. The gain control signal is fed to the transponder transmitting amplifier, which is a part of the transponder equipment rack.

18.2 THE GRUPPA EQUIPMENT

The Gruppa equipment, like the Gradient-N equipment described above, utilizes the FDMA principle, but in contrast to the Gradient-N equipment, not an individual telephone channel, but a group of channels is transmitted on a separate carrier frequency. There are two possible ways of transmitting a channel group: frequency modulation of the carrier by a standard 12-channel group (with FDM in the 12–60-kHz band), and phase modulation of the carrier by a 512-kb/s digital stream, generated as a result of analog-digital conversion on the basis of pulse-code modulation of eight standard audio frequency channels.

Main Specifications of the Gruppa Equipment

Minimum separation between carriers, MHz	1.35
Number of carriers	up to 24
Effective deviation per channel, transmitting a 12-channel analog group, kHz	125
Effective frequency band of one FM carrier, MHz	2.2
Signal-to-noise ratio necessary for achieving a signal-to-noise ratio on an audio frequency channel of >44 dB (in the 2.2-MHz band), dB	9.5 (in the analog mode)

Information transmission rate on one carrier, kb/s	512
Probability of incorrect reception (per bit) for P_s/P_n = 9.5 dB in 1.2-MHz band, not more than	10^{-6}
Pilot signal frequency, MHz	69, 72

The Gruppa equipment is manufactured in two different modifications: a discrete modem rack, consisting of six discrete modems and transponder equipment, and an SOP rack, consisting of two analog modems, two discrete modems and transponder equipment. The operating principle of the Gruppa equipment is examined by way of example of the SOP rack (Figure 18.5). The input of the SOP rack receives the information signal from an analog-digital converter as a 512-kb/s binary data stream. Relative phase modulation (PM) of the 6 MHz carrier takes place in the modulator. In the individual transmitting converter, the relative PM signal spectrum is transferred to 44.8 MHz. Then this signal is switched by a special switch to any of the mixers, where the relative PM signal spectrum is transferred from 44.8 MHz to any frequency of the following frequency net: 53.4, 54.75, 56.1, 57.45, 58.8, 60.15, 61.5, 62.85, 64.2, 65.55, 66.9, 68.25, 70.95, 72.3, 73.65, 75.0, 76.35, 77.7, 79.05, 80.4, 81.75, 83.1, 84.45. This transfer is accomplished with the aid of crystal oscillators. A phase-shift signal on the carrier frequency, allocated for the given communications link, is thus generated at the output of the transmitting converter. This signal, after being amplified additionally and filtered, then goes to the transmitter of the station. The signal is received in the reverse order. The receiver output signal at an earth station passes in the 70 ±17-MHz IF band to the input of the Gruppa rack, where it is first subjected to preamplification and filtering. Then the relative PM signal is transferred in the receiving converter with the aid of the very same crystal oscillator unit, used in the transmitting part of the equipment, to 44.8 MHz. This signal then can be connected through a switchboard to the input of any of the individual converters, where the relative PM signal spectrum is transferred to 6 MHz, on which detection itself takes place.

A coherent demodulator is used as a relative PM signal demodulator. The analog signal demodulator is designed as a synchronous phase detector. It provides a threshold gain of the order of 4–4.5 dB.

The Gruppa equipment is equipped with automatic receiver heterodyne frequency control and received signal gain control systems. These systems operate on a pilot signal from the central station of the network, and the frequency and level of this signal are used as reference signals at out stations.

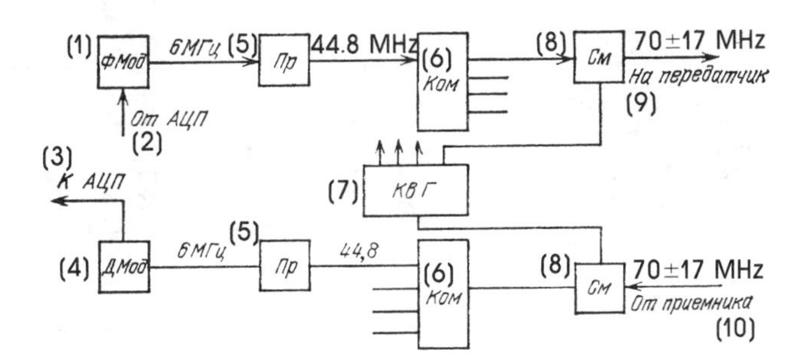

Key: (1) Phase modulator
(2) From analog-digital converter
(3) To analog-digital converter
(4) Demodulator
(5) Converter
(6) Switchboard
(7) Switch
(8) Mixer
(9) To transmitter
(10) From receiver

FIGURE 18.5. Schematic diagram of the Gruppa equipment.

18.3 THE MDVU-40 TIME DIVISION MULTIPLE ACCESS EQUIPMENT

The MDVU-40 equipment is based on the time division multiple access principle with a 40-Mb/s line digital stream transmission rate in a satellite. The equipment is intended for joint operation with terminal equipment that performs analog-digital conversion either of a standard 60-channel frequency division multiplex group (the binary stream rate is 5.12 MB/s), or of eight standard audio frequency channels (the binary stream rate is 512 kb/s). In either case, pulse-code modulation is used for analog-digital conversion.

Main Specifications of the MDVU-40 Equipment

Kind of modulation	Relative PM-4
Nominal incorrect reception probability	10^{-6}
Signal-to-noise-ratio (in 25-MHz band), dB	16
Line rate, Mb/s	40, 96

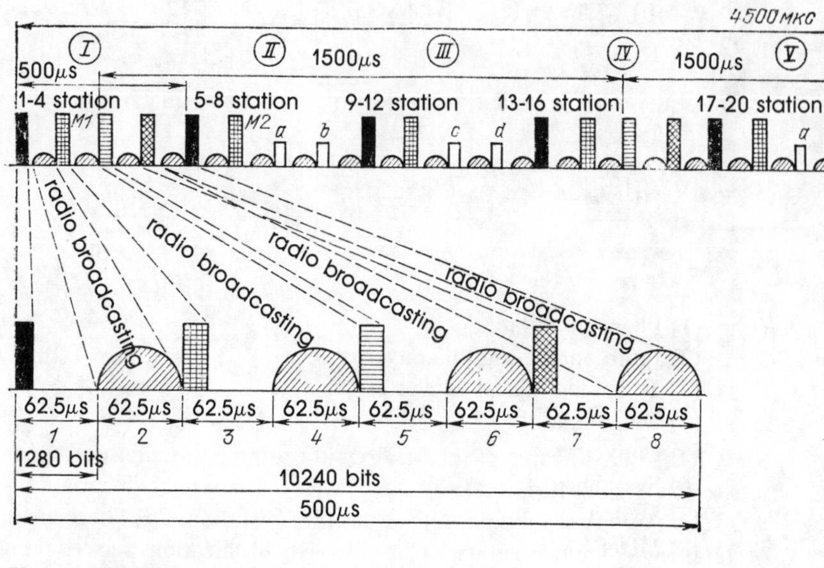

Key: **(a) Windows:**
(b) entry in subframe 1, T_r = 500 μs;
(c) service communications I in subframe 5, T_r = 1500 μs;
(d) hold in subframe 3, T_r = 500 μs;
(e) service communications II in subframe 7, T_r = 1500 μs
(f) vacant

Number of packets per frame at input digital signal rate, MB/s:	
5.12	8
0.512	72
Information frame duration, ms, at input signal rate, Mb/s:	
5.12	0.5
0.512	4.5
Frame duration of synchronization system, ms	18

A schematic diagram of the MDVU-40 equipment is shown in Figure 18.6. When the MDVU-40 equipment is used in conjunction with Orbita-RV equipment (see Chapter 14), the carrying capacity of the transponder is divided equally between the indicated systems: all even subframes (2, 4, 6, . . ., 288) are intended for transmitting audio broadcasting signals, and the signals of the MDVU-40 equipment are transmitted in the odd subframes (1, 3, 5, . . ., 287), and the maximum repetition period is 18 ms.

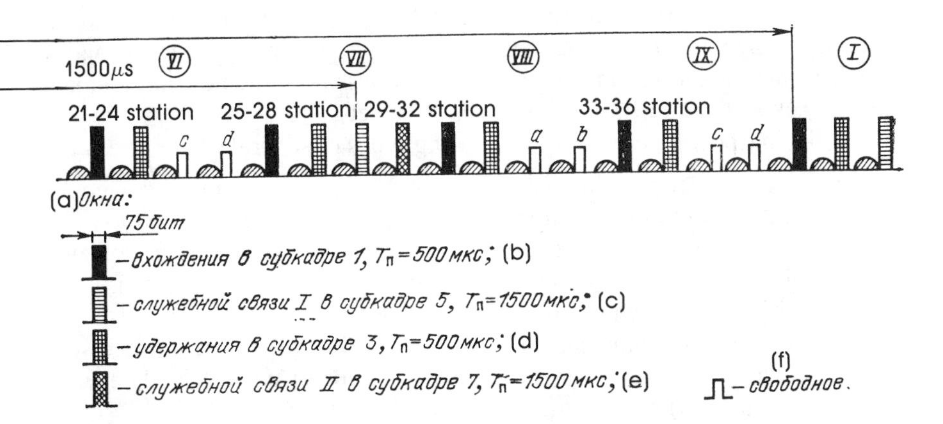

FIGURE 18.6. A frame of the MDVU-40 equipment: Length of window 75 bits (clock period duration f_{c1} = 20.48 MHz). Entry pack duration T = 500 μs. Total number of hold packets or number of stations 36. There are 36:9 = 4 stations per hold position. The hold synchronization packet repetition period per station is $T = 4.5 \cdot 10^{-3} \cdot 4 = 18$ μs. Service communications windows I and II are repeated at T_r = 1500 μs. The carrying capacity per channel V = 75 bits; $15 \cdot 10^{-8}$ = 50 kbaud. V = 16 kbaud per channel in consideration of losses to the preamble and synchronized pulse. Direct and answer channels are organized in two positions (subframes 5 and 7): $V_1 = V_2$ = 16 kbauds.

The separate synchronization frame principle is used in the equipment for simplifying in one transponder earth stations with different carrying capacities, and a frame is made up of packets (subframes) of a fixed duration. The carrying capacity of an earth station is changed by changing the repetition period of its information packet. For instance, the packet repetition period of an earth station with terminal equipment that performs analog-digital conversion of a standard 60-channed group (the digital stream rate is 5.12 Mb/s) is 500 μs, and when eight-channel terminal equipment is used, which performs analog-digital conversion of eight audio frequency channels (the digital stream rate is 512 kb/s), this period increases to 4500 μs. The packet duration is always constant and is 62.5 μs, which corresponds to 1280 symbols at a clock frequency of 20,480 kHz. The beginnings of every odd subframe with a duration of 75 symbols are time positions of the separate frame synchronization system. The information packets of the MDVU-40 equipment contain an entry part, consisting of the preamble of the demodulator, intended for entering the standard mode of the carrier and the clock frequency regenerating circuits, and a synchronizing signal, which identifies the boundary of the information part of a packet (code of the beginning of a packet).

By purpose, there are the following types of separate time positions:

- "entry windows," intended for receiving earth station entry signals, located in the first subframe with a repetition period of 500 μs;
- "hold windows," intended for transmitting frame synchronizing signals of all earth stations, operating in the network, located in the third subframe with a repetition period of 500 μs;
- "service communications windows I and II," intended for processing service communications channels, located in the fifth and seventh subframes with a repetition period of 1500 μs;
- "auxiliary communications windows," located in the thirteenth, fifteenth, twenty-first, and twenty-third subframes with a repetition period of 1500 μs, which can be used for telephone-telegraph service communications with individual terminal equipment.

The frame synchronizing signals of all earth stations have the identical duration and are transmitted in "hold" windows, and the positions of the central station and of a backup station are reserved for the third and eleventh subframes. "Entry windows" are used for receiving signals in them during the process of primary synchronization of earth stations for transmission. For entry an earth station transmits a short 500-μs long entry packet, a fragment of which, coinciding with an entry window on the receiving end, is used for computing the phase error and for establishing the necessary time relations on the transmitting end.

The MDVU-40 equipment consists of two racks: transponder equipment rack (MDVU-SOO) containing 40 Mb/s modems, frame synchronization, and service communications equipment; and an individual equipment rack (MDVU-SIO) containing five time companders (a compressor-expander system). The frame synchronization signals contain 24 symbol long demodulator preamables and the synchronizing signal itself (a 31-symbol long pseudorandom sequence). The kind of pseudorandom sequence that is transmitted is determined by the functional purpose of an earth station, which can be the central station, backup station or an out station. Accordingly, one of three possible pseudorandom sequences is transmitted: M1, M2, or M3, respectively. The frame synchronization signals are transmitted in the two-phase mode, so that frame synchronization of earth stations operating at 40 and 20 Mb/s can be synchronized.

When assembling a network of stations, the central station radiates its signal No. 1 first with a repetition period of 18 ms. The detection of this synchronizing signal on the receiving end enables all earth stations to synchronize the receiving part of the MDVU-40 equipment (receiver synchronization) by tuning the frequency and phase of the receiving clock generator with the aid of a phase-locked loop. The passband of the phase-locked loop is 2 Hz.

In the second step of synchronization, the backup station and all out stations enter pulse synchronism. In the method used for entering pulse synchronism, the entry signal causes the shortest possible interference with the information signals of the other earth stations. The phase uncertainty of the signal of an out station on the transmitting end is eliminated after primary entry. This occurs as follows.

A brief entry signal of an out station is transmitted one time by a command of a central processor, which is a part of the synchronization equipment. This signal is a 500-μs long pseudorandom sequency (10,240 symbols), the beginning of which coincides with the beginning of a frame on the transmitting end. At the repeater input, this signal is superimposed on the information signals of other earth stations. When the duration of the pseudorandom sequence is 500 μs, part of the entry signal must enter the "entry window," and this part of the signal will be received without distortions, because there are no other signals in the entry window. After the phase of the received fragment of the pseudorandom sequence is determined relative to its beginning, the necessary phase correction of the time of transmission of the packet is computed at the given earth station. After correction, the frequency uncertainty is eliminated on the transmitting end by a signal from the central processor; the earth station radiates an M3 synchronizing signal (or an M2 synchronizing signal at a backup station) in the middle of the subframe reserved for it. The repetition period of the M3 signal during the transition process is 4.5 ms. Frequency uncertainty on the transmitting end is eliminated by a phase-locked loop of a variable transmission clock generator. The passband of this phase-locked loop is 0.4 Hz. At the end of the transition process, the out station switches to the hold mode by a command from the central processor. After that, the out station continues to generate the M3 signal for transmission, but it now transmits it in a separate "hold window" with a repetition period of 18 ms; after the transient process is over, the backup station radiates an M2 signal in its assigned "hold window."

The status of frame synchronization is monitored, and the backup system is controlled with the aid of special test and control instruments and central processor. For instance, a synchronization failure at an out station during reception will cause the transmitter to be turned off to prevent interference with other stations, and at a backup station the same thing will result in automatic backup of the central station. On-line reliability monitoring is accomplished by counting missed M1, M2, or M3 frame synchronizing signals, and the equivalent digital stream rate of the reliability monitor channel is approximately 1.7 kb/s.

The equipment has instruments for processing the service telephone communications channel. This channel is processed with the aid of a delta-codec (the rate is 16 kb/s) and time compressors, which generate a service communications packet. A packet includes the preamble of the demodulator, "code of the beginning of a packet" signal and an information part. A call to any of 36 correspondents (the maximum number of stations in the network) is addressed by keying the phase of the "code of the beginning of a packet" signal in a particular way.

Bibliography

General

1. *Reglament radiosvyazi (Radio Communications Regulations).* Moscow: Svyaz′, 1975, 824 pp.
2. Borodich, S. V. *Iskazheniya i pomekhi v mnogokanal′nykh sistemakh radiosvyazi s ChM (Distortions and Interference in Multichannel FM Radio Communications Systems).* Moscow: Svyaz′, 1976, 256 pp.
3. Mashbits, L. M. *Zony obsluzhivaniya sistem sputnikovoj svyazi (Service Areas of Satellite Communications Systems).* Moscow: Radio i svyaz′, 1982, 169 pp.
4. Kantor, L. Ya., V. P. Minashin, and V. V. Timofeev. *Sputnikovoe veshchanie (Satellite Broadcasting).* Moscow: Radio i Svyaz′, 1981, 232 pp.
5. Chernyavskij, G. M. and V. A. Bartenev. *Orbity sputnikovoj svyazi (The Orbits of Satellite Communications).* Moscow: Svyaz′, 1978, 240 pp.
6. Final Acts, WARC-1977. Geneva: ITU, 1977. 146 pp.
7. Final Acts, SARC-1979. Geneva: ITU, 1979. 984 pp.
8. CCIR, XVth Plenary Assembly, Vol. IV-1. Geneva: ITU, 1982, 501 pp.
9. CCIR, XVth Plenary Assembly, Vol. IV/IX-2. Geneva: ITU, 1982, 162 pp.
10. CCIR, XVth Plenary Assembly, Vol. XI. Geneva: ITU, 1982, 377 pp.
11. CCIR, XVth Plenary Assembly, Vol. XII. Geneva: ITU, 1982, 263 pp.
12. CCITT, *The Orange Book,* Vol. III-1. Moscow: Svyaz′, 1980, 352 pp.
13. CCITT, *The Orange Book,* Vol. III-2. Moscow: Svyaz′, 1980, 248 pp.
14. CCITT, *The Orange Book,* Vol. IV-1. Moscow: Svyaz′, 1980, 240 pp.
15. CCITT, *The Orange Book,* Vol. IV-2. Moscow: Svyaz′, 1980, 177 pp.
16. Kantor, L. Ya. and V. M. Dorofeev, *Pomekhoustojchivost′ priema ChM signalov (Noise Immunity in FM Signal Reception).* Moscow: Svyaz′, 1970, 335 pp.

17. Fortushenko, A. D. (Editor). *Osnovy tekhnicheskogo proektirovaniya sistem svyazi cherez ISZ (Technical Planning Principles of Satellite Communications Systems).* Moscow: Svyaz′, 1970, 331 pp.
18. ———. *Osnovy tekhnicheskogo proektirovaniya apparatury sistem svyazi s pomoshch′yu ISZ (Technical Planning Principles of Satellite Communications Systems).* Moscow: Svyaz′, 1972, 344 pp.
19. Borodich, S. V. (Editor). *Spravochnik po radiorelejnoj svyazi (Handbook on Radio Relay Communications).* Moscow: Radio i svyaz′, 1981, 415 pp.
20. Krivosheev, M. I. *Osnovy televizionnykh izmerenij (Fundamentals of Television Tests).* Moscow: Svyaz′, 1976.

Chapter 1

1.1 GOST 24375-80 (State Standard), Radio Communications. Terms and Definitions.
1.2 Mashbits, L. M. "Zones of Guaranteed and Regulated Signal Levels of Satellite Communications Systems." *Radiotekhnika (Radio Engineering),* 1977, Vol. 32, No. 12.
1.3 Talyzin, N. V. and L. Ya. Kantor. "On the Use of Communications Satellites for Transmitting One-way (Simplex) Programs." *Radiotekhnika (Radio Engineering),* 1967, No. 6, pp. 1-7.
1.4 Special Report of Interim Working Party Plen. 2, CCIR. Geneva: ITU, 1977, 256 pp.
1.5 CCIR, Interim Working Party Plen. 3, Doc. IWP PLEN/3-13-E, October 1981, Ch. 4, System Synthesis and Cost Optimization.
1.6 Talyzin, N. V., L. Ya. Kantor, E. A. Manyakin, and Yu. M. Payanskij. "On the Cost Effectiveness of a Satellite Communications System." *Radiotekhnika (Radio Engineering),* 1969, No. 11, pp. 5-13.
1.7 Talyzin, N. V., L. Ya. Kantor, and M. Z. Tsejtlin. "The Orbita Earth Station for Receiving TV Programs from Earth Satellites." *Ehlektrosvyaz′ (Telecommunications),* 1967, No. 11, pp. 5-8.
1.8 Talyzin, N. V., L. Ya. Kantor, and Yu. M. Payanskij. "The Economic Aspects of Satellite Communications." *Radiotekhnika (Radio Engineering),* 1970, No. 1, pp. 3-8.
1.9 Kantor, L. Ya. and A. G. Pauk. "Choosing a Design Version of Satellite Broadcasting Communications Systems." *Ehlektrosvyaz′ (Telecommunications),* 1981, No. 2, pp. 24-27.
1.10 Roginskij, V. N. (Editor). *Teoriya setej svyazi (Theory of Communications Networks).* Moscow: Radio i svyaz′, 1981, 192 pp.
1.11 CCITT, *The Green Book,* Vol. VI-2. Geneva: MSE, 1973, pp. 233-412.

1.12 CCITT, *The Orange Book,* Vol. VI-3. Moscow: Svyaz′, 1980, 121 pp.
1.13 CCITT, *The Orange Book,* Vol. VI-2. Moscow: Svyaz′, 1980, 133 pp.
1.14 CCITT, *The Orange Book,* Vol. VI-1. Moscow: Svyaz′, 1979, 69 pp.

Chapter 2

2.1 *Pravila tekhnicheskoj ekspluatatsii sredstv veshchatel′nogo televideniya (Rules of Technical Operation of Broadcast TV Equipment),* Moscow: Radio i Svyaz′, 1981, 142 pp.
2.2 Nikonov, A. V. and L. Z. Papernov. *Izmeriteli urovnej zvukovykh signalov (Acoustic Signal Level Meters).* Moscow: Radio i svyaz′, 1981, 112 pp.
2.3 Ivanov, K. V. and V. G. Khodataj. "Variable Companding of Broadcast Signals," *Ehlektrosvyaz′ (Telecommunications),* 1976, No. 5, pp. 13-17.
2.4 Tsbulin, M. K. *Ehkhozagraditel′nye ustrojstva na setyakh svyazi (Echo Suppressors in Communications Networks).* Moscow: Svyaz′, 1979, 88 pp.
2.5 Rombro, V. A. and Yu. D. Farber. *Izmereniya kharakteristik mnogokanal′nykh sistem svyazi (Measurements of the Characteristics of Multichannel Communications Systems).* Moscow: Svyaz′, 1977, 272 pp.
2.6 Abramov, V. A., G. A. Emel′yanov, and A. A. Koshevoj. "Measurement of the Phase Jitter of a Digital Signal in Lines of Digital Transmission Systems." *Ehlektrosvyaz′ (Telecommunications),* 1982, No. 3, pp. 23-24.
2.7 Evnevich-Chekan, O. V. "Calculations of the Cumulative Interference and Distortions in TV Signal Transmission Lines." *Ehlektrosvyaz′ (Telecommunications),* 1974, No. 10, pp. 28-30.
2.8 Khvorostenko, N. P. *Statisticheskaya teoriya demodulyatsii diskretnykh signalov (Statistical Theory of Discrete Signal Demodulation).* Moscow: Svyaz′, 1968, 335 pp.

Chapter 4

4.1 Kalinin, A. I. *Raschet trass radiorelejnykh linij (Calculation of the Routes of Radio Relay Lines).* Moscow: Svyaz′, 1964. 243 pp.
4.2 Hogg, D. and R. A. Semplak. "The Effect of Rain and Water Vapor on Sky Noise at Centimeter Wavelengths." *BSTJ,* 1961, Sept., Vol. 40, pp. 1331-1348.
4.3 Tsejtlin, N. M. *Primenenie metodov radioastronomii v antennoj tekhnike (The Use of Radio Astronomy Techniques in Antenna Technology).* Moscow: Sovietskoe radio, 1979, 256 pp.

4.4 CCIR, Report 719, Kioto, 1978, Vol. V, pp. 97-107.
4.5 Kalinin, A. I. "The Influence of Rain and the Attenuation of Radio Waves in Earth-Satellite Links." *Ehlektrosvyaz' (Telecommunications),* 1976, No. 5, pp. 12-15.
4.6 CCIR, Doc. IV/41-E, 19 January 1962.
4.7 Petrovich, N. T. and E. F. Kamnev. *Voprosy kosmicheskoj radiosvyazi (Problems of Space Radio Communications).* Sovietskoe radio, 1965, 307 pp.
4.8 Hogg, D. and Ta-shing Chu. *Proc. IEEE,* 1975, No. 9, pp. 1308-1314.
4.9 Bartolome, P. Zh. "Experiments on Radio Wave Propagation on 11/14-GHz Frequencies within Framework of the Program of the Development of European Communications Satellite." *TIIER,* 1977, Vol. 65, No. 3, pp. 232-234.
4.10 Krejn, R. I. "A Forecast of the Influence of Precipitation on Satellite Communications Systems." *TIIER,* 1977, Vol. 65, No. 3, pp. 210-216.
4.11 Oguchi, T. *J. Radio Res. Lab.* (Japan), 1975, Vol. 22, No. 107, pp. 165-211.
4.12 Pokras, A. M. *Antennye ustrojstva zarubezhnykh linij svyazi cherez iskusstvennye sputniki Zemli (The Antenna Systems of Foreign Satellite Communications Links).* Moscow: Svyaz', 1965, 168 pp.
4.13 Bykov, V. L., V. A. Borovkov, and V. N. Kobylin. "Receiving Signals from Geostationary Satellites on High Latitudes." *Trudy NIIR,* 1980, No. 3, pp. 10-15.

Chapter 5

5.1 Suggestions of the French Administration of Communications to WARC-1979, Doc. No. 82, France, 1979, 19 March, 11 pp.
5.2 Concerning the Maximum Permissible Interference in Single-Channel-per-Carrier Transmissions in Networks of the Fixed Satellite Service, Doc. P/98-E, Special Preparatory Meeting, Geneva, 1979, 4 pp.
5.3 Borovkov, V. A. and M. G. Lokshin. "The Electromagnetic Compatibility of TV Broadcasting Services." *Ehlektrosvyaz' (Telecommunications),* 1979, No. 7, pp. 1-4.

Chapter 6

6.1 Kantor, L. Ya. "On an Evaluation of the Limiting Carrying Capacity of a Geostationary Orbit." *Radiotekhnika (Radio Engineering),* 1979, No. 4, pp. 5-12.
6.2 Undrou, B., D. Glover, D. Makkul, *et al.* "Adaptive Interference Compensators. The Principles of Design and Application." *TIIER,* 1975, Vol. 63, No. 12, pp. 69-98.

6.3 Kantor, L. Ya. "A New Approach to the Evaluation of the Effectiveness of the Utilization of a Geostationary Orbit." *Ehlektrosvyaz' (Telecommunications),* 1976, No. 1, pp. 5-11.
6.4 "Evaluation of Efficiency in the Use of the Geostationary Satellite Orbit by a Spot-Beam Satellite System." *Telecommunication Journal,* 1982, Vol. 49, No. 1, pp. 25-28.

Chapter 7

7.1 Sunde, E. D. "Intermodulation Distortion in Multicarrier FM Systems." *IEEE Internat. Conv. Record,* 1965, pp. 2, 22–26, March, pp. 130–146.
7.2 Davenport, W. B. "Signal-to-Noise Ratios in Bandpass Limiters I." *Applied Physics,* 1953, Vol. 24, No. 6, pp. 162-171.
7.3 Bykov, V.L., V. S. Rabinovich, A. L. Senyavskij, *et al.* "The Influence of the Amplitude Responses of a Limiter on the Performance of Correlators." *Trudy IV Vsesyuznogo simpoziuma. Metody predstavleniya i apparaturnyj analiz sluchajnykh protsessov i polej (Transactions of the Fourth All-Union Symposium. Methods of Representation and Hardware Analysis of Random Processes and Fields).* Leningrad, 1971, pp. 87–88.
7.4 Kotel'nikov, V. A. "On the Influence of the Cumulative Sinusoidal Voltages on Nonlinear Impedances." *Nauchno-tekhnicheskij sbornik LEhIS (Scientific-Technical Handbook of the Leningrad Electrotechnical Institute of Communications)* M. A. Bonch-Bruyevich (Editor), 1936, No. 4, pp. 23-30.
7.5 Westcott, R. I. "Investigation of Multiple FM/FDM Carriers Through a Satellite TWT Operating near Saturation." *Electronics Record,* 1967, Vol. 114, No. 5, pp. 726-740.
7.6 Schaft, R. D. "Hard Limiting of Several Signals and Its Effect on Communication System Performance." *IEEE Internat. Conv. Record,* Pt. 2, 1965, pp. 103-112.
7.7 Abolits, A. K. "Energy Relations in FDM Transmission Through a Nonlinear Repeater." *Ehlektrosvyaz' (Telecommunications),* 1967, No. 3, pp. 27-32.
7.8 Pustovojtov, E. P. "Crosstalk Interference in the Simultaneous Amplification of Several FM Signals in One Traveling Wave Tube." *Trudy MEhIS (Transactions of the Moscow Electrotechnical Institute of Communications),* 1968, pp. 36-40.
7.9 Chapman, R. C. and I. B. Millard. "Intelligible Crosstalk Between Frequency-Modulated Carriers Through AM-PM Conversion." *IEEE Transactions on Communication Systems,* 1964, June, Vol. GS-12, No. 2, pp. 159-166.

7.10 Bykov, V. L., V. A. Borovkov, and S. M. Khomutov. "Energy Analysis of FDM Multiple Access." 1967, No. 1 (46), pp. 58-63.
7.11 Kantor, L. Ya. and V. I. D′yachkov. "On the Optimum Planning of Satellite Communications Systems with Frequency Division Multiple Access." *Ehlektrosvyaz′ (Telecommunications),* 1969, No. 3, pp. 17-21.
7.12 Pritchard, W. L., N. MacGregor, and V. S. Military. "Commercial Comsat Design." *Astronaut and Aeronaut,* 1964, Vol. 2, No. 10, pp. 210-216.
7.13 Puente, Shmidt, and Vert. "Methods of Multiple Station Operation of Commercial Satellites." *TIIER,* 1971, No. 2, pp. 117-123.
7.14 Stein, S. and J. Jones. *Printsipy sovremennoj teorii svyazi i ikh primenenie k peredache diskretnykh soobshchenij (Principles of the Modern Theory of Communications and Their Application to Discrete Message Transmission).* Moscow: Svyaz′, 1971, 376 pp.
7.15 Khvorostenko, N. P. *Statisticheskaya teoriya demodulyatsii diskretnykh signalov (Statistical Theory of Discrete Signal Demodulation).* Moscow: Svyaz′, 1968, 335 pp.
7.16 Viterbi, Eh. D. *Printsipy kogerentnoj svyazi (The Principles of Coherent Communications).* Moscow: Sovietskoe radio, 1970.
7.17 Fink, L. M. *Teoriya peredachi diskretnykh soobshchenij (A Theory of Discrete Message Transmission).* Moscow: Sovietskoe radio, 1970, 727 pp.
7.18 Dorofeev, V. M. and E. A. Miroshnikov. "The Influence of Line Characteristics on PM Signal Transmission Reliability." *Trudy NIIR,* 1979, No. 1, pp. 20-23.
7.19 Gonorovskij, I. S. *Radiotekhnicheskie tsepi i signaly (Electronic Circuits and Signals).* Moscow: Sovietskoe radio, 1977, 607 pp.
7.20 Rabiner, L. and B. Fould. *Teoriya i primenenie tsifrovoj obrabotki signalov (The Theory and Application of Digital Signal Processing).* Moscow: Mir, 1978, 848 pp.
7.21 Grebel′skij, M. D., G. Kh. Pan′kov, and V. M. Tsirlin. "Optimization of Satellite Communications Links for Transmitting High-Rate Digital Streams." *Ehlektrosvyaz′ (Telecommunications),* 1979, No. 12, pp. 17-20.
7.22 "Time Division Multiple Access Equipment for Transmitting Digital Information Through a Satellite," Grebel′skij, M. D., G. Kh. Pan′kov, M. I. Rozenbaum, *et al., Ehlektrosvyaz′ (Telecommunications),* 1979, No. 11, pp. 14-16.
7.23 Simonov, M. M. "Synchronization Techniques in Time Division Multiple Access Systems." *Zarubezhnaya tekhnika svyazi. Ser. Radiosvyaz′, radioveshchanie, televidenie (Foreign Communica-*

tions Technology. Radio Communications, Radio Broadcasting and Television Series). Moscow: Ehkspress-informatsiya, Nos. 11, 12, pp. 3-32.

Chapter 8

8.1 Rudenko, V. M., D. B. Khalyapin, and V. R. Magnushevskij. *Maloshumyashchie vkhodnye tsepi SVCh priemnykh ustrojstv (Low-noise Input Circuits of SHF Receivers).* Moscow: Svyaz', 1971, 279 pp.

8.2 Kulikov, V. V. *Sovremennye sistemy besprovodnoj dal'nej svyazi (Modern Wireless Long-Range Communications Systems).* Moscow: Nauka, 1968, 229 pp.

8.3 Kantor, L. Ya., V. A. Polukhin, and N. V. Talyzin, "The New Orbita-2 Satellite Communications Stations." *Ehlektrosvyaz' (Telecommunications),* 1973, No. 5, pp. 1-8.

8.4 Petrovskij, Yu. B. "Low-noise Transistor SHF Amplifiers for Radio Relay Links." *Trudy NIIR,* 1973, No. 2, pp. 14-20.

8.5 Kantor, L. Ya., S. P. Kurilov, K. G. Trakhtenberg, and I. S. Tsirlin. "The New 'Lotos' Earth Station of the Satellite Communications System." *Trudy NIIR,* 1982, No. 1, pp. 11-17.

Chapter 9

9.1 Tverizovskij, V. N. *Kosmodrom (A Space Port).* Moscow: Mashinostroenie, 1976, 159 pp.

9.2 Spilker, Dzh. *Tsifrovaya sputnikovaya svyaz' (Digital Satellite Communications).* Moscow: Svyaz', 1979, 592 pp.

9.3 Solodov, A. V., Doctor of Engineering Sciences (Editor). *Inzhenernyj spravochnik po kosmicheskoj tekhnike (An Engineering Handbook on Space Technology).* Moscow: Voenizdat, 1977, 430 pp.

9.4 Kaganov, V. I. *SVCh poluprovodnikovye peredatchiki (Semiconductor SHF Transmitters).* Moscow: Radio i svyaz', 1981, 400 pp.

9.5 Kantor, L. Ya. (Editor). *Sputnikovaya svyaz' (Satellite Communications).* Thematic publication of *TIIER,* 1977, Vol. 65, No. 3, 256 pp.

9.6 Klehmpit, L. (Editor). *Moshchnye ehlektrovakuumnye pribory SVCh (SHF Electronic Power Instruments).* Moscow: Mir, 1974, 134 pp.

9.7 Botavin, A. P. and I. S. Tsirlin. "The Reliability of Disposable Transmitters, Designed on the Power Combining Principle." *Trudy NIIR,* 1982, No. 1, pp. 22-27.

9.8 *Zarubezhnaya tekhnika svyazi. Ehkspress-informatsiya. Ser. Radiosvyaz', radioveshchanie, televidenie (Foreign Communications Technology. Express Information. Radio Communications, Radio*

Broadcasting and Television Series). Moscow: Ehkspress-informatsiya, 1979, No. 4, 20 pp.

9.9 Rudenko, V. M., D. B. Khalyapin, and V. R. Magnushevskij. *Maloshumyashchie vkhodnye tsepi SVCh priemnykh ustrojstv (Low-noise Input Circuits of SHF Receivers).* Moscow: Svyaz′, 1971, 279 pp.

9.10 Raber, M. S. and I. S. Tsirlin. "Test Equipment for Earth Control of Satellite Television Broadcasting Space Repeaters." *Trudy NIIR,* 1980, No. 1, pp. 9-13.

Chapter 10

10.1 Model′, A. M., V. I. Krutikov, V. A. Stuzhin, *et al.* "The Waveguide of a Space Communications Station in the Centimeter Band." *Ehlektrosvyaz′ (Telecommunications),* 1975, No. 7, pp. 51-55.

10.2 Fel′dshtejn, A. L., L. R. Yavich, and V. P. Smirnov. *Spravochnik po ehlementam volnovodnoj tekhniki (Handbook on Components of Waveguide Technology).* Moscow: Sovietskoe radio, 1967, 651 pp.

10.3 Model′, A. M. and A. G. Kurashov. A Waveguide, Soviet Patent No. 807951. *Bulletin of Inventions,* 1982, No. 24.

10.4 Model′, A. M. *Fil′try SVCh v radiorelejnykh sistemakh (SHF Filters in Radio Relay Systems).* Moscow: Svyaz′, 1967, 352 pp.

10.5 Kharvej, A. F. *Tekhnika SVCh (SHF Engineering),* Vol. 1. Moscow: Sovietskoe radio, 1965, 783 pp.

10.6 Kuznetsov, V. D. "Frequency Division Multiplexing of Antenna Feed Lines without Resonators." *Ehlektrosvyaz′,* 1970, No. 7, pp. 48-52.

10.7 Kuznetsov, V. D., A. M. Model′, and V. A. Stuzhin. A Device for Combining Signals on Different Frequencies, Soviet Patent No. 492956. *Bulletin of Inventions,* 1975, No. 43.

10.8 Barkov, L. N. and A. M. Model′. "Semireflective Structures in a Transmission Link." *Trudy NIIR,* 1976, No. 1, pp. 41-48.

10.9 Barkov, L. N., V. D. Kuznetsov, A. M. Model′, and V. A. Stuzhin. "A Plan for Combining Signals of Different Frequencies with Low Losses on a Semireflective Structure." *Ehlektrosvyaz′ (Telecommunications).* 1976, No. 3, pp. 61-65.

10.10 Model′, A. M., V. A. Stuzhin, and Yu. S. Shkarinov. "Devices for Combining Wideband Signals of Different Frequencies." *Ehlektrosvyaz′ (Telecommunications),* 1979, No. 3, pp. 17-20.

10.11 Frenkel′, K. S. and I. I. Tsimbler. "Waveguide Traps." *Trudy NIIR,* 1969, No. 2, pp. 169-171.

10.12 Model′, A. M. and V. A. Stuzhin. *A Device for Combining Signals of Different Frequencies.* Soviet Patent No. 418146. *Bulletin of Inventions,* 1974, No. 27.

10.13 Model′, A. M., I. Z. Berlyavskij, V. N. Lychkin, and A. G. Ajzenberg. *A Waveguide Polarizer.* Soviet Patent No. 671648. *Bulletin of Inventions,* 1982, No. 22.

Chapter 11

11.1 Ajzenberg, G. Z., V. G. Yampol′skij, and O. N. Tereshin. *Antenny UKV (UHF Antennas),* Part 1. Moscow: Svyaz′, 1977, 381 pp.

11.2 Pokras, A. M. "Antennas of Earth Satellite Communications Stations." *Radiotekhnika (Radio Engineering),* 1979, Vol. 34, No. 12, pp. 9-18.

11.3 Zucker, H. and W. Jerly, "Computer Aided Analysis of Cassegrain Antennas." *BSTJ,* 1968, Vol. 47, No. 6, pp. 897-932.

11.4 Ajzenberg, G. Z., V. G. Yampol′skij, and O. N. Tereshin. *Antenny UKV (UHF Antennas),* Part 2. Moscow: Svyaz′, 1977, 288 pp.

11.5 Claydon, B. *The Marconi Review,* 1967, Vol. XXX, No. 165, pp. 98-115.

11.6 Bandukov, V. P. and A. M. Pokras. "Some Results of Analysis of an Antenna with Reflectors of a Modified Shape." *Radiotekhnika (Radio Engineering),* 1974, No. 2, pp. 38-45.

11.7 Basilaya, I. Sh. and A. M. Pokras. "A Nonaxially Symmetric Antenna, Equipped with a Light Guide." *Ehlektrosvyaz′ (Telecommunications),* 1979, No. 9, pp. 6-9.

11.8 Pokras, A. M., V. M. Tsirlin, and G. N. Kudeyarov. *Sistemy navedeniya antenn zemnykh stantsij sputnikovoj svyazi (Antenna Steering Systems of Earth Satellite Communications Stations).* Moscow: Svyaz′, 1978, 153 pp.

11.9 Kuznetsov, V. D., V. V. Lyalikov, N. L. Maksimova, A. M. Pokras. *A Conical Scanning Device.* Soviet Patent No. 424509. *Bulletin of Inventions,* 1976, No. 35.

11.10 Kvitko, A. G. and A. M. Pokras. "An Antenna of a Station." *Ehlektrosvyaz′ (Telecommunications),* 1981, No. 1, pp. 61-64.

11.11 Kuznetsov, V. D., V. K. Paramonov, and N. V. Soshnikova. "Receiving Antennas of the Ehkran System." *Ehlektrosvyaz′ (Telecommunications),* 1977, No. 5, pp. 19-23.

11.12 Pokras, A. M., G. G. Tsurikov, and M. A. Shifrin. "Pointing the Beam of a Two-Reflector Antenna at a Stationary Satellite by Rocking the Counterreflector." *IV nauchno-tekhnicheskaya konferentsiya po antennam i fidernym traktam dlya radiosvyazi, radioveshchaniya i televideniya (Fourth Scientific-Technical Conference on Antennas and Feed Lines for Radio Communciations,*

Radio Broadcasting and Television). Moscow: Svyaz', 1977, 4-7 January, Theses of Reports, p. 35.

11.13 Erokhin, G. A., V. G. Kocherzhevskij, and A. M. Pokras. "Calculation of Radio Transparent Foundations of Antenna Structures by Means of Synthesis." *Trudy NIIR,* 1974, No. 1, pp. 110-114.

11.14 Technical Exhibition on the Occasion of the 14th Plenary Assembly of CCIR, 1978. Kyoto: The ITU Association of Japan, Inc., 106 pp.

11.15 Brain, K. R. and I. A. Gill, "An 11/14-GHz 19-m Satellite Station Antenna." IEE Conf. Public. No. 169, Pt. 1, Antennas, 28-30 November, 1978, London, pp. 384-388.

11.16 Glezerman, E. G., B. A. Remizov, and A. V. Shishlov, "Second Generation Space Antennas." *Zarubezhnaya radioehlektronika (Foreign Radio Electronics).* 1981, No. 8, pp. 93-106.

11.17 van Tress, H. L., *et al.* "Planning for the Post-1985 Intelsat System." AIAA 7th Communications Satellite Systems Conference (24-27 April 1978, San Diego, California). 1978, pp. 43-54.

11.18 Makarov, I. S. "Planning of the Intelsat System for the Period after 1985." *Ehkspress-informatsiya. Zarubezhnaya tekhnika svyazi. Ser. Radiosvyaz', radioveshchanie, televidenie (Express Information. Foreign Communications Technology. Radio Communications, Radio Broadcasting and Television Series).* TsNTI Informsvyaz', 1979, No. 5, pp. 1-5.

11.19 ———. "Investigations in Satellite Communications Technology Abroad during 1977–1979." *Ehkspress-informatsiya (Obzor po materialam zarubezhnoj pechati). Zarubezhnaya tekhnika svyazi. Ser. Radiosvyaz', radioveshchanie, televidenie (Express Information (A Survey of Foreign Press Materials). Foreign Communications Technology. Radio Communications, Radio Broadcasting and Television Series).* TsNTI Informsvyaz', 1980, No. 3, pp. 1-19.

11.20 Ohm, E. A. "System Aspects of a Multibeam Antenna for Full Use Coverage." *ICC '79.* Boston and N.Y.: 1979, pp. 49.2.1–49.2.5.

11.21 Makarov, I. S. "A Satellite System with a Multibeam Antenna." *Ehkspress-informatsiya. Zarubezhnaya tekhnika svyazi. Ser. Radiosvyaz', radioveshchanie, televidenie (Express Information. Foreign Communications Technology. Radio Communications, Radio Broadcasting and Television Series).* TsNTI Informsvyaz', 1981.

11.22 Belyanskij, P. V. and B. G. Sergeev. *Upravlenie nazemnymi antennami i radioteleskopami (The Control of Terrestrial Antennas and Radio Telescopes).* Moscow: Sovietskoe radio, 1980, 280 pp.

11.23 Solodukho, Ya. Yu., R. Eh. Belyavskij, S. N. Plyakanov, *et al.*

Tiristornyj ehlektroprovod postoyannogo toka (A DC Thyristor Electric Drive). Moscow: Ehnergiya, 1971, 103 pp.

11.24 Zimin, E. N., V. L. Katsevich, and S. K. Kozyrev. *Ehlektroprivody postoyannogo toka (A DC Electric Drive).* Moscow: Gosehnergoizdat, 1981, 109 pp.

Chapter 12

12.1 Gnedenko, B. V., *et al. Matematicheskie metody v teorii nadezhnosti (Mathematical Methods in Reliability Theory).* Moscow: Nauka, 1965, 524 pp.

12.2 Kozlov, B. A. and I. A. Ushakov, *Spravochnik po raschetu nadezhnosti apparatury radioehlektroniki i avtomatiki (Handbook on the Reliability Calculation of Radio Electronic and Automation Equipment).* Moscow: Sovietskoe radio, 1975, 472 pp.

12.3 Rakov, A. I. *Nadezhnost' radiorelejnykh i sputnikovykh linij peredachi (The Reliability of Radio Relay and Satellite Transmission Links).* Moscow: Radio i svyaz', 1981, 160 pp.

12.4 Intelsat System Status Report, June 1981, Addendum No. 1 BG-47-4E, w/9/81, 14 August 1981.

Chapter 13

13.1 "The Number of Communications Satellites is Increasing/A Company Coworker Group." *Ehlektronika (Electronics),* 1980, No. 20, pp. 66-72.

13.2 Borodich, S. V., V. L. Bykov, L. Ya. Kantor, *et al.* "Intersputnik—The International Satellite Communications System." *Ehlektrosvyaz' (Telecommunications),* 1977, No. 11, pp. 7-11.

13.3 Domingo, J. "The European Regional System ECS and Its Implication for the Planning of the European Communication Network." *Telecommunication Journal,* 1980, Vol. 47, No. 12, pp. 740-745.

13.4 Fawcettle, James. "Bold Move Planned for Italy's Satcom." *Microwaves System News,* 1980, Vol. 10, No. 3, pp. 124-126.

13.5 Tokaaki Kikuchi, Toshio Suzuki, and Shigeru Kimura, "Tunnel Relay System for Automobile Telephone System." *Japan Telecom. Review,* 1980, Vol. 22, No. 2, pp. 157-159.

13.6 Borovkov, V. A. and M. G. Lokshin. "Problems of the Electromagnetic Compatibility of TV Broadcasting Services." *Ehlektrosvyaz' (Telecommunications).* 1979, No. 7, pp. 1-4.

13.7 Zaslavskij, S. A. and V. G. Khodataj. *A Controllable Compander* (Soviet Patent No. 482904). *Bulletin of Inventions,* 1975, No. 32.

Chapter 14

14.1 Goron, I. E. *Radioveshchanie (Radio Broadcasting).* Moscow: Svyaz′, 1979, 368 pp.

14.2 D′yachkova, M. N. and V. I. D′yachkov. "On the Influence of the HF Circuit of Radio Relay Links on the Noise Immunity of an Extra Channel, Transmitted on a Subcarrier." *Trudy NIIR,* 1970, No. 2, pp. 47-53.

14.3 Gurevich, V. Z., Yu. G. Lopushnyan, and G. V. Rabinovich, *Impul′snokodovaya modulyatsiya v mnogokanal′noj telefonnoj svyazi (Pulse-Code Modulation in Multichannel Telephony).* Moscow: Svyaz′, 1973, 336 pp.

14.4 Talyzin, N. V., L. Ya. Kantor, V. I. D′yachkov, *et al.* "Facsimile Transmission in the Orbita System." *Ehlektrosvyaz′ (Telecommunications),* 1969, No. 5, pp. 1-7.

14.5 Kantor, L. Ya. and Eh. Chekhovskij. "Satellite Facsimile Transmission." *Zemlya i vselennaya (The Earth and the Universe),* 1979, No. 3, pp. 20-23.

Chapter 15

15.1 Kantor, L. Ya., A. M. Model′, V. M. Zaev, *et al.* "The Gradient Satellite Communications Receiving-Transmitting Complex." *Ehlektrosvyaz′ (Telecommunications),* 1975, No. 1, pp. 1-6.

15.2 Kantor, L. Ya., S. P. Kurilov, K. G. Trakhtenberg, and I. S. Tsirlin, "The New Lotos Earth Station of the Satellite Communications System." *Trudy NIIR,* 1982, No. 1, pp. 11-17.

Chapter 16

16.1 Kantor, L. Ya., V. A. Polukhin, and N. V. Talyzin. "The New Orbita-2 Satellite Communications Stations." *Ehlektrosvyaz′ (Telecommunications),* 1973, No. 5, pp. 1-8.

Chapter 17

17.1 D′yachkov, V. I., N. M. Zevelev, L. Ya. Kantor, *et al.* "The Ehkran Receiver Systems." *Ehlektrosvyaz′ (Telecommunications),* 1977, No. 5, pp. 15-19.

17.2 Shamshin, V. A. "From the Ehkran to the Tele-Ehkran." *Radio,* 1977, No. 5, pp. 1-3.

17.3 Prospekt mezhdunarodnoj vystavki "Svyaz′-81" (Prospectus of the International Exhibition "Svyaz′-81"). Moscow, Soviet Exposition of the USSR, 1981.

Index